Proof, Computation and Agency

SYNTHESE LIBRARY

STUDIES IN EPISTEMOLOGY, LOGIC, METHODOLOGY, AND PHILOSOPHY OF SCIENCE

VOLUME 352

For further volumes:
http://www.springer.com/series/6607

Proof, Computation and Agency

Logic at the Crossroads

Edited by

Johan van Benthem
ILLC, University of Amsterdam
The Netherlands and Stanford University, USA

Amitabha Gupta
Indian Institute of Technology Bombay, India

and

Rohit Parikh
Brooklyn College and The Graduate Center
City University of New York, USA

Springer

Editors
Prof. Johan van Benthem
University of Amsterdam
Institute for Logic, Language
and Computation (ILLC)
Science Park, P.O. Box 94242
1090 GE Amsterdam
The Netherlands
johan@science.uva.nl

Prof. Amitabha Gupta
Adi Shankaracharya Marg
503 Whispering Woods
Powai Vihar, Bldg. 3
700076 Powai, Mumbai
India
agcg503@gmail.com

Prof. Rohit Parikh
City University of New York
Brooklyn College and the CUNY Graduate Center
Computer Science, Mathematics and Philosophy
365 Fifth Avenue
New York, NY 10016
USA
rparikh@gc.cuny.edu

ISBN 978-94-007-0079-6 e-ISBN 978-94-007-0080-2
Set ISBN 978-94-007-0920-1
DOI 10.1007/978-94-007-0080-2
Springer Dordrecht Heidelberg London New York

Library of Congress Control Number: 2011923788

Printed on acid-free paper

Springer is part of Springer Science+Business Media (www.springer.com)

Foreword

This book is all about new links in logic. The "First Indian Conference on Logic and its Relationship with Other Disciplines" took place in Mumbai at IIT Bombay, from January 8 to 12, 2005, as an initiative of the Mumbai logic circle, bringing together Indian logicians from various disciplinary backgrounds and different locations with a group of open-minded and active international colleagues.

The conference took place over 6 days: two of them devoted to tutorials and four to advanced talks. Tutorials as well as advanced talks were given by Indian logicians and by visitors from abroad. The visitors responding to the Call for the Mumbai event came from Australia, the Czech Republic, Finland, Great Britain, Italy, Israel, Japan, the Netherlands, the USA, and other nations. Together, they formed a distinguished galaxy seldom found even with conferences in the West. The papers emphasized novel approaches and perspectives emanating from research by some of the masters of logic and its interfaces with surrounding disciplines.

The Conference venue were the outstanding logistical and scenic facilities at IIT Bombay, with smooth organizational efforts and crucial support provided by faculty, and especially also by student volunteers. The geese, the pond, and the view of the lake provided most people with the needed inspiration to think of deeper matters.

Talks ranged from reflections on the range of mathematical proof and definability to recent developments in computational logic, as well as new interfaces between logic, information dynamics, and games. In addition, there was a wide range of presentations on schools of Indian logic. One term used nowadays for this broad view of logic is "intelligent interaction". The Mumbai Conference took this term in the double sense of both information exchange and community formation, and indeed both processes were in evidence. Accordingly, the "Second Indian Conference on Logic and its Relationship with Other Disciplines" was held from January 9 to 11 in 2007 at IIT Bombay, Mumbai, with equal success. Besides these two Conferences, a first Indian Winter School with the same topic and title was held at IIT Bombay, Mumbai from January 1 to 15, 2006, primarily for students and researchers.

The current volume contains a broad selection of material from the first Conference, while further volumes will document the follow-up. We see this publication venture as providing information, but also as a means of shaping a community. Our aim is to present an Indian audience with what we see as a representative sample of modern logic, in the hands of some of its best practitioners. This volume is not a textbook, however, accessible throughout to absolute beginners. But we do expect that the papers collected here will give a sense of vitality, strength and direction, and that, through the story-lines and references, our chapters will open further doors to the field as the reader gets inspired.

We also hope that this volume will help the Indian logic community take shape and flight, and indeed, there are some very encouraging signs. Since this book was put together, an 'Association for Logic in India' has been formed that has taken the IIT Bombay initiative as well as other congenial events in Kolkata 2007 under its aegis. The website http://ali.cmi.ac.in/ shows its lively past and current activities. Thus the Mumbai founding events have found a natural continuation in a national series of conferences and schools.

Finally, on a still larger scale, we hope that the present publication initiative will maintain the ties between the Indian community and its friends and colleagues worldwide. In particular, we see the papers on Indian Logic in this volume as reaching out to more classical communities in the field while informing their modern colleagues. One day, we envision that logic will have one history, with Gotama and Gangesa along with Aristotle and Boole. Be that as it may be, our book may be a first gentle push toward joining forces between historically different, but maybe in the final analysis, not all that different strands in our fascinating field.

The editors of this volume have had an easy task in some ways. The positive response and generous cooperation from both Indian and international colleagues has been overwhelming. Editing is easy with authors of this energy and quality, while serious referees were easy to find. Nevertheless, producing this volume has also meant solving a host of non-trivial logistic problems, which could only have been done by our Mumbai experts. Special thanks are due to Mr. Abhisekh Sankaran, who has provided vital technical support to our efforts in putting the act together. We must not fail to register here our appreciation to Allied Publishers, especially Mr. Sharad Gupta, for their concern, persistence and endurance to see the project through in its first Indian instalment. We are grateful that we now have a chance to bring this book to an even wider audience through the medium of Springer Publishers.

We would like to record here the support and advice given to the organizers of the First Conference by several faculty members from various departments and authorities at IIT Bombay, Mumbai, as well as the Institute of Mathematical Sciences, Chennai and TIFR, Mumbai. Furthermore, generous financial support was received, for the Conference and this publication, from Infosys Technologies Ltd., the Department of Science and Technology, Government of India, the Indian Council of Philosophical Research, New Delhi and Microsoft, India. All this is hereby gratefully acknowledged.

Mumbai *Amitabha Gupta*
New York *Rohit Parikh*
Amsterdam *Johan van Benthem*

Preface

Logic: A View from the Cross-roads

Logic stands at the cross-roads of many disciplines and cultures. The discovery that human reasoning and argumentation contains stable patterns that can be studied as such has been made independently in Antiquity in the Greek, Indian, and Chinese traditions, showing a remarkable unity to human thought. Ever since those days, logic has had a special relationship with disciplines where pure reasoning is of the essence, in particular, philosophy and mathematics.

Modern logic as we have today began with developments in the 19th century due to Frege, Boole, Peirce, Cantor and others concerned with logic as well as with the foundations of mathematics. They were followed by early 20th century figures like Russell, Wittgenstein, and Ramsey. A sense of crisis provoked by Russell's and other paradoxes eventually led to Hilbert's program and to deep technical results in the first half of the 20th century on provability, computability, truth, and definability found by Hilbert, Herbrand, Gödel, Gentzen, Skolem, Tarski, Turing, and others. Since that time, mathematical notions and techniques have pervaded the field, and this book contains several chapters devoted to the great themes from that mathematics-oriented tradition.

But modern logic is also a meeting ground of many further disciplines. The study of decidability and undecidability of mathematical questions led to the foundations of computing and eventually, to the birth of computers. Thus, nowadays, there is a wide family of interfaces between logic and computer science, ranging from the study of programming languages to the design of databases or intelligent multi-agent systems, and sometimes even straight into practical matters of information technology. This trend, too, is well-represented in various chapters of this volume, and 'computational logic' broadly conceived may indeed be the bulk of all logic research today.

Still, this is by no means the complete picture of current developments. Logic is being used extensively in the study of topics beyond narrower proof or computation, with a particular emphasis on the behaviour of rational agents which have knowledge, beliefs, preferences, and obligations, and need to adjust these in a world that changes over time. All this has traditionally been the domain of "philosophical logic" at the interface of logic and philosophy, with ties to epistemology, philosophy of action, philosophy of science, and ethics. But over the last decades, this stream of research has started converging with computational logic, as computer systems become more and more sophisticated, alone or in networked societies, thus becoming more and more like us. This is why AI, in particular, has been a large scale "consumer" of logic.

Finally, logic in the last century has had extensive contacts with linguistics, since its account of meaning and definability provides intriguing interfaces with natural language in all its facets, from syntax to semantics and pragmatics. Frege, with his distinction between "sense" and "reference" already gave us the morning star – evening star problem central to both semantics and the philosophy of language. But more recent influences have been those of Lewis with his somewhat controversial notion of common knowledge, and Grice with his notion of implicature which has made pragmatics deeply embedded into semantics. Another important figure has been Saul Kripke whose semantics of modal logic and whose notion of rigid designator have played central roles in recent developments.

Right now, it seems fair to say that the discipline of logic also finds *itself* at a crossroads in the other sense, that of choosing its future directions. This book documents what are arguably the major ones.

First, the study of rational agency as a broader goal beyond computation and deduction naturally leads to the logical study of intelligent interaction and social behaviour. Let's not forget that logic probably arose in the first place as a study of legal, philosophical and political debate! An inspiring paradigm for this is "Social Software", the study of patterns of social interaction by means of techniques from logic and computer science. In this way, logic also joins forces with the social sciences such as game theory in economics, or social choice theory, or even the Law, as our most famous historical example of rational social engineering.

Next, even though logic has traditionally viewed itself as normative, and not as a description of actual behaviour, connections between logic and empirical psychology and cognitive science are on the rise, as more facts about actual behaviour and even about the human brain are becoming available for inspection. Logical theories suddenly find themselves at the forefront of theorizing about human reasoning skills. In a sense, this was already implicit in earlier studies of logic and natural language, where actual facts of meaning and inference in human practice have long served as a constraint on theory.

Finally, the future is sometimes opened up by paying proper attention to one's past. There is a rising interest in logic in other traditions than the Western one, including the independent origins in the Indian and Chinese traditions (a fact which by itself should suffice for shaking up some received cultural stereotypes), but also, its transformation in the golden age of early medieval Arabic culture. Some first explorations have already exposed fascinating vistas.

It was in this lively current setting that a group of Indian logicians decided to start a series of meetings in Mumbai devoted to the full width of logic today, and its modern and ancient roots in Indian cultural life. One purpose of these meetings was to restore India to its rightful place in the logical world. While India has been one of the traditional homes of logic, in recent days, Indian contributions in logic have remained a small part of Indian contributions to science or even to mathematics and philosophy. One could say that the Indian genius has been inadequately harnessed in this important area.

As has already been pointed out that the current volume contains a representative selection of material from the first Conference, while the planned further volumes in this book series will document the follow-up.

Here is a brief guide to the chapters in this volume.

The first two chapters in Part I, entitled "Views of Logic Today," provide a grand setting to what is to follow. John Crossley gives a masterful survey of mathematical logic, which highlights the essential and far-ranging insights that have accumulated by now into proof, computation, and definability via formal systems. Rohit Parikh then follows up with an extended view of logic as a way of studying patterns and procedures in social agency, showing how the perspective of "social software" can systematize this field, while bringing out interesting aspects of different cultures.

Part II comprising a group of three papers on "Logic and Mathematics" demonstrates the continued vitality of mathematical logic and foundations of mathematics. John Crossley examines what is a mathematical proof, a question which continues to inspire new logical theory. Petr Hajek develops a rigorous treatment of fuzzy logic, once thought to be a "soft" subject, but by now fully respectable mathematically. Finally, Wilfrid Hodges takes us back to the formative years of model theory and recursion theory, and Tarski's seminal results joining expressive power and complexity in arithmetical settings.

As a counter-point to this grand tradition in logic, we have placed in Part III some papers from another grand tradition providing "Perspectives from Indian Logic". K. Ramasubramanian presents an overview that will help Western readers understand the rich tapestry of this field, while even teaching Indian readers a thing or two. Moreover, the same author adds a concrete sample, in the form of an exposition of the subtlety of a specific theme in Indian Logic, viz. the concept of *hetvabhasa* in the *Nyāya-sastra*. Finally, Sundar Sarukkai positions Indian logic in its connections with philosophical epistemology and the philosophy of science, showing how topics ran naturally into each other. This interface seems especially relevant in understanding logic/philosophy interfaces today.

The next groups of papers spread over three sections document some major streams in contemporary logic. We start with Part IV, viz., "Logic and Computation," an area which is richly represented in India today. John Crossley examines the entangled and complementary notions of "proof" and "program", and charts their marriage in current theories, including Combinatory Logic and Hoare Logic. Yuri Gurevich and Andreas Blass discuss the state of the art concerning Zero-One Laws, the mysterious statistical regularities underlying the expressive power of many logical systems that were first brought to light in the 1970s. Ron van der Meyden develops another strand, the penetration of systems from philosophical logic, such as epistemic and temporal logic, into the study of computational systems and network security. Daniele Mundici and and Ferdinando Cicalese then take up the topic of coding theory, and its deep and surprising connections with many-valued logics which came originally from the Polish School in the early 20th century. Finally, Noson Yanofsky gives a masterful yet light-footed survey of the new paradigm of quantum computing, which both students and experts will find both informative and delightful.

It may be said with some exaggeration that modern computing is rational agency in societies of interacting agents, which pass information, correct themselves, and engage in purposeful strategic behaviour. Thus, connections between logic, philosophy of action, computer sciences, and social sciences come to the fore. The next part, i.e. Part V on "Logic, Agency, and Games" has some samples of this recent trend. Johan van Benthem presents an extensive survey of ways in which games have been used to model logical core activities such as model checking, model building, or argumentation, and then relates this interactive view of logic to standard notions from game theory such as strategies, equilibrium, and imperfect information. Eric Pacuit tackles one specific issue in social software, viz. the correctness and scope of the algorithm of "Adjusted Winner" in fair division, and shows how logical techniques provide general viewpoints and more reliable conclusions about what such procedures achieve. Next, Krister Segerberg discusses the more general phenomenon of belief revision in the light of dynamic logics of computation and action, and shows how this systematic stance can solve the problem of "iterated belief revision" which has plagued belief revision theory in AI and philosophy for a long time. Finally, G. Venkatesh develops systems of temporal preference logic which can describe players' goals and evaluations over time, as different parts of a game tree become available when successive moves are played. This points the way toward rich logics of games as a paradigm for intelligent interaction over time.

While social phenomena suggest some more careful attention to empirical facts than what has been the norm for logicians, game-theoretic analyses may still be normative flashes of brilliance, rather than the outcome of genuine painstaking observation of behaviour. Our final part, Part VI, is on "Logic, Language and Cognition". It moves closer to the realities of at least one major empirical phenomenon, the "sea of discourse" that we all live in, viz. Natural Language. D.B. Acharya and Shalini Joshi start by exploring the scope and limits of discrete mathematical models (most logical systems are) for the behavioral, cognitive, and social sciences. Then Wilfrid Hodges gives a masterful analysis of the Principle of Compositionality and the emergence of meaning from sentence understanding. This theme has been crucial to understanding the recursive learnability of language since, not just Frege and the modern semanticists, but even, as the author shows, all the way back into medieval Arabic Logic, another relevant historical and cultural tradition.

In their totality, these contributions offer a view of reasoning, computation, and rational agency in a wide modern sense, with logic acting as a common language, and indeed, an intellectual catalyst.

Amsterdam — *Johan van Benthem*
New York — *Rohit Parikh*
Mumbai — *Amitabha Gupta*

Contents

Part I Logic Today: Some Reflections

1 **What Is Mathematical Logic? A Survey** 3
John N. Crossley
1.1 Introduction 3
1.2 Syntax and Semantics 5
1.3 The Completeness Theorem and Model Theory 7
1.4 Intuitionist or Constructive Logic 8
1.5 Recursive Functions 10
1.6 Set Theory 11
1.7 Proof Theory 13
1.8 Conclusion 14
References 15

2 **Is There a Logic of Society?** 19
Rohit Parikh
2.1 Introduction 19
2.2 Logic vs Rationality 19
2.3 Plans 21
2.3.1 Proving Computer Programs Correct 21
2.4 Communication and Knowledge 23
2.5 Incentives and Preferences 24
2.6 Co-ordination and Conflict 25
2.6.1 The Two Horsemen 26
2.7 The Free-Rider Problem 28
2.7.1 Akbar and Birbal 28
2.8 Culture and Tradition 29
References 30

Part II Logic and Mathematics

3 **What Is a Proof?** ... 35
John N. Crossley
3.1 Prelude ... 35
3.2 Introduction ... 35
3.3 A Little History ... 36
3.3.1 Aristotle and Euclid ... 36
3.3.2 Leibniz to Boole ... 39
3.3.3 Frege and Hilbert ... 40
3.4 Differing Proof Systems ... 41
3.5 Applied Logic ... 46
3.6 Comprehending and Constructing Roofs ... 48
3.7 Conclusion ... 49
References ... 50

4 **A Visit to Tarski's Seminar on Elimination of Quantifiers** ... 53
Wilfrid Hodges
4.1 Tarski's Seminar ... 62
4.2 Presburger Arithmetic ... 62
4.3 The Algorithm ... 63
4.4 The Definition of Truth ... 64
References ... 65

5 **Deductive Systems of Fuzzy Logic** ... 67
Petr Hájek
5.1 Introduction ... 67
5.2 Fuzzy Connectives ... 68
5.3 Basic Fuzzy Propositional Logic ... 69
5.4 Basic Fuzzy Predicate Logic ... 72
5.5 Probability ... 74
5.6 Sorites Paradox ... 74
5.7 The Big Family of Fuzzy Logics ... 75
5.8 Making Fuzzy Description Logic More General ... 76
References ... 78

Part III Logic and Computation

6 **What Is the Difference Between Proofs and Programs?** ... 81
John N. Crossley
6.1 Introduction ... 81
6.2 The Lambda Calculus and the Curry-Howard Correspondence ... 82
6.3 Strong Normalization and Program Extraction ... 85
6.4 Beyond Traditional Logic ... 87
6.4.1 Algebraic Specifications ... 87
6.4.2 Imperative Programming ... 90

6.5 Programs and Proofs 95
6.6 Conclusion 96
References 97

7 Zero-One Laws: Thesauri and Parametric Conditions 99
Andreas Blass and Yuri Gurevich
References 114

8 Recent Developments of Feedback Coding and Its Relations with Many-Valued Logic 115
Ferdinando Cicalese and Daniele Mundici
8.1 Basic Facts in Feedback Coding 115
8.1.1 Introduction 115
8.1.2 Symmetric Coding 116
8.1.3 Asymmetric Coding: Basic Facts 120
8.2 Part Two: The Logic of Feedback Coding 123
8.2.1 Asymmetry and Multichannels 123
8.2.2 Search Space, Questions, Answers 124
8.2.3 The Truth-Value 124
8.2.4 The Partially Ordered Monoid of States of Knowledge 125
8.2.5 Multichannel Games vs. Hájek Basic Logic 126
Further Reading 129
References 130

9 Two Applications of Epistemic Logic in Computer Security 133
Ron van der Meyden
9.1 Introduction 133
9.2 The Logic of Knowledge and Time 134
9.3 Model Checking 137
9.4 Verifying the Dining Cryptographers Protocol 137
9.5 Synthesis from Epistemic Specifications 140
9.6 A Strategic Notion of Deducibility 141
9.7 Conclusion 142
References 144

10 An Introduction to Quantum Computing 145
Noson S. Yanofsky
10.1 Intuition 145
10.1.1 Classical Deterministic Systems 146
10.1.2 Classical Probabilistic Systems 149
10.1.3 Quantum Systems 153
10.1.4 Combining Systems 160
10.2 Basic Quantum Theory 163
10.2.1 States 163
10.2.2 Dynamics 165
10.2.3 Observables 165

10.3 Architecture 166
10.3.1 Bits and Qubits 167
10.3.2 Classical Gates 171
10.3.3 Quantum Gates 172
10.4 Deutsch's Algorithm 173
10.4.1 First Attempt 175
10.4.2 Second Attempt 177
10.4.3 Deutsch's Algorithm 178
References 180

Part IV Logic, Agency and Games

11 Logic Games: From Tools to Models of Interaction 183
Johan van Benthem
11.1 From Products to Activities: Logic in Games 183
11.2 The Basic Logic Games 184
11.2.1 Gamification 184
11.2.2 Argumentation 184
11.2.3 Obligation 185
11.2.4 Model Checking 185
11.2.5 Model Construction 186
11.2.6 Model Comparison 187
11.2.7 Other Logic Games 188
11.3 The Unifying Role of Strategies 188
11.3.1 Strategies as a Unifying Notion 188
11.3.2 $\exists$-sickness, and Its Cure 189
11.3.3 Strategies: Actions or Powers? 190
11.3.4 From Logic Games to Game Theory 191
11.4 Game Equivalences and Game Languages 191
11.4.1 When Are Two Games the Same? 191
11.4.2 Game Equivalences and Game Languages 192
11.4.3 Logic and Games: The Plot Thickens 193
11.5 Players' Powers and Modal Forcing Languages 193
11.5.1 Determinacy 193
11.5.2 Powers and Representation 194
11.5.3 Modal Forcing Languages 195
11.6 Extensive Games and Modal Action Logics 196
11.6.1 Extensive Games as Modal Process Models 196
11.6.2 Dynamic Logic as Strategy Calculus 197
11.7 Game Constructions 198
11.7.1 Logical Game Operations 198
11.7.2 Game Algebra of Sequential Operations 199
11.7.3 Excursion: A Complete System 200
11.7.4 Dynamic Game Logic 200
11.7.5 Logics of Parallel Game Operations 201

11.8 Finite Versus Infinite Games 201
11.8.1 The Importance of Infinite Runs 201
11.8.2 Extending the Game Logic Perspective 201
11.9 From Game Logics to Logic Games 203
11.9.1 Questioning Game Equivalence 203
11.9.2 The Outcome Perspective in Logic Games 203
11.9.3 Fine-Structure: The Action Level in Logic Games 204
11.9.4 Digression: Strategy Calculus in Type-Theoretic Format . 205
11.9.5 Operations on Logic Games 206
11.9.6 Finite Versus Infinite 208
11.10 From Game Theory to Logic Games 209
11.10.1 Preferences 209
11.10.2 Solving Games 211
11.10.3 Imperfect Information 211
11.10.4 More Agents and Coalitions 213
11.11 Conclusions 214
References 214

12 In memory of Jasu Magan Bhana Panchia (1963–1991): Iterated Belief Revision in Dynamic Doxastic Logic 217
Krister Segerberg
12.1 Introduction to DDL 217
12.2 Basic DDL 218
12.3 Iteration 220
12.4 Two Examples 221
12.5 The Lattice of Fallback Candidates 224
12.6 Concluding Remarks 226
References 227

13 Towards a Logical Analysis of *Adjusted Winner* 229
Eric Pacuit
13.1 Introduction 229
13.2 The *Adjusted Winner* Procedure 230
13.3 Towards a Logical Analysis 232
13.3.1 Relevant Details 233
13.3.2 Formalizing *AW* 235
13.3.3 Discussion 236
13.4 Conclusion 238
References 239

14 Temporal Logic with Preferences and Reasoning About Games 241
G. Venkatesh
14.1 Introduction 241
14.1.1 Related Work 242
14.1.2 Organisation 242
14.2 Games 243

14.2.1 Strategic Form Games 243
14.2.2 Extensive Form and Repeated Games 243
14.3 Preference Orderings in Propositional Logic 244
14.3.1 Semantics 245
14.3.2 Preference Ordering of Models 245
14.3.3 Preferred Models and Non-monotonic Consequence 246
14.3.4 Modeling Preferences Using Theories 246
14.3.5 Game Theoretic Consequence Relation 247
14.4 Preference Orderings in Propositional Temporal Logic 247
14.4.1 Propositional Temporal Logic 247
14.4.2 Using PTL to Model Games 248
14.4.3 Semantics of PTL 248
14.4.4 Eventually Periodic Models 249
14.4.5 Preference Ordering 249
14.4.6 Ordering EP Models 250
14.4.7 Modeling Preferences Using Theories 251
14.4.8 Nash Equilibrium for Repeated Games 252
14.5 Temporal Logic with Preferences (TLP) 252
14.5.1 The Language 252
14.5.2 Models of TLP 253
14.5.3 Game Theoretic Consequence 254
14.6 Modeling Games in TLP 254
14.6.1 Strategic Form Games 254
14.6.2 Extensive Form Games 255
14.7 Expressive Power of TLP 256
14.8 Conclusions and Possible Extensions 257
References 257

Part V Logic, Language and Cognition

15 From Sentence Meanings to Full Semantics 261
Wilfrid Hodges
15.1 Mathematical Theory 262
15.1.1 Equivalence of Frames 266
15.1.2 The Case Where *X* Is Not Cofinal 266
15.2 Commentary 267
15.2.1 The Freedom to Choose a Grammar 267
15.2.2 Freedom in the Choice of $|e|_\mu$ 268
15.2.3 The Centrality of Sentence Meanings 269
15.2.4 Splitting Syntax From Semantics 270
15.2.5 The Indeterminacy of Translation 273
15.2.6 Tarski's Definition of Truth 273
15.2.7 Tarski-Style Semantics for Other Languages 275
References 276

16 Some Reflections on Discrete Mathematical Models in Behavioral, Cognitive and Social Sciences . . . 277
B. D. Acharya and Shalini Joshi
16.1 Cybernetics and Systems . . . 277
16.2 Combinatorial Models . . . 279
16.3 Opinion Influence Processes . . . 281
16.4 Simmelian Triads . . . 283
16.5 Roberts' Energy Use Sidigraph Model . . . 285
16.6 Structural Balance . . . 288
16.7 Theory of Balance in Social Systems . . . 293
16.8 Intergroup Relations . . . 297
16.9 Societal Networks . . . 299
16.10 Conclusions and Scope . . . 302
References . . . 304

Part VI Perspectives from Indian Logic

17 History and Development of Indian Logic: An Overview . . . 311
K. Ramasubramanian
17.1 Introduction . . . 311
17.2 Place of Logic in the Scheme of Inquiry and Its Purpose . . . 312
17.3 Rise and Development of Logic in India . . . 314
17.3.1 Phase 1 . . . 314
17.3.2 Phase 2 . . . 315
17.3.3 Phase 3 . . . 321
17.4 Motivation Behind the Development of Logic in India . . . 326
17.5 Nature of Indian Logic . . . 328
17.6 Concluding Remarks . . . 330

18 Indian Logic and Philosophy of Science: The Logic-Epistemology Link . . . 333
Sundar Sarukkai
18.1 The Logical and the Empirical in Indian Logic . . . 333
18.2 Logic-Epistemology Link in Indian Logic . . . 336
18.3 Indian logic and philosophy of science . . . 340
18.4 Indian Logic, Science and Semiotics . . . 341
18.5 Signs, Symbols and Theory . . . 343
18.6 Science and Semiotics . . . 347
18.7 Sign-Signified Relation in Applied Mathematics . . . 351
References . . . 353

19 The Concept of Hetvābhāsa in Nyāya-śāstra . . . 355
K. Ramasubramanian
19.1 Introduction . . . 355
19.2 Means and Types of Cognition in Nyāyaśāstra . . . 356
19.2.1 Means of Knowledge . . . 357

19.2.2 Irreducibility of Knowledge ... 359
19.3 Anumāna and Nyāya-prayoga ... 360
19.3.1 Justification for Using Five-Membered Structure ... 361
19.4 Hetvābhāsa ... 362
19.4.1 Definition of Hetvābhāsa ... 363
19.4.2 Import of the Word *yathārtha* in the Definition ... 364
19.4.3 Falsifier is Valid Knowledge and Not Fact ... 365
19.5 Classification of Hetvābhāsa ... 366
19.5.1 *Savyabhicāra* (The Straying Reason) ... 366
19.5.2 *Viruddha* (Contradicting Reason) ... 367
19.5.3 *Satpratipakṣa* (Opposable Reason) ... 368
19.5.4 *Asiddha* (Unestablished Reason) ... 369
19.5.5 *Bādhita* (Stultified Reason) ... 369
19.6 Summary and Conclusion ... 370

About the Authors

B. D. Acharya
B. D. Acharya joined the Department of Science & Technology, Government of India, in 1992 as a Director and became an Advisor in 1997. He is Head of the Earth System Science Division of DST. His main current interest is the application of discrete mathematics to social and behavioural sciences.
Homepage: http://dst.gov.in/scientific-programme/bd-acharya-bio.htm

Johan van Benthem
Johan van Benthem is a University Professor of Logic in Amsterdam, and Henry Waldgrave Stuart Professor of Philosophy at Stanford University. He has worked in modal logic, logic and natural language, and interfaces between logic and philosophy. His main current interests are dynamic logics of information and games.
Homepage: http://staff.science.uva.nl/~johan

Andreas Blass
Andreas Blass is a Professor of Mathematics at University of Michigan, and Visiting Researcher at Microsoft Research. His primary interest is mathematical logic particularly, set theory, and applications to algebra, computer science, and other areas. Other areas: finite combinatorics, topos theory.
Homepage: http://www.math.lsa.umich.edu/~ablass/

Ferdinando Cicalese
Ferdinando Cicalese is an Associate Professor of Computer Science and Applications at the University of Salerno. His areas of interest are combinatorial search algorithms, and their applications in bio-informatics.
Homepage: http://www.dia.unisa.it/c̃icalese/

John N. Crossley
John N. Crossley is Emeritus Professor, Clayton School of Information Technology, Monash University. His main areas of interest are Mathematical Logic and Theoretical Computer Science.
Homepage: http://www.csse.monash.edu.au/∼jnc/

Yuri Gurevich
Yuri Gurevich is Head of the Foundations of Software Engineering group at Microsoft Research, Washington and Professor Emeritus, University of Michigan. He started his career as an algebraist, later became a logician, and then moved to computer science, where his main projects have been Abstract State Machines, Average Case Computational Complexity, and Finite Model Theory.
Homepage: http://research.microsoft.com/∼gurevich/

Petr Hájek
Petr Hájek is Professor at the Institute of Computer Science, Academy of Sciences, Prague, the Czech Republic. Areas of interest: Mathematics - mathematical logic; Computer Science and Artificial Intelligence, and the mathematics of Fuzzy Logic.
Homepage: http://www.uivt.cas.cz/∼hajek/

Wilfrid Hodges
Wilfrid Hodges is an Emeritus Professor of Mathematics at Queen Mary, University of London and is known for his work in model theory. His areas of interest are: Mathematical logic, Foundations, and Model Theory.
Homepage: http://www.maths.qmul.ac.uk/∼wilfrid/

Shalini Joshi
Shalini Joshi was in the Department of Studies in Mathematics, University of Mysore, before she moved to the Department of Psychology, University of Allahabad. Her current interest is in the application of discrete mathematics to social and behavioural and cognitive sciences.

Ron van der Meyden
Ron van der Meyden is a Professor in the School of Computer Science and Engineering, University of New South Wales, Sydney, Australia. His areas of interest are Logic in Computer Science, Logic of Knowledge and Belief, Temporal Logic, Theory of Distributed Systems, Computer Security, Electronic Commerce

Infrastructure, Deductive Databases and Logic Programming.
Homepage: http://www.cse.unsw.edu.au/~meyden/

Daniele Mundici
Daniele Mundici is Professor of Mathematical Logic in the Department of Mathematics, University of Florence. His areas of interest are: Mathematics and Computer Science, with particular reference to many valued logic, lattice-ordered groups, coding with feedback, polyhedra and AF C*-algebras.
Homepage: http://homes.dsi.unimi.it/~mundici/

Eric Pacuit
Eric Pacuit is currently a resident fellow at the Tilburg Center for Logic and Philosophy of Science. Previously, he was a Postdoctoral Researcher at the Institute for Logic, Language and Computation at the University of Amsterdam and Stanford University. His research interests include modal logic, game theory and its foundations, and social choice theory.
Homepage: http://ai.stanford.edu/~epacuit/

Rohit Parikh
Rohit Parikh is a Distinguished Professor of Computer Science, Mathematics and Philosophy, City University Graduate Center, and Department of Computer and Information Science, Brooklyn College, New York. Areas of current interest: Social Software, Reasoning about Knowledge, Belief Revision, Game Theory and Philosophy of Language. His earlier research was in Recursive Function Theory, Proof Theory, Formal Languages, Non-Standard Analysis, and Dynamic Logic.
Homepage: http://www.sci.brooklyn.cuny.edu/cis/parikh/

K. Ramasubramanian
K. Ramasubramanian is an Associate Professor, Cell for Indian Science and Technology in the Department of Humanities and Social Sciences at Indian Institute of Technology Bombay. His areas of current interest are: Nonlinear Dynamics, and Indian Astronomy and Mathematics. He is also interested in Indian Sciences and Physics. He is a scholar in Advaita-vedanta-sastra.

Sundar Sarukkai
Sundar Sarukkai is a Professor, School of Humanities, Centre for Philosophy, in the National Institute of Advanced Studies, Indian Institute of Science Campus, Bangalore. His main research interests are: Philosophy of Science and Mathematics,

Postmodernism, Phenomenology.
Homepage: http://www.iisc.ernet.in/nias/sspro.htm

Krister Segerberg
Krister Segerberg is Professor Emeritus, Uppsala University, Sweden. His current areas of interest are: Modal Logic, Doxastic Logic and Belief Revision.
Homepage: http://www.filosofi.uu.se/personal/kristerse.htm

G. Venkatesh
G. Venkatesh is currently a board member of Sasken Communication Technologies Ltd. Bangalore and the Corporate Chief Technology and Strategy Officer. He is also a Visiting Professor at IIM Bangalore. His areas of interest are: Applications of game theory to strategic thinking in the technology industry, specifically the telecom and semiconductor industries, temporal logic, functional/logic programming, and applications of logic and languages to VLSI design.
Homepage:
http://www.icst.org/?page=about&site=scientificcouncil&subsite=gvenkatesh

Noson Yanofsky
Noson Yanofsky is an Associate Professor, Department of Computer and Information Science, Brooklyn College, New York. His research interests are: Theoretical Quantum Computing, Applied Category Theory, Quantum Complexity Theory, Categorical Universal Algebra, and Categorical Logic.
Homepage: http://www.sci.brooklyn.cuny.edu/~noson/

List of Contributors

B.D. Acharya
Department of Science and Technology, Government of India,"Technology Bhawan", New Delhi 110 016, India, e-mail: bdacharya@yahoo.com

Johan van Benthem
Institute for Logic, Language and Computation (ILLC), University of Amsterdam, Amsterdam, The Netherlands and Spring Quarters: Department of Philosophy, Stanford University, Stanford, CA 94305, USA, e-mail: johan@science.uva.nl

Andreas Blass
Mathematics Department, University of Michigan, Ann Arbor, MI 48109–1043, USA, e-mail: ablass@umich.edu

Ferdinando Cicalese
AG Genominformatik, Technische Fakultät, Universität Bielefeld, D-33594 Bielefeld, Germany, e-mail: nando@cebitec.uni-bielefeld.de

John N. Crossley
School of Information Technology, Monash University, Clayton, VIC 3800, Australia, e-mail: John.Crossley@infotech.monash.edu.au

Yuri Gurevich
Microsoft Research, One Microsoft Way, Redmond, WA 98052, USA, e-mail: gurevich@microsoft.com

Petr Hájek
Institute of Computer Science, Academy of Sciences of the Czech Republic, 182 07 Prague, Czech Republic, e-mail: hajek@cs.cas.cz

Wilfrid Hodges
Queen Mary, University of London, London, UK, e-mail: w.hodges@qmw.ac.uk

Shalini Joshi
Department of Psychology, University of Allahabad, Allahabad 211 002, India, e-mail: shalinijoshi2000@yahoo.com

Ron van der Meyden
School of Computer Science and Engineering, University of New South Wales, Sydney 2052, Australia, e-mail: meyden@cse.unsw.edu.au

Daniele Mundici
Department of Mathematics "Ulisse Dini", University of Florence, 50134 Florence, Italy, e-mail: mundici@math.unifi.it

Eric Pacuit
Tilburg University, Tilburg Institute for Logic and Philosophy of Science, Warandelaan 2, 5037 AB Tilburg, The Netherlands, e-mail: e.j.pacuit@uvt.nl

Rohit Parikh
City University of New York, New York, NY 10016, USA, e-mail: rparikh@gc.cuny.edu

K. Ramasubramanian
Cell for Indian Science and Technology in Sanskrit, Department of HSS, IIT Bombay, Mumbai 400 076, India, e-mail: kramas@iitb.ac.in

Sundar Sarukkai
National Institute of Advanced Studies, Indian Institute of Science Campus, Bangalore 560012, India, e-mail: sarukkai@nias.iisc.ernet.in

Krister Segerberg
Uppsala University, Uppsala, Sweden, e-mail: krister.segerberg@filosofi.uu.se

G. Venkatesh
Indian Institute of Management, Bangalore 560076, India

Noson S. Yanofsky
Department of Computer and Information Science, Brooklyn College, CUNY, Brooklyn, NY 11210, USA; Computer Science Department, The Graduate Center, CUNY, New York, NY 10016, USA, e-mail: noson@sci.brooklyn.cuny.edu

Part I
Logic Today: Some Reflections

Chapter 1
What Is Mathematical Logic? A Survey

John N. Crossley

1.1 Introduction

What is mathematical logic? Mathematical logic is the application of mathematical techniques to logic.

What is logic? I believe I am following the ancient Greek philosopher Aristotle when I say that logic is the (correct) rearranging of facts to find the information that we want.

Logic has two aspects: formal and informal. In a sense logic belongs to everyone although we often accuse others of being illogical. Informal logic exists whenever we have a language. In particular Indian Logic has been known for a very long time.

Formal (often called, "mathematical") logic has its origins in ancient Greece in the West with Aristotle. Mathematical logic has two sides: syntax and semantics. Syntax is how we say things; semantics is what we mean.

By looking at the way that we behave and the way the world behaves, Aristotle was able to elicit some basic laws. His style of categorizing logic led to the notion of the *syllogism*.

The most famous example of a syllogism is

$$\frac{\begin{array}{c}\text{All men are mortal}\\ \text{Socrates is a man}\end{array}}{\text{[Therefore] Socrates is mortal}}$$

Nowadays we mathematicians would write[1] this as

$$\frac{\begin{array}{c}\forall x(\mathit{Man}(x) \rightarrow \mathit{Mortal}\ (x))\\ \mathit{Man}(S)\end{array}}{\mathit{Mortal}\ (S)} \tag{1.1}$$

John N. Crossley
School of Information Technology, Monash University, Clayton, VIC 3800, Australia,
e-mail: John.Crossley@infotech.monash.edu.au

[1] A list of the symbols used is included in the appendix.

J. van Benthem et al. (eds.), *Proof, Computation and Agency*, Synthese Library 352,
DOI 10.1007/978-94-007-0080-2_1, © Springer Science+Business Media B.V. 2011

One very general form of the above rule is

$$\frac{A \qquad (A \to B)}{B} \tag{1.2}$$

otherwise know as *modus ponens* or *detachment*: the A is "detached" from the formula $(A \to B)$ leaving B. This is just one example of a logical rule.

This rule, and other rules of logic such as

$$\frac{A \quad B}{(A \wedge B)} \quad \text{or} \quad \frac{(A \wedge B)}{A}$$

where $\wedge$ is read "and", have obvious interpretations. These rules come from observing how we use concepts.

This analytic approach was that taken by George Boole, an Irish mathematician in the 19th century. It is from his work that we have *Boolean algebra* or *Boolean logic* or, as it is often known today, *propositional calculus*.

Later in the 19th century Gottlob Frege developed a small suite of logical laws that are with us today and suffice for all of mathematics. These are the rules of the *predicate calculus*. The rules relate only to the syntax. Although they are abstracted from the way we talk and think, the meaning, the *semantics*, is something quite separate.

The most familiar example of semantics is given by *truth-tables* such as the one for conjunction (or "and", $\wedge$):

$\wedge$	T	F
T	T	F
F	F	F

Here we are given the *truth-values* of the formulae A and B and we work out the truth of the conjunction $(A \wedge B)$ by taking the value for A at the side and the value for B at the top and finding the value where column and row intersect. In general, determining the truth of a syntactic expression (formula) A requires looking carefully at its constituents.

At this point we should pause because we are now dealing with two completely different things, two different aspects of languages.

On the one hand we have the rules for working with strings of symbols – rules such as *modus ponens* above; on the other hand we are looking at meanings. This latter is the study of *semantics*. The former is *syntax* – the study of the way that language is put together from symbols. In English this latter includes the way we put words together to make a sentence.

Consider something we, in essence, wrote earlier (in (1.1)), or rather a slight variant of it.

$$\forall x(M(x) \to D(x)) \tag{1.3}$$

If $M(x)$ is interpreted as "x is a man" and $D(x)$ as "x will die", then this formula is obviously true under this interpretation. However, if we interpret $M(x)$ as "x is an animal" and $D(x)$ as "x has at most two legs", then it is obviously false under this different interpretation.

Note that the formula itself is neither true nor false, it is a *formula* – a piece of syntax.

1.2 Syntax and Semantics

The first concern of mathematical logic is the relation between syntax and semantics.

First let us consider syntax a little more closely.

Any logic starts from certain basic symbols. In ordinary mathematical logic, and I will say what "ordinary" means later in Section 1.4, we have symbols (letters) for variables: $x, y, z, \ldots$ and letters for predicates or properties, such as M and D above. We also have letters for relations and functions. For example, $L(x, y)$ might arise in a context: "there is a line between the points x and y", or in a context "x is married to y". We use small letters for functions, $f_1, f_2 \ldots$. Such arise in arithmetic where we may use f_1 for $+$, or have a function i, say, in group theory that is intended to denote the inverse of an element. Applying these function letters (perhaps repeatedly) gives rise to (individual) *terms*.

Together all of these give us *atomic formulae* – basic formulae such as $L(x, y)$ or $M(f_3(x, f_1(y_2)))$ or $D(y)$.

Atomic formulae can the be joined together by *logical connectives* to form more complicated *formulae*, sometimes called "*well-formed formulae*", such as

$$(M(x) \wedge D(y)) \qquad \text{or} \qquad (M(x) \rightarrow D(x)) \qquad \text{or} \qquad \forall x(M(x) \rightarrow D(x)).$$

These have been joined using *connectives*: *propositional connectives* $\wedge$ (read as "and"), $\vee$ (read as "or"), $\rightarrow$ (read as "implies"), $\neg$ (read as "not"); and the *quantifiers*: $\forall$ (read as "for all") and $\exists$ (read as "there exists").[2] The precise rules for constructing formulae can be found in any logic textbook, such as [22].

We then have rules for generating more formulae. We have already met *modus ponens* in (1.2). This rule is also know as *implication elimination*, ($\rightarrow$-E): the $\rightarrow$ above the line is eliminated below it.

In Fig. 1.1 we give the rules[3] that were discovered in the late 19th century by Frege (though Leibniz knew many of them long before in the 17th century). The formulation here is known as *Natural Deduction* since the rules (with two exceptions) look very close to our actual practice in reasoning. In this formulation we do not have axioms but we may have hypotheses, i.e. formulae that we assume.

---

[2] Some people avoid using negation, $\neg$. They employ a constant $\perp$ for the *false* formula. Then they use the formula $(A \rightarrow \perp)$ instead of $\neg A$.

[3] The phrase "x is free/not free in [some formula]" is a technical condition that avoids misunderstandings.

$$\frac{}{\vdash A}\ (\text{Axiom-I}) \qquad \frac{}{A \vdash A}\ (\text{Ass-I})$$

$$\frac{\Delta, A \vdash B}{\Delta \vdash (A \to B)}\ (\to\text{-I}) \qquad \frac{\Delta \vdash A \quad \Delta' \vdash (A \to B)}{\Delta, \Delta' \vdash B}\ (\to\text{-E})$$

$$\frac{\Delta \vdash A}{\Delta \vdash \forall x.A}\ (\forall\text{-I}) \qquad \frac{\Delta \vdash \forall x.A}{\Delta \vdash A[c/x]}\ (\forall\text{-E})$$

x is free in A, not free in Δ

$$\frac{\Delta \vdash P[a/y]}{\Delta \vdash \exists y.P}\ (\exists\text{-I}) \qquad \frac{\Delta_1 \vdash \exists y.P \quad \Delta_2, P[x/y] \vdash C}{\Delta_1, \Delta_2 \vdash C}\ (\exists\text{-E})$$

where x is not free in C

$$\frac{\Delta \vdash A \quad \Delta' \vdash B}{\Delta, \Delta' \vdash (A \wedge B)}\ (\wedge\text{-I})$$

$$\frac{\Delta \vdash (A_1 \wedge A_2)}{\Delta \vdash A_1}\ (\wedge\text{-E}_1) \qquad \frac{\Delta \vdash (A_1 \wedge A_2)}{\Delta \vdash A_2}\ (\wedge\text{-E}_2)$$

$$\frac{\Delta \vdash A_1}{\Delta \vdash (A_1 \vee A_2)}\ (\vee\text{-I}_1) \qquad \frac{\Delta \vdash A_2}{\Delta \vdash (A_1 \vee A_2)}\ (\vee\text{-I}_2)$$

$$\frac{\Delta \vdash (A \vee B) \quad \Delta_1, A \vdash C \quad \Delta_2, B \vdash C}{\Delta_1, \Delta_2, \Delta \vdash C}\ (\vee\text{-E})$$

$$\frac{\Delta \vdash \bot}{\Delta \vdash A}\ (\bot\text{-E})$$

$A[a/x]$ is read "A with a for x" and denotes the formula A with a substituted for the variable x.

Fig. 1.1 The basic rules of predicate calculus.

Here an expression such as $\Delta, A \vdash B$ is read "from the formulae in Δ and the formula A we can prove the formula B".

The rules should be easy to read with at most two exceptions. These are ($\vee$-E) and ($\exists$-E). The former corresponds to proof by cases. We can paraphrase it as follows: "If, from A we can prove C and from B we can prove C, then we can prove C from $(A \vee B)$". Likewise ($\exists$-E) can be paraphrased as: "If we can prove C from some particular x such that P (with x replacing y), and we can also prove that there is some y such that P, then we can prove C".

It is obvious that these rules are "good" rules if we just interpret them in an intuitive way. That is to say, when so interpreted they lead from true assertions to other true assertions.

We build proofs by repeatedly applying the rules. Since some of the rules have two *premises* (top lines) we actually get a tree. The tree is a *proof* of the formula at the root of the tree.

1.3 The Completeness Theorem and Model Theory

It should be quite surprising that when we analyze the sort of language we use in mathematics, and elsewhere too, that the few basic rules, given above and originally due to Frege, suffice to yield all those syntactic expression that are always true – and no others. This is the most dramatic result that mathematical logic produced in the first half of the twentieth century:

Theorem 1.1 (The Completeness Theorem) *There is a finite (small) set of logical rules which will yield all, and only, those formulae which are true under any interpretation.*

What is an *interpretation*? We have given a hint when discussing (1.3) above.

First we have to establish what domain we are talking about, e.g. people, or natural numbers. Then we have to interpret the predicates in the language and these will usually be interpreted as relations, one such example in the case of numbers is the relation $\leq$. Here is a different example. If we have a language involving $P(x,y)$ and we consider interpreting this predicate P as "x divides y" and we only allow x,y, etc. to be interpreted as natural numbers, then we can see that

$$(P(x,y) \wedge P(y,z)) \rightarrow P(x,z)$$

is always true.

On the other hand

$$(P(x,y) \vee P(y,x)) \tag{1.4}$$

is sometime true and sometimes false, while

$$\forall x \neg P(x,x) \tag{1.5}$$

is false since every number divides itself.

However if we interpret $P(x,y)$ as "y is strictly greater than x", then (1.4) is sometimes true and sometimes false and (1.5) is true.

If we go to a completely different interpretation and let the variables range over human beings and now interpret $P(x,y)$ as "x is married to y" then we get different results.

The actual formal definition of "true in an interpretation" is quite complicated (see e.g. [22]), so we omit it here.

Those formulae, which are true under all possible interpretations, are called *universally valid* or, sometimes, simply "valid". One example is any formula of the form $(A \vee \neg A)$.

Formally we say that an interpretation, $\mathcal{M}$, is a *model* of a formula A (or a set of formulae Δ) if A is true in $\mathcal{M}$ (if every formula in Δ is true in $\mathcal{M}$).

In [17], Kreisel pointed out that the Completeness Theorem actually catches our intuitive notion of truth. The argument is simple. The Completeness Theorem says that every formula true in all (formal, mathematical) interpretations is provable. Clearly, anything provable is true in all interpretations (including informal ones).

Finally anything true in all interpretations is true in all formal interpretations. Thus these three classes: A: provable in predicate calculus, B: true in all interpretations, and C: true in all formal interpretations, are such that $A \subseteq B \subseteq C \subseteq A$ and therefore A, B and C all coincide.

One half of the Completeness Theorem is more powerful in the following form.

Theorem 1.2 (The Compactness Theorem) *If a set, Σ, of formulae is consistent, then it has a model, i.e. an interpretation in which all formulae in Σ are true.*

Here *consistent* means, as you would expect, that we cannot prove a contradiction (such as $(A \wedge \neg A)$) from Σ.

The idea of model gives rise to *Model Theory*. This got a great impetus from the Completeness Theorem. Model Theory is the study of interpretations of a set of sentences of logic. The area is basically a grand exploitation of the semantics of logic. Surprisingly, one can obtain significant results simply by looking at the style of the sentences. At its simplest level it gives us beautiful results such as the following.

Theorem 1.3 *If a formula $\forall x_1 \forall x_2 \ldots \forall x_n A(x_1, \ldots, x_n)$ with no quantifiers inside the formula A is true in an interpretation then it is true in any (non-empty) sub-interpretation.*

A simple application of this is the following. If we write down axioms for a group involving the inverse function (as mentioned above) and a function m, say for group multiplication, then all these axioms can be written in the form $\forall x_1 \forall x_2 \ldots \forall x_n A(x_1, \ldots, x_n)$ with no quantifiers inside the formula A. It follows that if $\mathscr{G}$ is a model of these axioms, i.e. a group, then any non-empty subset of the elements of $\mathscr{G}$ closed under inverses and multiplication, is actually a sub-group.[4]

But Model Theory has much more powerful results too and allows us to do calculus in a new way: the *Non-standard analysis* of Abraham Robinson [23]. It has given us deep insights into group theory and into models of set theory, and has become an autonomous discipline (see [1]).

1.4 Intuitionist or Constructive Logic

Now there is not just one "true" logic. At the beginning of the 19th century Brouwer questioned the law of the excluded middle and gave a new interpretation to the syntax that we use in mathematics. He regarded a proof as telling us how to make a construction. For Brouwer, a proof of $(A \vee \neg A)$ was only acceptable if one could give *either* a proof of A *or* a proof of $\neg A$. In the ordinary mathematics of that time, as used by Hilbert, $(A \vee \neg A)$ was trivially true. This was, roughly speaking, based

[4] Some people take this as the definition of a sub-group but other examples can be given, see [21] or [5].

on the idea that A was either true or not, even if we do not know which: there was no middle alternative. This was unacceptable, indeed meaningless, for Brouwer.

Now the ordinary logic that is (still!) commonly used by mathematicians and most philosophers[5] relies on the law of the excluded middle:

$$(A \vee \neg A)$$

or, equivalently, the law of double negation:

$$\frac{\neg\neg A}{A}$$

Both of these are equivalent to our last rule in Fig. 1.1:

$$\frac{\Delta \vdash \bot}{\Delta \vdash A}\ (\bot\text{-E})$$

For many computer scientists and philosophers, it is better *not* to use this rule.[6] This gives so-called *constructive* or *intuitionist* logic.

When I was a student this kind of logic was regarded as odd.

In the 1960s Saul Kripke, then a young student, produced a different kind of interpretation in order to prove a Completeness Theorem for intuitionist logic. His "interpretations" were not single models in the sense we met above, they were families of such interpretations with relations between them. They are know as *possible world* interpretations. Intuitively speaking, a formula is then provable if (and only if) it is true in all possible worlds. For a detailed treatment see [8, 19] or [2].

Nowadays there are lots of different logics that all have their value and application. These logics include various kinds of modal logics. Modal logics have *modalities* which are indicated by new symbols. Thus we may have $\Diamond$ for *possibly* and $\Box$ for *necessarily*. One of the most famous of these is the logic called $S5$. This is a propositional logic in which the formulae are built up from *propositional variables* $p, q, \ldots$ using the propositional connectives (see above Section 1.2) and also allowing prefixing of formulae by $\Box$ or $\Diamond$. (Thus $(p \vee \Diamond(q \wedge \Box r))$ is a formula of $S5$. It has all the true formulae of ordinary propositional calculus as axioms together with the rules shown in Fig. 1.2.

Many of these logics also have completeness theorems analogous to Theorem 1.1. The proofs of these have similarities with the proof of the completeness of intuitionist logic. Indeed, Kripke proved completeness results for modal logics first [18] and only subsequently used his ideas there to prove the completeness of intuitionist logic. Details of such theorems may be found in [8].

---

5 But not as much by computer scientists.

6 This is not a question of the rule being right or wrong, it is a question of what one can say about what computers do. There are certainly problems which a compute cannot decide (see below, Section 1.5, so the computer does not necessarily "know" whether A is true or $\neg A$ is true.

If $\vdash A$ then $\vdash \Box A$	$\vdash \Box(A \to B) \to (\Box A \to \Box B)$
$\Box A \vdash A$	$\Diamond A \to \Box\Diamond A$

Fig. 1.2 The rules for the modal logic *S*5.

1.5 Recursive Functions

When one turns to specific domains, for example the natural numbers, $0,1,2,\ldots$, the power of the logic changes. In order to study the natural numbers, the theory of formal arithmetic was described by axioms. These axioms are known as the Peano axioms but they are really due to Richard Dedekind (see [13]).

Kurt Gödel showed, in 1931, that there is no finite system of axioms that will give you all the true statements of arithmetic.[7] This is his famous *Incompleteness Theorem.*

So logic, in one sense, fails. But this weakness is also a strength. In proving the theorem Gödel developed the notion of *recursive* or *computable* function.[8] These functions are those that are *representable* in formal arithmetic. That is to say there are predicates that exactly mirror these functions. Later, however, it was found that there are many other ways of describing exactly the same functions.

Alan Turing [27] showed that the functions that can be computed on his idealized machines, now known as *Turing machines*, are exactly the same as Gödel's recursive functions. All our computers developed from this notion of Turing's.[9] It is now almost universally believed that a function is computable if, and only if, it can be programmed on some computer and, for every type of computer so far developed, it has been shown that those functions that can be computed are amongst Turing's computable functions. So recursive functions are *exactly* the functions we can compute. This work also showed that some functions can*not* be computed. In particular it is not possible to compute whether a given Turing machine with a given program will stop or not. This is the *Halting problem.*

The development of recursion theory has spawned a whole sub-industry of mathematical logic in which I have played a part.

In advancing the theory of computable functions several different approaches have been adopted including the *Turing machines* mentioned above and the *Lambda calculus* which we shall treat below (see Section 1.7).

---

7 In fact he even showed that there is no finite complete system of axiom schemes for formal arithmetic.

8 To be precise, attention actually focussed on *partial* functions, those that may not be defined for all arguments.

9 At least as far as we can tell. It seems obvious that John von Neumann used Turing's ideas but there is no record of him admitting to that! See Martin Davis [11].

$$Z \to 0$$

$$Z \to Z+1$$

$$\text{If } Z \neq 0 \text{ GO TO } L$$

Z stands for an arbitrary register name.
Instructions are labelled with *labels* $L, L_1, L_2, \ldots$.

Fig. 1.3 The programming instructions for Shepherdson-Sturgis machines.

One of the nicest approaches is that of *Shepherdson-Sturgis machines* or *Unlimited Register Machines* (see [16]). An excellent description of these can be found in Davis, Sigal and Weyuker [12]. We consider an abstract machine, very much like a modern computer. It has a finite number of *registers* in which natural numbers are stored. The registers are identified by variables $X_1, X_2, \ldots, Y, Z_1, Z_2, \ldots$. The X_i are the input registers. There is also an output register, where the answer will be found. The Z_j are auxiliary ones. Initially each register has zero in it. The whole programming language is very simple and is shown in Fig. 1.3. All recursive functions can be computed using programs in this language!

It was some time before people were able to prove that there are different *degrees* of computability. However this has now become a large industry, strongly developed by Sacks in the latter half of last century. (See Barwise [1].)

One of the areas which still requires considerable development, in my view, is that of *higher order* recursive functions. Although a great deal of work was done last century by Kleene, Sacks and others, our understanding of the kinds of functions that will take a program as an argument and always yield another program is not very well understood (even though such higher order functions are extensively used).

1.6 Set Theory

Besides arithmetic logic has also been applied to set theory. Indeed, one of the major impetuses for mathematical logic was the problem, in the nineteenth century, of the foundations of analysis, including, in particular, the infinitesimal calculus.

Cantor [3, 4] started investigating Fourier series and came across difficulties. The first thing was that one could count beyond infinity. This had been remarked long ago by Galileo in his *Discorsi e dimostrazioni matematiche intorno a due nuove scienze* (1638) (which I have not seen) where he showed there are the same number of square numbers as there are numbers. Here is an even simpler example. Suppose we arrange all the even numbers first and then all the odd numbers, then we get a list

$$0, 2, 4, \ldots, 1, 3, 5, \ldots.$$

So if we start counting we count $1,2,3,\ldots$ and then run out of counting numbers before we get to the end. Cantor introduced infinite (ordinal) numbers to be able to count this far. He counted

$$1,2,3,\ldots,\omega,\omega+1,\omega+2,\ldots.$$

And this process could be carried much further to $\omega\times\omega$ or ω^{ω} or even to $\omega^{\omega^{\omega^{\omega^{\cdot^{\cdot^{\cdot}}}}}}$ which is now known as ε_0 and even that was not the end. There is no end.

In developing his theories of large cardinal and ordinal numbers he introduced set theory. Sets can be thought of as being formed in two different styles. The first is by building them up, e.g. by taking unions of smaller sets, etc. The other is by defining them by what they comprehend: defining a set as the set of x such that $A(x)$. The *naïve axiom of comprehension* says that $\{x : A(x)\}$ always exists. This gives rise to *Russell's paradox* (1.6):

If R is the set of x such that $x \notin x$, then we have both $x \in R$ and $x \notin R$.

The basic axioms of set theory were not troublesome and were formulated by Zermelo and Fraenkel. The axiom of comprehension, however, had to be circumscribed.

Thanks to the formulations available through the mathematization of logic, set theory has developed enormously.

Nevertheless this (finite) set of axiom schemes did not succeed in resolving all the open questions. Not merely was it incomplete (as it had to be because it was possible to develop arithmetic in set theory, see my remarks on incompleteness above in Section 1.5) but it did not resolve the status of the Axiom of Choice. The Axiom of Choice says that, given a non-empty set of non-empty sets, there is a set that has exactly one element from each of those sets. Russell [24] gives the nice example: given an infinite set of pairs of socks, how do you pick one sock from each pair?

The first problem was the problem of consistency. This has been attacked by building models of set theory. Gödel was the first to do this in spectacular fashion by building models of set theory within set theory, see [14]. However, such models do not solve all the problems.

Now Gödel [14] had established the consistency, but not the independence, of both the Axiom of Choice and the Continuum Hypothesis[10] back in 1940.

About the time I obtained my PhD (1963) Paul Cohen [6, 7] showed the independence of the Axiom of Choice (and the Continuum Hypothesis) from the other axioms of set theory. He applied a kind of model theory which we now know is closely related to the model theory that Saul Kripke used for modal and intuitionist logic. The technique is known as *forcing* and depends on being able to create possible worlds that behave in unusual ways. Since that time very many other statements

[10] The Continuum Hypothesis says that there are no infinite cardinal numbers between the smallest infinite cardinal number, that of the set of natural numbers and the cardinal number of the set of all subsets of the natural numbers.

with mathematical import have been proved independent of the other axioms of set theory.

The search for “nice” axioms for set theory continues. Although the concept of set appears simple from an intuitive point of view, we have no precise conception of what a set is. Moreover, since about 1970, the field of Set Theory has become extremely complicated and difficult. It is perhaps not surprising that even its practitioners use words such as “morass”.

1.7 Proof Theory

Finally there is *Proof Theory*. This studies the formal proofs in mathematical logic in the same way that one studies the real numbers, for example. Originally this was done by Gerhard Gentzen in the 1940s when he tried to prove that the Peano axioms for arithmetic are consistent.

Gödel had not only proved the Incompleteness Theorem (see above Section 1.5) but had also shown that it was impossible to prove the consistency of formal arithmetic in the theory itself. Gentzen devised a new way of presenting proofs, in fact closely related to the system of natural deduction we used in Section 1.2. He was then able to show that one could simplify certain proofs.

For example, suppose we are given two proofs: where the first is a proof of A:

$$\begin{array}{c}\vdots\\ A\end{array}\tag{1.6}$$

and the second is a proof of B from A from which we get a proof[11] of $A \to B$:

$$\dfrac{\begin{array}{c}[A]\\ \vdots\\ B\end{array}}{(A \to B)}\tag{1.7}$$

Then suppose that we use *modus ponens* to remove the A from $(A \to B)$ thus:

We now have a proof:

$$\dfrac{\begin{array}{c}\vdots\\ A\end{array}\qquad \dfrac{\begin{array}{c}[A]\\ \vdots\\ B\end{array}}{(A \to B)}}{B}\tag{1.8}$$

But if instead we had simply put the first proof, (1.6) of A, on top of the second proof, (1.7), of $(A \to B)$, then we no longer need the hypothesis $[A]$ in the second

[11] The square brackets indicate that A can be discharged, i.e. is not needed for the proof of B, though it is for the proof of B, of course.

proof in order to get a proof of B. Previously we had made an unnecessary detour since we already had a proof of B.

That is to say, we can *reduce* the proof in (1.8) to a simple proof of B of the form

$$\begin{array}{c} \vdots \\ A \\ \vdots \\ B \end{array}$$

This removal of unnecessary detours is known as *cut elimination*.

Gentzen [26] showed, by using a special form of induction, transfinite induction,[12] that there could be no proof of a contradiction, i.e. that arithmetic was consistent. The actual technique was to assume that there was a proof of a contradiction and then to reduce that proof, by cut elimination, until it was, in fact, of a very simple form. From there it was obvious that there could be no such proof.

Gentzen's techniques have been greatly developed by Feferman in the USA and Schütte and his group in Germany. These people have extended his results to much more complicated systems of logic than simple arithmetic.

However, the work started by Gentzen about 60 years ago has started a new and perhaps surprising industry in computer science. Although Gentzen was aware of the information contained in a proof, it was not until Bill Howard showed in his [16] that propositional calculus (which is even simpler than predicate calculus and can be, at least partially, identified with Boolean algebra) reflects the lambda calculus. Thereby the usefulness of Gentzen's work for producing computer programs was realized.

This is despite the fact that the lambda calculus was one of the ways that recursive functions were developed. It was an alternative to Turing machines. More recently lambda calculus has formed the basis for the programming language, LISP, and in fact one can obtain programs directly from formal proofs by developing Howard's ideas further. For more about this you may care to look at my lecture in this conference [9]: *What is the difference between proofs and programs?* which is devoted to this topic of extracting programs from proofs.

1.8 Conclusion

Over little more than a century mathematical logic has developed from nothing to a very multi-faceted subject. It has thrown a great deal of light on many areas of philosophy: particularly through modal logics; mathematics: especially through set theory; and computer science: through the analysis it has permitted and the (correct) programs it has allowed to be generated. It has also shown the limits of computability.

[12] In fact he only needed transfinite induction up to ε_0, see Section 1.6, in order to prove his result.

At present, mathematical logic encompasses model theory, set theory, recursion theory and proof theory. Although modal logics have long been used, especially by philosophers, in my lifetime I believe that the most important change in mathematical logic has been the development of many, many other kinds of logics, which have supplemented the standard or classical one used in mathematics. I have touched on but a few of these. Some of the others are the subjects of other lectures in this conference. There is still more work to be done and I hope I have encouraged you to find out what there is to do in the areas in which you yourselves are interested.

References

1. Barwise J., editor. *Handbook of Mathematical Logic*, North-Holland Pub. Co., Amsterdam, 1977.
2. Blackburn P., van Benthem J., and Wolter F., editors. *Handbook of Modal Logic*, 3, (*Studies in Logic and Practical Reasoning*), vol. 3. Elsevier, Amsterdam, 2006.
3. Cantor G. Über einen die trigonometrischen Reihen betreffenden Lehrsatz, *Journal f. reine und angew. Math.*, 72: 130–138, 1870.
4. Cantor G. Über die Ausdehnung eines Satzes aus der Theorie der trigonometrischen Reihen, *Math. Annalen*, 5: 123–132, 1872.
5. Chang C. C., and Keisler H. J. *Model Theory*, 3rd ed., North-Holland Pub. Co., Amsterdam 1973, 1990.
6. Cohen P. J. The independence of the continuum hypothesis, *Proc. Nat. Acad. Sci. USA*, 50: 1143–1148, 1963.
7. Cohen P. J. The independence of the continuum hypothesis, II, *Proc. Nat. Acad. Sci. USA*, 51: 105–110, 1964.
8. Cresswell M. J., and Hughes, G. E. *A New Introduction to Modal Logic*, Taylor & Francis Inc., London, 1996.
9. Crossley J. N. What is the difference between proofs and programs? Lecture at the First International Conference on Logic and and its application to other disciplines, IIT Bombay, 2005, submitted for publication.
10. Crossley J. N., Brickhill C., Ash(†) C. J., Stillwell J. C., and Williams N. H. *What is Mathematical Logic?* Oxford University Press, New York, NY, 1972, Latest edition, Dover, 1990.
11. Davis M. *Engines of Logic: Mathematicians and the Origin of the Computer*, W.W. Norton, New York, NY, 2001.
12. Davis M. D., Sigal R., and Weyuker E. J. *Computability, Complexity, and Languages: Fundamentals of Theoretical Computer Science*, Academic Press, Harcourt, Brace, Boston, MA, 2nd edition, 1994.
13. Dedekind R. The nature and meaning of numbers. In *Essays on Theory of Numbers*, Dover, New York, NY, 1963. Originally published in 1901. Translation of *Was sind und was sollen die Zahlen?*
14. Gödel K. *The Consistency of the Continuum Hypothesis*, Princeton University Press, Princeton, NJ, 1940.
15. van Heijenoort J., editor. *From Frege to Gödel*, Harvard University Press, Cambridge, MA, 1967.
16. Howard W. The formulae-as-types notion of construction. In J. Roger Hindley and J. Seldin, editors, *To H.B. Curry: Essays on Combinatory Logic, Lambda Calculus, and Formalism*, pages 479–490. Academic Press, New York, NY, 1969.

17. Kreisel G. Mathematical logic. In T. L. Saaty, editor, *Lectures on Modern Mathematics*, vol. 3, pages 95–195. Wiley, New York, NY, 1965.
18. Kripke S. A completeness theorem in modal logic, *J. Symb. Logic*, 24: 1–14, 1959.
19. Kripke S. Semantical analysis of intuitionistic logic I. In J. N. Crossley and M. A. E. Dummett, editors, *Formal Systems and Recurive Functions*. North-Holland, Amsterdam, 1965.
20. Lemmon E. J. *Beginning Logic*, Nelson, London, 1971.
21. Lyndon R. C. Properties preserved under homomorphism, *Pacific J. Math.*, 9: 143–154, 1959.
22. Mendelson E. *Introduction to Mathematical Logic*, 4th ed., Chapman & Hall, 1997.
23. Robinson A. *Non-Standard Analysis*, North-Holland Pub. Co., Amsterdam, 1966.
24. Russell B. A. W. *Introduction to Mathematical Philosophy*, G. Allen and Unwin, London, 1970. First published in 1919, thirteenth impression, 1970.
25. Shepherdson J. C., and Sturgis H. E. Computability of recursive functions, *J. Assoc. Comput. Mach.*, 10: 217–255, 1963.
26. Szabo M. E., editor. *The Collected Papers of Gerhard Gentzen*, North-Holland Pub. Co., Amsterdam, 1969.
27. Turing A. M. Computability and lambda-definability, *J. Symb. Logic*, 2:153–163, 1937.

Appendix: A Brief Guide to the Symbols

Symbol	*Read*
$\forall$	forall
$\rightarrow$	implies
$\dfrac{A \quad (A \rightarrow B)}{B}$	From A and $(A \rightarrow B)$infer B
$\wedge$	and
T	True
F	False
$\vee$	or
$\neg$	not
$\exists$	there exists
($\rightarrow$-E)	$\rightarrow -$elimination
($\vee$-E)	$\vee -$ elimination
($\wedge$-E)	$\wedge -$ elimination
($\exists$-E)	$\exists -$ elimination
($\forall$-E)	$\forall -$ elimination
($\rightarrow$-I)	$\rightarrow -$introduction
($\vee$-I)	$\vee -$ introduction
($\wedge$-I)	$\wedge -$ introduction
($\exists$-I)	$\exists -$ introduction
($\forall$-I)	$\forall -$ introduction
$\bot$	False(the false proposition)
$\Box$	necessarily
$\Diamond$	possibly
$\in$	is a member of
$\notin$	is not a member of
$\begin{matrix} \vdots \\ A \end{matrix}$	a proof of A
$\begin{matrix} A \\ \vdots \\ B \end{matrix}$	a proof of B from A

Chapter 2
Is There a Logic of Society?

Rohit Parikh

2.1 Introduction

Continuing the discussion above, we came to identify "valid proof" with "proof in first order logic," proved complete by Gödel. And we came to identify effective procedures with procedures which could be carried out (or somehow simulated) by Turing machines.

But reasoning precedes first order logic; in India and Greece, as well as in China, Logic has existed for several thousand years. And the notion of algorithm is implicit in so many things which happen in everyday life. We humans are tool-making creatures (as are chimps to a somewhat smaller extent) and both individual and social life are over-run with routines, from cooking recipes to elections.

So it would seem that here is another area, dealing with people rather than computers, crying out for development and study. Such a study seems to have been neglected until recently, but the work of [11, 14, 16, 19] has constituted a start.

But let us start first with logic.

2.2 Logic vs Rationality

When we come to the question of why we want an argument to be logically rigorous, one answer is that a logically correct argument leads from true premises to true conclusions. This criterion may be modified somewhat and we may require, as Adams [1] does, that an argument should lead from premisses true with high probability to conclusions also true with high probability. But the point stands.

But what is important in social contexts is not just that one should be logical but also that one should be rational. The two requirements are related, but of course

Rohit Parikh
City University of New York, New York, NY, USA, e-mail: rparikh@gc.cuny.edu

J. van Benthem et al. (eds.), *Proof, Computation and Agency*, Synthese Library 352,
DOI 10.1007/978-94-007-0080-2_2, © Springer Science+Business Media B.V. 2011

not identical. And the requirement of rationality enters into Grice's defense of the material conditional. The material conditional $A \supset B$ of two formulas A and B is defined to be equivalent to $\neg A \vee B$ or $\neg(A \wedge \neg B)$. The material conditional is truth functional in that the truth value of $A \supset B$ depends *only* on the truth values of A and of B. It is true if A is false, or if B is true. No connection between A and B is needed. But is $A \supset B$ the *same* as "If A then B"?

Dorothy Edgington [20] shows that *if* "If P then Q" is truth-functional, then it must coincide with the material conditional. But then she gives a nice argument against interpreting "if A then B" as $A \supset B$. For consider the following statement, X = *If God does not exist, then it is not the case that if I pray, my prayers will be answered.* Surely this seems right. Now there are two "if"s in X. If we formalize both the "if"s as material conditionals, we get $\neg G \supset \neg(P \supset A)$. Suppose now that I don't pray. Then P is false, and hence $P \supset A$ is true. Thus $\neg(P \supset A)$ is false. Thus if X is true, which we granted, $\neg G$ must be false, and hence G is true. I can prove the existence of God simply by not praying!

But surely the existence of God cannot be proved so simply and it would seem that "if A then B" cannot always be represented as $A \supset B$.
Can the material conditional be defended in spite of such examples? The best known defense of material conditionals is due to Paul Grice [6] who introduced his notion of *implicature* to this end.
Grice brings in rationality considerations in his celebrated defense of the material conditional. The material conditional $A \supset B$ is true if A is false or B is true. But then why do we resist that interpretation for *if A then B*?
Grice's answer is that it is a social convention that when making a statement one should supply the maximum relevant information. If I know $\neg A$ then it is *misleading* to say $A \supset B$ because the person who hears me will assume that I did not know $\neg A$; otherwise why not just say *that* rather than the longer formula $A \supset B$? Similarly, if I already know B, then it is *true* according to Grice, that "if A then B". But it is misleading to say that.

Coming back to our God example, suppose I say to someone, "Every time I pray to God, my prayers are answered." This is true if I do not pray. But the hearer assumes that this simple sentence "I don't pray" could not be what I intended to say because I have said something more complex instead. So he assumes that it was necessary to make the more complex statement, that I do pray, and that there is a connection between my praying and my prayers being answered. If I don't pray, then my statement, "Every time I pray to God, my prayers are answered" is true, but misleading.

Thus Grice proposes to defend the material conditional by saying that when it appears to differ in sense from "if A then B", that is because of the conventions of conversation.

Grice's analysis of conditionals, although brilliant, is not generally accepted because there are other problems (which we shall not discuss here) with his analysis. But the notion of *implicature* which he introduced has proved to be of great value. The implicature of a statement is something which is conveyed by the act of making that statement and goes beyond the mere truth of it. If a child says, "I am hungry," and I say, "Your mother will be back in a few minutes," I have implicated but not said that the mother will bring food. The child infers this from the fact that if the mother was not going to bring food, then my remark, even if true, is pointless.

This view of Grice has led to new developments in our understanding of conversation, that it does not consist merely of exchanging true assertions, but that it is an activity with a social purpose which is usually helpful (but sometimes not so). If a professor is asked for a letter of recommendation for a student and the professor says, "She always came regularly to class," and nothing else, then the professor has not *said* but he has implicated that there is nothing good about the student which is of relevance to the application.

We have seen how the logic which goes on within a social context is different from the logic of textbooks. But there is also a logic *about* social situations to which I shall now turn.

2.3 Plans

Suppose that I am going from my apartment to a hotel in Chicago. Then my schedule (or program) will consist of several steps. Taking a cab to Laguardia airport, checking in, boarding the plane, and then in Chicago, taking a cab to the hotel. Each step needs to be tested for correctness. For instance, if I do not have a picture ID, I cannot board the plane, and it is irrelevant that the other steps would have gone well.

The schedule I described is a simple *straight line* program. There could also be decision points like, *take a taxi if available, otherwise take a bus.* A plan with many such decision points would look like a tree rather than a line. But the entire schedule does have to be checked for correctness – something we do informally.
But other social procedures like organizing a conference or a wedding can be far more complex. We can invite aunt Betsy only if her ex-husband Eric has said he cannot come. Formal instruments can come in useful when the complexity surpasses some modest level. And we have not developed such instruments for *social* contexts. A logic is needed.

The paradigm of that logic comes (in part) from the logic of computer programs. A social procedure, to be useful, should serve some purpose, and it should be designed so that in normal circumstances the purpose is indeed fulfilled. The same holds for computer programs, and there is a well developed field devoted to proving that they work.

2.3.1 Proving Computer Programs Correct

Let us look quickly at an example of the use of Hoare logic for proving a program correct [8].

Consider the program α_g for computing the greatest common divisor $gcd(u,v)$ of two positive integers u,v. We want this program to have the property

$$\{u > 0, v > 0\}\ \alpha_g\ \{x = gcd(u,v)\}$$

This property has three components: the *pre-condition* $u > 0, v > 0$, the desired *post-condition* $\{x = gcd(u,v)\}$ and the program α_g itself. We want that if we start with two integers u,v which are both positive, then the program α_g when it ends will have set x to the gcd of u,v. One such program α_g, *Euclid's algorithm*, is given by

$$(x := u); (y := v); (\text{while } (x \neq y \text{ do} \\ (\text{if } x < y \text{ then } y := y - x \text{ else } x := x - y))).$$

The program sets x to u and y to v respectively, and then repeatedly subtracts the smaller of x, y from the larger until the two numbers become equal. If u, v are positive integers then this program terminates with $x = gcd(u,v)$, the greatest common divisor of u, v.

The reason is that after the initial $(x{:=}u);(y{:=}v)$, clearly $gcd(x,y) = gcd(u,v)$. Now it is easy to see that both the instructions $x := x - y$ and $y := y - x$ leave the gcd of x,y unchanged. Thus if B is $gcd(x,y) = gcd(u,v)$, and β is ($if\ x < y\ then\ y := y - x\ else\ x := x - y$), then $\{B\}\beta\{B\}$ holds. β *preserves* B.

Thus if the program α_g terminates, then by Hoare's rule for "*while*," $x \neq y$ will be false, i.e. $x = y$ will hold, and moreover B will hold. Since $gcd(x,x) = x$, we will now have $gcd(u,v) = gcd(x,y) = gcd(x,x) = x$.

Hoare's rules allow us to derive the properties of complex programs from those of simpler ones. Given program β let α ="while A do β". A,B are first order formulas. Then the Hoare rule says that if we know $\{B\}\beta\{B\}$ i.e. that β preserves the truth of B,[1] then we can conclude $\{B\}\alpha\{B \wedge \neg A\}$. That is that α will also preserve B and will moreover falsify A.

$$\frac{\{B\}\beta\{B\}}{\{B\}while\ A\ do\ \beta\{B \wedge \neg A\}}$$

In other words, provided that β preserves the truth of B and that B holds when α begins, then B will still hold when α ends and moreover, A will be false. This rule allows us to predict that when the gcd program terminates, $gcd(u,v) = gcd(x,y)$ will still hold and $x \neq y$ will be false, i.e. x will equal y.

A social algorithm may have a similar proof of correctness, which essentially amounts to showing that provided the steps are performed correctly, the result will also be correct. Since social algorithms consist of a series of steps organized in some logical sequence, logical methods similar to those used for computer programs will sometimes work. Reference [14] shows how formal methods can be used to prove the correctness of the Banach-Knaster algorithm for fairly dividing a cake.
However, this sort of logic works only for sequential computer programs. When it comes to people taking part in social algorithms, four other factors enter. They are

[1] Indeed it is sufficient if β, applied once, yields B under the precondition that both A and B hold.

- Communication and Knowledge
- Preferences and Incentives
- Co-ordination and Conflict
- Culture and Tradition

Of these four, the first enters already in computer science, i.e., in distributed computing. When *several* processors are co-operating on a task, then each needs to know what the others are doing so that actions can be co-ordinated [15]. Reference [17] provides a model of knowledge acquired in the course of activity and relies on a Kripke model based on histories to develop a theory of the knowledge of each agent, whether a human or a computer.

2.4 Communication and Knowledge

Here is a discussion from [13]. The following examples illustrate the type of situations we have in mind.

Example 1: Uma is a physician whose neighbour is ill. Uma does not know and has not been informed. Uma has no obligation (as yet) to treat the neighbour.

Example 2: Uma is a physician whose neighbour Sam is ill. The neighbour's daughter Ann comes to Uma's house and tells her. Now Uma does have an obligation to treat Sam, or perhaps call in an ambulance or a specialist.

The difference between Uma's responsibilities in examples 1 and 2 is that in the second one she has knowledge of a situation which requires action on her part. In the first case, none of us would expect her to address a problem whose existence she does not know of. Thus any decent social algorithm must allow for the provision of requisite knowledge. However, in the example 3 below, it is the agent's own responsibility to *acquire* the proper knowledge.

Example 3: Mary is a patient in St. Gibson's hospital. Mary is having a heart attack. The caveat which applied in case (1) does not apply here. The hospital cannot plead ignorance, but rather it has an obligation to *be aware* of Mary's condition at all times and to provide emergency treatment as appropriate.

In all the cases we mentioned above, the issue of an obligation arises. This obligation is circumstantial in the sense that in other circumstances, the obligation might not apply. Moreover, the circumstances may not be fully known. In such a situation, there may still be enough information about the circumstances to decide on the proper course of action. If Sam is ill, Uma needs to know that he is ill, and the nature of the illness, but not where Sam went to school.

To see that such matters can have serious political consequences, consider the following from the *New York Times*.

By CLAUDIO GATTI and JAD MOUAWAD

Published: May 8, 2007

> Chevron, the second-largest American oil company, is preparing to acknowledge that it should have known kickbacks were being paid to Saddam Hussein on oil it bought from Iraq as part of a defunct United Nations program, according to investigators.

> The admission is part of a settlement being negotiated with United States prosecutors and includes fines totaling $25 million to $30 million, according to the investigators, who declined to be identified because the settlement was not yet public.
> The penalty, which is still being negotiated, would be the largest so far in the United States in connection with investigations of companies involved in the oil-for-food scandal.

[13] contains a detailed discussion of such cases relying on the semantics from [17] which is used to calculate the obligations.

Such knowledge issues arise all the time in real life. Suppose a shy student is in your office and you wonder if it is time for your next appointment. If you look at your watch, then you will know the time, but the student will also realize that you *wanted* to know the time, and may, being shy, leave even though he need not.

But we turn now to the other two aspects.

2.5 Incentives and Preferences

This aspect is of course what decision theory is all about. In decision theory, an agent is faced with some uncertainty as to how things are, and needs to make a decision. Suppose for instance that a die is to be tossed and the agent has to choose between (A) getting $30 if a six shows up, and (B) getting $12 if a six does *not* show up. Now the probability of a six is of course 1/6 (assuming that the die is fair) and that of a non-six is 5/6. Thus the expected value of the first choice is $30/6 or $5. That of the second choice is $12(5/6) or $10. It seems that the second choice is better and the theory says that that is what the agent should choose.

This situation was clear enough, but when an agent needs to choose between two jobs or two men who want to marry her, then numbers in the form of value or probabilities in the form of reals in [0,1] may be hard to come by. Still, it can be argued that some of the theory which we applied with the die will continue to apply. Savage [22] following on ideas of Ramsey and de Finetti shows that if the choices made by an agent (both in simple situation and in probabilistic bets) satisfy certain axioms, then there exists a utility function for that agent, *and* a subjective probability over the outcome space, such that the agent's actions can be seen as seeking to maximize the expected utility of an action.

We saw how an agent who believed that the die was fair and whose utility was linearly proportional to the amount of money would choose option (B) above. If the agent chooses (A) instead, then we would conclude that either the agent's subjective probability for a six was more than 1/6, i.e., the agent thought the die was not fair, or perhaps the utility of $30 to the agent was more than 2.5 times the utility of $12. The latter could happen if the agent needed to get to her home town, and the busfare was $27. Then a 1/6 chance of getting $30 would be better than a 5/6 chance of getting $12. Yet another possibility is that the agent did not know probability theory [18].

Sen [24] points out that our reliance on utility theory is a bit naive, for people want many things which may not be comparable.

> I argue in favour of focusing on the capability to function, i.e., what a person can *do* or can *be* and argue against the more standard concentration on *opulence* (as in 'real income' estimates) or on *utility* (as in traditional 'welfare economic' formulations). Insofar as opulence and utility have roles (and they certainly do), these can be seen in terms of their indirect connections with well-being and advantage, in particular, (1) the *causal* importance of opulence, and (2) the *evidential* importance of utility (in its various forms, such as happiness, desire-fulfilment and choice)

Thus Sen emphasizes *abilities* rather than utilities. However, of course he did not have (I assume) access to the relevant material from computer science like analysis of algorithms, or logic of programs. Naturally he does not offer a detailed theory of abilities.

But such a detailed theory is crucial. For instance, when a bus line from town A to town B going through towns C,D,E is established, then the presence of the bus line increases the abilities of people living in these towns. Or alternately, when cellphones are invented and services established, then that too increases the abilities of people. Such things enter into the GNP, but only in an indirect way, via what people are willing to pay for a cellphone and not in terms of how it actually improves the abilities of people. This is why an expensive treatment for cancer or an expensive divorce turns out to have enormous (positive) implications for the GNP – not what we would intuitively expect.

Now decision theory is a theory of a *single* agent facing uncertainty.

2.6 Co-ordination and Conflict

Game theory extends this to a theory of an agent trying to maximize his utility in a situation where others are also trying to maximize theirs. Since the agent wants to maximize his own utility which depends on the actions of others, he needs a theory of what *they* want. The earliest major contributions here are due to von Neumann and Morgenstern, and to John Nash [10, 26].

A Nash equilibrium (for two players) is a pair of choices (a,b) by them which has the property that given that one agent is choosing a, the other's best bet is to choose b and vice versa.

Suppose for instance that a husband and wife want to go to a music concert and she prefers Mozart, whereas he prefers Stravinsky. Still each would rather go with the other than go alone.

Thus, putting the wife's location (or choice) first in the pairs below, their orderings are,
for the wife: $(M,M) > (S,S) > (M,S) > (S,M)$, and
for the husband, $(S,S) > (M,M) > (M,S) > (S,M)$.
Then there will be two Nash equilibria, one where they both go to Mozart and are both happy but the wife is happier. She gets her top choice and the husband gets his second choice. But if she *is* going to Mozart, then M is his best choice, For now he can only choose between (M,M) and (M,S) and the second possibility, (M,S) is worse for him. The other Nash equilibrium is where they both go to Stravinsky, and this time the husband is the happier one.

In so called zero sum games, a profit to one agent is a loss to the other, and each tries to outsmart the other. In such a case, pure strategy Nash equilibria like (M, M) do not usually exist, but mixed strategies do exist which are equilibria.
Suppose Jack and Ann are playing the game of matching pennies and Jack is the matcher and Ann is the mismatcher. Thus Jack gets both pennies if the pennies match and Ann gets them both if they do not. Then Jack wants to show heads if Ann does and tails if she shows tails. Ann wants the opposite and so each tries to guess what the other is going to do. There is no (pure) equilibrium here as there was in the Mozart-Stravinsky scenario.

The following story of either Indian or Iranian origin is very instructive of the game theoretic situation.

2.6.1 The Two Horsemen

Suppose we want to find out which of two horses is faster. This is easy, we race them against each other. The horse which reaches the goal first is the faster horse. And surely this method should also tell us which horse is *slower*, it is the other one. However, there is a complication which will be instructive.

Two horsemen are on a forest path chatting about something. A passerby M, the mischief maker, comes along and having plenty of time and a desire for amusement, suggests that they race against each other to a tree a short distance away and he will give a prize of \$100. However, there is an interesting twist. He will give the \$100 to the owner of the *slower* horse. Let us call the two horsemen Bill and Joe. Joe's horse can go at 35 miles/h, whereas Bill's horse can only go 30 miles/h. Since Bill has the slower horse, he should get the \$100.

The two horsemen start, but soon realize that there is a problem. Each one is trying to go slower than the other and it is obvious that the race is not going to finish. There is a broad smile on the canny passerby's face as he sees that he is having some amusement at no cost. Figure 2.1, below, explains the difficulty. Here Bill is the row player and Joe is the column player. Each horseman can make his horse go at any speed upto its maximum. But he has no reason to use the maximum. And in Fig. 2.1, the left columns are dominant (yield a better payoff) for Joe and the top rows are dominant for Bill. Thus they end up in the top left hand corner, with both horses "going" at 0 miles/h.

However, along comes another passerby, let us call her S, the problem solver, and the situation is explained to her. She turns out to have a clever solution. She advises the two men to switch horses. Now each man has an incentive to go fast, because by making his competitor's horse go faster, he is helping his own horse to win! Figure 2.2 shows how the dominant strategies have changed. Now Joe (playing row) is better off to the bottom, and Bill playing column is better off to the right – they are both urging the horse they are riding (their opponents' horse) as fast as the horse can go. Thus they end up in the bottom right corner of figure II. Joe's horse, ridden by Bill comes first and Bill gets the \$100 as he should.

	0	10	20	30	35
0	0, 0	100, 0	100, 0	100, 0	100, 0
10	0, 100	0, 0	100, 0	100, 0	100, 0
20	0, 100	0, 100	0, 0	100, 0	100, 0
30	0, 100	0, 100	0, 100	0, 0	100, 0

Fig. 2.1 Bill and Joe race their own horses.

	0	10	20	30	35
0	0, 0	0, 100	0, 100	0, 100	0, 100
10	100, 0	0, 0	0, 100	0, 100	0, 100
20	100, 0	100, 0	0, 0	0, 100	0, 100
30	100, 0	100, 0	100, 0	0, 0	0, 100

Fig. 2.2 Bill and Joe switch horses.

2.7 The Free-Rider Problem

When many agents participate in an activity where the benefit is to all, the so called *free rider problem* can arise. If all households are asked to use less water to avoid a shortage, then someone who surreptitiously used more water will benefit. Whether there is water shortage or not will depend on the actions of thousands of others, and his own action will not decide whether he has to endure a water shortage. On the other hand by taking long showers he will get more pleasure, and if the others save, he will *also* get the benefit of more water in the future.
Some readers may be familiar with the **Tragedy of the Commons**. In [7], Hardin introduces a hypothetical example of a pasture shared by local herders. The herders are assumed to wish to maximize their yield, and so will increase their herd size whenever possible. The utility of each additional animal has both a positive and negative component:

- **Positive:** the herder receives all of the proceeds from each additional animal
- **Negative:** the pasture is slightly degraded by each additional animal

Crucially, the division of these components is unequal: the individual herder gains all of the advantage, but the disadvantage is shared among all herders using the pasture. Consequently, for an individual herder weighing up these utilities, the rational course of action is to add an extra animal. And another, and another. However, since all herders reach the same conclusion, overgrazing and degradation of the pasture is its long-term fate.

2.7.1 Akbar and Birbal

There is a wonderful Indian story about the Mughal emperor Akbar[2] and his minister Birbal [2] about the way in which incentives (and knowledge) affect a social algorithm. Birbal had asserted to the emperor that all wise people think alike and had been challenged.

Then at Birbal's instance, all men in Agra, the capital, were ordered to come at night to the palace grounds, and pour one potful of milk into the pool on the palace grounds. The punishment for not doing so was severe, so one by one, all the residents came at night and poured a potful into the pool which was covered by a white sheet. When the sheet was removed in the morning, it turned out that the pool was entirely full of water! Birbal explained, "Your majesty, each man thought that if he, and he alone were to pour a potful of water into the pool, it would not make much difference, and no one would notice. So, since all wise men think alike, and of course all of your subjects are wise, they all did the same thing and the pool is full of water."

---

[2] Akbar was the grandfather of Shah Jehan who built the Taj Mahal in memory of his wife Mumtaz.

Akbar was a emperor of India during the second half of the 16th century, so this fable predates William Forster Lloyd's 1833 parable of the Tragedy of the Commons by some 300 years. However, Aristotle, who said, "That which is common to the greatest number has the least care bestowed upon it," predates Birbal by almost two thousand years!

There is a similar situation which we might call the *Tragedy of the Beer*. Suppose that ten people go out to eat dinner and there is a convention that the total bill is to be split equally. Many restaurants in fact encourage this convention by refusing to give itemized bills. Suppose that a beer costs $7 which is a bit high. But if you, as a diner, are thinking whether to have that extra beer, its cost to *you* will only be 70 cents (plus tax and tip). Again, the game theoretic situation is the same as before, the benefit is to the individual and the cost is to the whole group; an individual who acts irresponsibly will gain.

2.8 Culture and Tradition

Finally we come to culture, which is our shorthand for inherited algorithms, so that we are using the word *culture* in a fairly wide sense.

One interesting example of the influence of culture is the ultimatum game, what the theory predicts about it and what actually happens in practice. Suppose I offer a hundred rupees to Laxmi and Ram in the following way. First Laxmi decides who is to get how much of the hundred each of them will get. Then Ram has to decide if he accepts her division. If he does, then the money is allocated as Laxmi decided. If he rejects, neither gets anything.

Suppose Laxmi decides that she gets 90 and Ram gets 10. Now it is Ram's turn to decide. According to conventional game theory, 10 is better than nothing which is what he will get if he rejects. So he should accept and take the 10. In practice, it is quite unlikely that Ram will accept less than 30. Moreover, in some cultures, Laxmi may offer 40:60 with 40 for herself and 60 for Ram, and Ram may reject this division as being excessively generous.

So actual behavior differs from theoretical prediction and seems to follow some pre-existing cultural pattern. Essentially our view is that in game theoretic situations, a complex calculation needs to be made and the correctness of the calculation depends on the other party making a corresponding calculation.

Theoretical considerations cannot work out these scenarios and in practice people use ready made solutions which are part of culture or tradition.
Thus both Lewis and Schelling [9, 23] refer to the notion of saliency. If two people in New York city get separated, how will they find each other? One solution is to say that they will make their way to the giant clock in Grand Central station which is a *salient* point.

But what makes that a salient point? Surely that is part of the culture. If the two people are Chinese they may go to the Buddhist temple in Chinatown instead!

Group membership is crucial here. People tend to use algorithms which they have learned or inherited from their group, which may be a group defined by nationality, or ethnicity, or religion, or even by some subfield of some area. Thus the same kind of object, a subset of some space W will be called a *proposition* by a philosopher and an *event* by an economist. When a philosopher talks to an economist, they may use different conventions, and there could be confusion.
in [3] Michael Bacharach suggests that identifying oneself as a member of a group is crucial to making choices when co-ordinated action is needed.

> What produces team reasoning? Team reasoning — in any of its versions — is unmysterious; it is an algorithm, or routine, for arriving at decisions. As many things could get people to execute it, as could get them to do long division. It could be done as a game, or as a mathematical exercise, or because someone in control of a group of people found it in her interest that they should. (p. 135)

Recently, when attending a logic conference at IIT Mumbai, I met the young graduate student Tithi Bhatnagar who was doing research in what was important to human beings. She has found that regardless of age, gender or class, the most important thing for people is relationship. This aspect of human beings, that we value relationships above most everything, is crucial to understanding human beings who have, for far too long, been thought of primarily as selfish individuals looking out only for ourselves.
In sum, we have suggested that a study of society and its social algorithms requires logic (especially the logic of programs), but also requires attention to the other three legs of knowledge, incentive, and culture. With all these things in hand a good theory can be worked out.

Acknowledgements Thanks to Johan van Benthem, Jongjin Kim and Eric Pacuit for comments.

References

1. Adams E. Probability and the logic of conditionals. In P. Suppes and J. Hintikka, editors, *Aspects of Inductive Logic*, Pages 265–316. North Holland, Amsterdam, The Netherlands, 1968.
2. Sarin A. *Akbar and Birbal*, Penguin India, Panchseel Park, New Delhi, 2005.
3. Bacharach M. In N. Gold and R. Sugden, editors, *Beyond Individual Choice: Teams and Frames in Game Theory*. Princeton University Press, Princeton, NJ, 2006.
4. Benthem J., van Eijck J., and Kooi B. Common knowledge in update logics. In R. der Meyden, editor, *Theoretical Aspects of Rationality and Knowledge*, pages 253–261, Proceedings of the Tenth Conference, National University of Singapore, 2005.
5. Gödel K., Die Vollständigkeit der Axiome des logischen Funktionen-kalküls, *Monatshefte für Mathematik und Physik*, 37: 349–360, 1930.
6. Grice P. *Studies in the Way of Words*. Harvard University Press, Cambridge, MA, 1989.
7. Hardin G. The tragedy of the commons, *Science*, 162: 1243–1248, 1968.
8. Kozen D., and Tiuryn J. Logics of programs. In J. V. Leeuwen, editor, *Handbook of Theoretical Computer Science*, vol. B, pages 789–840. North Holland, Amsterdam, 1990.

9. Lewis D. *Convention: A Philosophical Study*. Harvard University Press, Cambridge, MA, 1969.
10. Nash J. Equilibrium points in N-person games, *Proc Natl Acad Sci USA*, 36: 48–49, 1950.
11. Pacuit E. Topics in Social Software: Information in Strategic Situations Doctoral dissertaion, City University of New York, 2005.
12. Pacuit E., and Parikh R. Social interaction, knowledge, and social software. In D. Goldin, S. Smolka, and P. Wegner, editors, *Interactive Computation: The New Paradigm*. Springer, Berlin, Germany, 2006.
13. Pacuit E., Parikh R., and Cogan E. The logic of knowledge based obligation, presented at Society of Exact Philosophy meeting in Maryland, and at DALT 2004, *Synthese*, 149: 311–341, 2006. Also in in Knowledge, Rationality & Action, 57–87, 2006.
14. Parikh R. The Logic of games and its applications, *Ann. Discrete Math.*, 24: 111–140, 1985.
15. Parikh R. Knowledge based computation (Extended abstract). In *Proc. AMAST-95 Montreal*, July 1995, Edited by Alagar and Nivat, Springer Lecture Notes in CS no. 936, pages 127–142.
16. Parikh R. Social software, *Synthese*, 132: 187–211, September 2002.
17. Parikh R., and Ramanujam R. A knowledge based semantics of messages, *J. Logic, Language and Information*, 12: 453–467, 2003.
18. Parikh R. WHAT do we know and what do WE know? *the proceedings of Theoretical Aspects of Rationality and Knowledge*, R. van der Meyden, editor, University of Singapore, June 2005.
19. Pauly M. Logic for Social Software, Ph.D. Thesis, University of Amsterdam. ILLC Dissertation Series 2001–10, ISBN: 90-6196-510-1.
20. Edgington D. 'Conditionals', http://plato.stanford.edu/entries/conditionals
21. Ramsey F.P. Truth and probability. In *The Foundations of Mathematics*, pages 156–198. Routledge and Kegan Paul, London, 1931.
22. Savage L.J. *The Foundations of Statistics*. Wiley, New York, NY, 1954.
23. Schelling T. *The Strategy of Confict*. Harvard University press, Cambridge, MA, 1960.
24. Sen A. *Commodities and Capabilities*. Elsevier Science, Amsterdam, 1985.
25. Turing A.M. On computable numbers, with an application to the Entscheidungsproblem, *Proc. London Maths. Soc.*, ser. 2, 42: 230–265, 1936–1937.
26. von Neumann J., and Morgenstern O. *Theory of Games and Economic Behavior*. Princeton University Press, Princeton, NJ, 1944.

Part II
Logic and Mathematics

Chapter 3
What Is a Proof?

John N. Crossley§

3.1 Prelude

> The term "form" [of a logical argument] ... is not easy to define but it is easy to illustrate. (Prior [56], p. 1.)

This paper is principally concerned with the logic of predicate calculus and its development, principally in the West. I leave it to others to discuss the development of other methods of proof, such as are found in, for example, Indian logic or modal logics. Moreover, I am much more concerned with the practicalities of proofs rather than with the philosophy of logic.

The paper starts off with quite concrete and simple systems. It then moves on to somewhat more complicated considerations that mirror the development of proofs in formal logics. The changing rôle of logics, so that in the twentieth century the distinction between proof systems and computational systems has become blurred, is brought into sharp focus. Finally the place of formal proofs and the changing natures of them are discussed.

3.2 Introduction

"What is a formal system?"

I was asked this as the very first question in my *viva voce* examination for my doctorate in 1963. It was, I was later told, intended as a friendly opening. I did

John N. Crossley
School of Information Technology, Monash University, Clayton, VIC 3800, Australia, e-mail: John.Crossley@infotech.monash.edu.au

§This paper has had a long gestation. A version was presented at the Australasian Association for Logic meeting in Christchurch, New Zealand in 1989, but was never published. The present version reflects changes in logic since that time.

J. van Benthem et al. (eds.), *Proof, Computation and Agency*, Synthese Library 352,
DOI 10.1007/978-94-007-0080-2_3,

not know how to answer the question but, after coaxing, I eventually trotted out an answer referring to an alphabet, predicate letters, logical connectives, etc.

I considered, even then, that this answer was not satisfactory. It gave an *example* of a formal system rather than a definition.[1] In a recent paper [17], I have made tentative steps towards a definition of what a (formal) logic is. In that paper I said that in addition to a syntax and semantics, one needed an inference mechanism. Even in 1989 it seemed clear to me that we logicians were not at all clear as to what a rule (of inference) was. Since rules of inference are inextricably involved with formal proofs, it follows that the idea of a proof is also not clear. I shall therefore begin by looking at history in Section 3.3. I begin with the ideas of axioms in Section 3.3.1, and move on to rules of inference. I stress the importance of the purpose of a proof in a logic, and its dependence on the human situation. After that, in Sections 3.3.2 and 3.3.3, I trace the development of formal proofs through Leibniz, Boole and Frege to Russell and Hilbert.

Next, in Section 3.4, I consider some different systems of logics, which provide different styles of proofs, and discuss some relations between them. This section also treats the blurring of the distinction between proof and computation, which is a subject that has intrigued me for a very long time. Things take a more practical turn in Sectionapplied. First comes the question of constraints on proofs, in particular those imposed by the finiteness of our lives. Penultimately, in Section 3.6 I look rather more at mathematicians' proofs, following Devlin in his differentiation between right- and left-wing proofs. I also raise some questions about how we know that actually have a proof, and the lack of formal certification for this. Finally I present a brief summing up of my argument in Section 3.7.

3.3 A Little History

3.3.1 Aristotle and Euclid

Mathematicians of the 20th century, especially the first half, have been particularly bad at explaining *why* they are doing things. So it is worthwhile stating explicitly that in my view logic is the study of the structure of arguments; that it derives from the study of human discourse; and that, to be effective, it must relate back to human activity. I therefore criticize those mathematicians who seize on an abstract structure and explore it for all it is worth, without caring about its connexions with anything else. To quote von Bertalanffy [9], p. 99 (from a slightly different context): "The danger is to consider too early the theoretical model as being closed and definitive."

So let us trace some developments in the history of proofs. It is also appropriate to bring in anthropological and linguistic considerations alongside the more

[1] Likewise Curry, in [25], pp. 14–15, does not give a definition. He says: The definition does not tell us what a formal system is, in the philosophical sense, but describes the nature of the conventions made in setting one up.

philosophical and mathematical ones. To quote von Bertalanffy again [9], p. 237: "Conceptualization is culture-bound because it depends on the symbolic systems we apply."

Logic, as discussed here and in [17], is an applied discipline.[2] There is a certain (varying) subject matter usually presented in linguistic form, whose structure we wish to model mathematically, or at lest abstractly. So my focus is more on symbolic logic and its proofs than on philosophical logic, but not exclusively so.

The earliest person I know of who produced proofs is Thales of Miletus (640–546 BC). According to Gow, [37], pp. 138–139, Thales is responsible for five theorems including "The circle is bisected by its diameter." The obvious way to prove this (if one thinks it needs a proof!)is to fold the circle along the diameter. Then, because every point on the circle is the same distance from the centre, the two halves fit exactly.[3]

Here there is no hint of a formal system until one starts to *define* a circle, for example as Euclid did, by fixing a point and saying that all points on the circle are an equal distance from the centre (Euclid I.17, see [38], vol. I, p. 131).

Another theorem of Thales was that if the base angles of a triangle are equal, then the sides are also equal. The natural method of proof[4] is again to fold the triangle over. Euclid I. 26 does not do this. He argues verbally, not visually, about lengths. Indeed, if you look carefully at the proof you will see that he makes no distinction between right and left, no reference to moving the triangle – a two-dimensional figure – through the third dimension. Euclid's argument considers only two dimensions. the form of the argument is quite precise and logical. The interpretation of the formal terms is up to the reader. And a reader who lived in a two-dimensional world would have no difficulty with Euclid's proof.

Exercise 3.1 Try and explain to someone who has no point of reference in common with you, what is left and right. Do not cheat by assuming your interlocutor has the same sort of body as yourself.

Here we see the importance of the distinction between syntax and semantics (cf. also [17]).

Aristotle, who remains of major importance in logic, introduces us to the natural history of logic in his *Prior Analytics*.

> We must first state the subject of our enquiry and the faculty to which it belongs; its subject is demonstration and the faculty that carries it out demonstrative science. ([5], 24a, p. 65.)

Aristotle does not investigate systems (plural), he looks at *the* system. His study of logic is restricted to classifying various kinds of propositions. The ground rules are laid down:

---

[2] Many in the West may dispute this; few who know anything at all about Navya-Nyāya would question it.

[3] Euclid did not give a proof of this result. However, Proclus (ca. 410–485) [57] appears to suggest it was proved by superposition.

[4] That is to say "natural" to one trained in a Western Euclidean tradition.

> Yet every sentence is not a proposition; only such are propositions as have in them either truth or falsity. ... Let us therefore dismiss all other types of sentences but the proposition. ([3], 17a, p. 42.)

> A premiss then is a sentence affirming or denying one thing of another. ([5], 24a, p. 65.)

In view of these remarks we are committed to a two-valued logic. This commitment has (almost completely) dominated Western logic ever since. It is only in the last few decades that there has been much exploration of other logics (cf. [21]).

With Aristotle we start with premises and infer a conclusion. Notice that Aristotle does not begin with axioms; each argument has its own beginning: the premises ([5], 65b, p. 95). Euclid and his successors, on the other hand, believed that their axioms were true. This explains why the advent of non-Euclidean geometry, where Euclid's fifth postulate is replaced by another that contradicts it, was so striking (cf. [10, 47]).

Logicians tend to look at axioms as merely selected statements and require nothing more than that they be consistent, i.e. do not permit the deduction of a contradiction. Mathematicians select axioms that they believe are true. Philosophers vary in their practice. John Lucas [48] claims that all that is necessary is for the interlocutors to agree on the premises. Imre Lakatos [44] beautifully demonstrates how axioms (and definitions) in the case of the proof of the formula $V + F - E = 2$ for polyhedra, that have seemed clear and definite at one point in time, may subsequently need further refinement, especially in the light of technical developments or new insights. In Indian logic the premises seem usually to be assumed to be true, and to reflect the world. Thus Matilal [49], p. 75, says: "There must be observed data and other principles which could be used as premises of the argument to follow."

It is an interesting exercise to go through the *Posterior Analytics* check that "necessary premises" and similar phrases can be replaced by "unquestioned, but assume true, premises" without disturbing Aristotle's analysis of arguments.[5] Aristotle is characterizing arguments in general, rather than in any particular subject matter.

Finally, of course, we have the formal descriptions of the figures of the syllogism ([4], 75b, pp. 121–122). When, however, we look for the justification of the figure we find that Aristotle is again doing natural history. He observes that some figures work, some do not. The "proofs" he adduces can be reduced to the study of the obtaining of contradictions. He simply asserts, he does not argue for, the validity of his arguments. Thus, for example, in *On interpretation* ([3], 17b, p. 44) we read: 'We *see* that in a pair of this sort both propositions cannot be true.' (Present author's emphasis.)

When we turn to Euclid, who was born about the same time as Aristotle (c. 320 BC) we find further refinements of Aristotle's approach, albeit in the restricted context of geometry and arithmetic.

Euclid's main contribution, a far as the present paper is concerned, is his delineation of axioms. He has three basic pronouncements: definitions, postulates and common notions.

---

[5] Thus Aristotle is detaching the logical arguments from the empirical data in a way that is not done in Navya-Nyāya logic.

Although his idea of a definition does not meet present-day standards, the aim is to define some object or property in terms of simpler, or previously known, objects and properties. Sometimes definitions and theorems get conflated, as in Euclid I.7: A diameter of the circle is any straight line drawn through the centre and terminate in both directions by the circumference of the circle, *and such a straight line also bisects the circle*. The emphasis is mine, and the italicized clause seems to present a theorem. The common notions are often what we would today call "axioms".

Finally the postulates are axioms specific to geometry. Some of these are, in effect, axioms that assert the existence of geometrical entities. For example "To describe a circle with any centre and distance." (Postulate I.3 [38] I, p. 154.) In modern terminology we would say: there *exists* a circle with given centre and radius. However, in Euclid's time it was necessary, having given the construction, to prove the existence of the figure (see [38], vol. I, pp. 121ff.).

It is hard for the modern reader to distinguish the three types: definitions, postulates and common notions. We tend to think in terms of definitions and axioms. However, since Euclid was concerned to *construct* geometrical figures, he needed to have axioms asserting that certain figures could indeed be constructed.

We do not have any precise classification of his axioms, but there are essentially two classes: those asserting general properties and existential axioms. In the modern setting these roughly correspond to logical and non-logical axioms.

Just how advanced Euclid's work was is suggested by the comparative ease with which Hilbert put Euclid's geometry into modern logical language (see [41]).

3.3.2 Leibniz to Boole

The formalization of mathematics in the sense of the development of a symbolic language using letters for elements, functions, and other objects, had begun in the sixteenth century with, for example, Maurolico [50], but Leibniz took it considerably further. He started his various attempts at formalization quite early in *De arte combinatoria*, published in 1666, when he was 20 years old. There he used numbers for concepts and tried to establish an arithmetic of concepts. However, his most interesting pieces – looked at from the point of view of formalization– are his *Non inelegans specimen* (see [46] or [16], pp. 239–243). In *Non inelegans* he presents what is generally regarded as a precursor of Boolean algebra. He represents intersections of classes and also uses "+" for exclusive "or". He had, in essence all the machinery needed for Boolean algebra. Why did he not achieve what Boole did? The answer is one that is quite usual for Leibniz: "It s that amongst lots of attempts and various projects he did not know how to discern what was best and then to adopt it and develop it systematically." ([15], present author's translation.) The most important thing that Leibniz did sort out was the idea of making a formal language that could be manipulated in a mechanical way ([30], vol. 7, p. 200, or [16]):

> ...when controversies arise, there will be no more dispute between two philosophers, than between two [human] computers. For it will suffice to take their pens in their hands and sit down at abaci, and say to each other (if it pleases his summoned friend): 'Let us calculate.'

So Leibniz's aim was to get a mechanical presentation of arguments or proofs. he made various attempts. When he was fifty he wrote a short letter to Professor Vagetius at Giessen ([29], vol. III, pp. 338–339) in which he laid down the basic construction for a formal language for mathematics/ He starts off with simple terms (*termini simplices*), which are almost the same as the formal terms we use today. Then come atomic statements (*enunciationes*). The latter include equalities, inequalities and proportions, for he is dealing with arithmetic. Then come rule *consequentiae*, which involve operations such as addition and composition. Finally the method *methodus* is the way of getting theorems or finding solutions.

It could be argued that the *consequentiae* are not quite the same as formal rules of inference, but it could equally well be argued that they are formal rules for manipulating the operations (such as addition). They therefore correspond to axioms for such operations, and include, for example, axioms of commutativity and associativity for addition and multiplication.

Thus we have terms, atomic formulae, axioms, rules of inference and formal theorems all foreshadowed.

3.3.3 Frege and Hilbert

Between Leibniz and Boole I have been unable to find any significant contribution to the notion of proof system. Boole completed the Leibnizian task of giving a mathematical formalization of a calculus of classes. However it appears that Boole was unaware of Leibniz's work.[6] Slightly later Frege developed what has become a standard node of treating formal languages: first order logic or predicate calculus. However Frege's system was inconsistent. This arises because it was, in fact, a higher order logic[7] that allowed in Russell's paradox, indeed Russell pointed this out to Frege as is well known (see e.g. [19]).

Frege's approach was axiomatic and this idea spilt over into other mathematical disciplines, for example, group theory (see, e.g., [11]). The aim of the mathematicians was to give ultimate descriptions, definitive accounts. It appears that Hilbert felt this had been achieved for formal systems. However, there was a fly in the ointment. Formalization did not mean the automatic avoidance of inconsistency. Hilbert [39], p. 130, says: "Rather, from the very beginning a major goal of the investigations into the notion of umber should be to avoid contradictions [such as that obtained by Russell in Frege's work] and to clarify these paradoxes."

---

[6] There is no concrete evidence that Boole knew of Leibniz's work, though this is not impossible (see Dipert [28]).

[7] Nevertheless, it is often referred to as a second order-logic in many publications.

Hilbert's way of dealing with the problem, that is, the formalist approach, was to attempt to provide a consistency proof to go alongside, or perhaps to cover, the formal logic. This immediately brings up questions of meaning. There are several difficulties here that I do not wish to dwell on. Brouwer and others have questioned the value of this formalist approach to mathematics. Formalism can be charged with distorting the meaning of mathematical assertions, since it was previously assumed that mathematical assertions were assertions about the world (even if it was an ideal or Platonic world) and now they seem to be only about, for example, marks on paper. Hilbert and the formalists' approach leaves the question of the interpretation of all these marks on paper unanswered.[8] What it did do, and this is a point of interest for the present paper, was to make mathematical proofs, albeit formalized ones, into objects of mathematical interest, which are open to investigation by the methods of mathematics.

In his second lecture on the foundations of mathematics ([40], p. 64), Hilbert says: "With this new [formalist] way of providing foundations for mathematics, which we may appropriately call a proof theory, I pursue a significant goal, for I should like to eliminate once and for all the questions regarding the foundations of mathematics, ..., by turning every mathematical proposition into a formula that can be concretely exhibited and strictly derived." Hilbert thereby opened the way not only to a study of proofs but to a study of systems that embrace proofs: proof systems.

Hilbert's method of producing a consistency proof, a method still commonly used today, is to transform a given proof (for example of $0 = 1$ in arithmetic), into another "simpler" proof of the same conclusion. The way this is done is by specifying methods for rewriting (parts of) the proof so as to "simplify" it. In addition to this we need also to prove that the process of simplification does actually come to a stop after a finite number of steps. If it does, we say that we have a *normal form* for the proof. How this is done depends on what kind of proof system we have.

3.4 Differing Proof Systems

We take the notions of "formula", etc. as given. Hilbert-style systems then comprise a set of axioms together with a couple of rules of inference of which the most important is *modus ponens*. (The other is generalization: from $A(x)$ infer $\forall x A(x)$.)

Natural deduction systems were introduced by Gerhard Gentzen (see [61]). These are aimed at mimicking the informal arguments that we use. For example, the rule for $\wedge$-introduction (and-introduction) is:

$$\frac{\Delta \vdash A \quad \Delta' \vdash B}{\Delta, \Delta' \vdash (A \wedge B)} \ (\wedge\text{-I})$$

[8] For further discussion of this see my earlier paper [17]).

where Δ and Δ' are sets of formulae. In addition to the introduction rules for the various connectives are the elimination rules. The one for $\wedge$-elimination on the left is:

$$\frac{\Delta \vdash (A_1 \wedge A_2)}{\Delta \vdash A_1} \; (\wedge\text{-E}_1)$$

Normalization for proofs in such a natural deduction formalization of logic proceeds by eliminating occurrences of an introduction rule that is immediately followed by an elimination rule for the same connective. *Strong normalization*, that is to say, normalization where it does not matter in what order the reductions take place, is provable for the predicate calculus and also for arithmetic (see e.g. Girard's book [32]).

Besides the natural deduction formalization Gentzen also introduced various *Sequent calculi*. These have rules similar to natural deduction but they allow the introduction of connectives at both sides of the provability symbol $\vdash$. In this case one does not need elimination rules.[9]

There is one major problem about proving strong normalization. Kurt Gödel showed, in his [35] of 1931, that there is no finite system of axioms that will give you all (and only) the true statements of arithmetic: his first incompleteness theorem.[10] The second incompleteness theorem showed that, if arithmetic is consistent, then one cannot prove that consistency within arithmetic, something further is required. Gentzen's proof involved using transfinite induction up to the ordinal ε_0. (We start counting $1, 2, 3, \ldots$ for finite ordinals. Cantor introduced infinite (ordinal) numbers to be able to count further, and he counted

$$1, 2, 3, \ldots, \omega, \omega + 1, \omega + 2, \ldots.$$

This process could be carried much further to $\omega \times \omega$ or ω^{ω} or even to $\omega^{\omega^{\omega^{\omega^{\cdot^{\cdot^{\cdot}}}}}}$ which is the infinite ordinal ε_0.) Thus a new proof-technique has been introduced.

Note that in the normalization process one does not deal with actual proofs but with schemes. We operate on proofs to get other proofs. In this way I believe we are getting closer to ordinary mathematical practice, a point to which I shall return later (see Section 3.6).

There is another way of formalizing logic. This is to use the lambda calculus of Church [7, 14, 33]. This calculus arose because Church wanted a non-set-theoretic version of mathematics and he therefore based the calculus on the notions of function and argument. (Shortly afterwards it was shown by Turing [62] that the functions that could be computed in the lambda calculus were exactly the (partial) recursive functions. See [20] for a survey.)

---

[9] Roughly speaking introduction of a connective on the right of the $\vdash$ corresponds to introduction in natural deduction, and introduction of a connective on the left to elimination.

[10] In fact he even showed that there is no finite complete system of axiom schemes for formal arithmetic.

Definition 3.1 (Lambda terms) The alphabet comprises variables $x_1, y_1, \ldots$, together with λ and ".", and the brackets (and).

The *lambda terms*, Λ, are formed as follows (in Panini or Backus-Naur notation):

$$T = x|\lambda x.T|(T_1 T_2)$$

The lambda calculus has two major constructions: *abstraction* and *application*. Consider the following:

What is the function denoted by x^y? We have several choices: as a function of two variables, as a function of x only with y held constant and as a function of y only with x held constant. These are usually denoted by

$$\lambda x \lambda y.x^y \text{(often written as } \lambda xy.x^y\text{)}, \ \lambda x.x^y \text{ and } \lambda y.x^y.$$

This is *abstraction.*

Application is written in a familiar way: thus $(T_1 T_2)$ denotes the application of the lambda term T_1 to the lambda term T_2. In particular fa is the application of the lambda term f to the lambda term a. (We omit brackets where there is no ambiguity.) These notations have the obvious interpretations.

In ordinary mathematics if we apply the function $\lambda x.f$ to a then we get $f[a/x]$, which is read "f with a for x". In the lambda calculus however this is *not* the same as the (application) term $(\lambda x.f)a$, i.e. $\lambda x.f$ applied to a. That is to say they are *syntactically* different. We therefore have to introduce the notion of *β-reduction*[11]

$$(\lambda x.f)a \rhd f[a/x]$$

(Here $\rhd$ is read "reduces to".)

Some time after Church's work, Curry [24, 25] and Howard [42] observed that ordinary propositional logic can also be viewed as a functional (i.e. a programming) language by making them into a lambda-calculus that has types. Thus programs are contained, in a certain sense, in proofs in mathematical logic. The underlying reason (in the present author's view) is because of the formal, that is to say, purely syntactic, similarities between logical rules and those of the lambda calculus.

The special kind of typed lambda calculus involves taking formulae of logic as the types. Now this is a strange idea to accept but it is easier to work with it if you just think of a type (formula) as the set of proofs of that formula. Instead, therefore, of variables, we use typed variables of the form $a : A$.

The rule of *modus ponens* then becomes:

$$\frac{a : A \quad g : (A \to B)}{(ga) : B} \tag{3.1}$$

where we have changed f to g to avoid confusion in what follows.

[11] α-reduction refers to the simple renaming of one variable by another (without clashes).

If we had a proof of B from A then we would get an expression $\lambda x : A.f : B$ by the rule of ($\rightarrow$-I) which has type $(A \rightarrow B)$. If the g in the expression (3.1) is actually of the form $(\lambda x : A.f : B) : (A \rightarrow B)$, then we get

$$\frac{a : A \qquad (\lambda x : a.f : B) : (A \rightarrow B)}{((\lambda x : A.f : B) : (A \rightarrow B))a : A) : B}$$

which is somewhat hard to read. However the bottom line has the formula B as its type, and the expression reduces to

$$f : B[a : A/x : A] \tag{3.2}$$

where the substitution of $a : A$ for $x : A$ takes place throughout the term $f : B$.

If we translate this back into proofs it means that the corresponding proofs look as follows. On the one hand we have the complicated proof:

$$\dfrac{A \qquad \dfrac{\begin{matrix}[A]\\ \vdots \\ B\end{matrix}}{(A \rightarrow B)}}{B} \tag{3.3}$$

and on the other hand, by putting the proof of A from the left on top of the proof of B, and *not* introducing the $\rightarrow$, we no longer need the hypothesis $[A]$ in the proof on the right in order to get a proof of B.

That is to say, we *reduce* the proof in (3.3) to a simple proof of B of the form

$$\begin{matrix}A\\ \vdots \\ B\end{matrix}$$

This corresponds in the lambda calculus to the reduction[12] that resulted in (3.2). So we have a direct correspondence between proofs and terms of our typed lambda calculus. This is called the *Curry-Howard correspondence*.[13] (For more details see e.g. [18] and Howard's original [42]).

Now Church's work was in a tradition initiated by Schönfinkel [59] and developed by Curry and Feys (see e.g. [25]).

Two special lambda terms interest us here. We denote the lambda term $\lambda x \lambda y.x$ by k and $\lambda x \lambda y \lambda z.(xz).yz)$ by S. Thus

$$\begin{aligned} Kxy &= x \\ Sxyz &= (xz)(yz). \end{aligned}$$

[12] This process of reduction is also called *cut elimination*.

[13] Some people use the term *isomorphism* but there are technical difficulties involved in making the correspondence one to one, so I prefer the weaker terminology.

The *combinators* K, S, as they are called, were first introduced by Schönfinkel [59]. His aim was simply to be able to represent all logical formulae (of predicate calculus) by means of one connective. At the time he wrote this connective as $|^x$ writing, for example, $A(x)|^x B(x)$. $|^x$ is a generalization of the Sheffer stroke and the above formula is equivalent to$\forall x(\neg A(x) \wedge \neg B(x)$. However, Schönfinkel needed to get any finite number of variables, not just one, so he reduced functions of several arguments to functions of one argument.[14]

He showed that by using the combinators K and S one gets a *combinatorially complete* algebra (see [8], p. 100), i.e. all functions that are combinations explicitly definable by means of application and substitution have a name built up out of S and K. Now all the terms of the lambda calculus can also be represented by combinations of the combinators K and S, since we can inductively define

$$\begin{aligned}\lambda x.x &= SKK\\ \lambda x.y &= K \text{ if } y \text{ is a variable different from } x\\ \lambda x.(uv) &= S(\lambda x.u)(\lambda x.v)\end{aligned}$$

(Note that SKK is the identity, usually denoted by I, since $SKKx = (Kx)(Kx) = x$.) Hence the lambda calculus and the theory of the combinators S and K, otherwise known as *combinatory logic* are equivalent. It turns out that, at times, it is more convenient to deal with one presentation than the other.

These combinators have very close connexions with logical calculi. We take such a lambda or combinatorial calculus with formulae as types. In usual mathematics we think of $\lambda x : \alpha.t : \beta$ as a function $f : \alpha \to \beta$. In this way, since $Kx : \alpha y : \beta = x : \alpha$, $Kx : \alpha$ maps $\beta \to \alpha$ and K maps α into $\beta \to \alpha$: the set of all functions from β into α. We may write this as $K : \alpha \to (\beta \to \alpha)$. A similar calculation shows that from $S(x : \alpha)(y : \beta)(z : \gamma)$ we get

$$S : (\gamma \to (\beta \to \alpha)) \to ((\gamma \to \beta) \to (\gamma \to \alpha)).$$

The two expressions in Greek letters and arrows are, of course, familiar as the first two axioms in a Hilbert-style presentation of propositional calculus (see e.g. Mendelson [52]).

Proceeding in this way we are able to replace the proof system by an equivalent computation system. The crucial point needed here is that the semantics should be correct.[15]

When we look more closely we see that in general we have been doing metamathematics when dealing with the various logical systems: we have usually not dealt with the formulae themselves but with place-holders for the formulae. Consider the simple rule of *modus ponens* once more. We do not usually think of spe-

---

[14] This process is nowadays known as "Currying" but should perhaps better be called "Schönfinkelling".

[15] A nice presentation of this is in the work of Lauchli [45], where he considers types as non-empty sets and "*A* is provable" is translatable as "the type *A* is non-empty".

cific formulae A and $A \to B$, we simply think of one formula, and then the same formula followed by an implication sign and a second formula.

This technique was often used in the literature. A nice example is in Schütte's treatment of positive and negative parts of formulae in his [60]. He uses "Formelvariablen" ("formula variables") as the place-holders. Another example is in Parikh [55] where they are essential in his analysis of proofs.

It is worth pointing out, however, that there is no "nice" algorithm for going between these systems. This is even true between the various kinds of system Gentzen devised. For example, although a proof in natural deduction can be transformed in a natural way into one in the two-sided Gentzen systems, nevertheless there is no similar natural way of going back. Of course there is an algorithm, since the two systems proved equivalent theorems. This algorithm, however, simply says: "Look for the first proof that proves the same result"!

There is also a positive side to these different ways of formalizing. In particular the lambda calculus teaches us to look at the same system in different ways at the same time since, in the typed form, it is both a proof system and a computation system.

3.5 Applied Logic

The Curry-Howard correspondence has also been important in facilitating the study of other logical systems. At this point I feel that the study of applied logic has really begun. "Applied" in the sense of finding out the logic of actual systems other than the logic of humans. Here, of course, it is very clear that if we find such logics then the justification of the logic is a purely inductive one:[16] it models the system accurately. Such logics are, for example, the logics of certain kinds of computers. This immediately leads us to the question of resources.

It is well known that some quite simply described functions take impractically long to compute. First define the auxiliary function $\alpha(a,0) = 0, \alpha(a,1) = 1, \alpha(a,n) = a$ for $n > 1$. Now the Ackermann function, ϕ, is defined by

$$\begin{aligned} \phi(a,b,0) &= a+b \\ \phi(a,0,n+1) &= \alpha(a,n) \\ \phi(a,b+1,n+1) &= \phi(a,\phi(a,b,n+1),n) \end{aligned}$$

and this is a function which is much more time consuming to compute than exponentiation. However, even exponentiation to base 2 is generally regarded as impractical to compute. The values rapidly become astronomical. The limit of functions which are considered (theoretically) feasible to compute are the *polynomial-time*

[16] "Inductive" is used here in the sense of scientific induction, as opposed to logical *de*duction.

computable functions, that is to say, computations that can be completed in a time bounded by a polynomial (of some fixed degree) in the length of the data.[17]

The functions that are representable in arithmetic(see [52]) are exactly the recursive functions. Parikh [54] looked at restricting the size of proofs in a standard logic. This work was subsequently developed into an elegant calculus in which the representable functions were exactly the polynomial-time computable functions by Sam Buss [12]. (See also [13] for the connexion between feasibility and the lengths of proofs.) The major difference between Buss's system of arithmetic and the familiar Dedekind-Peano one is a simple limitation on the form of the induction axiom. The induction schema becomes:

$$(A(0) \wedge \forall x(A(\lfloor x/2 \rfloor) \rightarrow A(x))) \rightarrow \forall x A(x)$$

where $\lfloor x/2 \rfloor$ is the last integer not exceeding $x/2$, i.e. x in binary notation with the last digit deleted [12], p. 27.

Other directions have been explored looking at different machines and different ways of computation (see e.g. [22, 23, 34, 53], the last of these being based on Girard's linear logic [31]). In [23] we attacked the question of how difficult it is to compute (or check) even atomic formulae. Little attention seems to have been paid to this question and it perhaps worth expanding upon. Suppose one has developed a theory with certain predicates, for example, for addition and multiplication, and then, in the light of further research one wishes to work instead with simpler predicates such as the successor predicate.[18] Then the atomic predicates become "split": they are replaced by longer formulae.

We may therefore think of atomic predicates that can, despite, the name, be split at any time. We can also look at the time required to read an atomic formula, or even a variable. In the context of a machine it does make sense to think that reading a variable $x\_1000000$ takes longer than to read x_1; and how long does it take to read "next x"? It is in fact possible to press this kind of analysis even further, but the only person I know who has done so is Yesenin-Volpin. In a lecture that I heard in SUNY, Buffalo, NY, in 1972, he was even looking at the question of recognizing that two occurrences of a symbol were indeed occurrence of the same symbol. (See also [64]).

Hilbert seems to have thought that there was a final description of a formal system. Yesenin-Volpin's approach questions this.

---

[17] This characterization of "feasible" seems too weak if one thinks in the terms of a practical engineer. For example, using the finite element method (see e.g. [26]) for a calculation that requires time proportional to the square of the number of data points, then if one increases the number of points considered by a factor of ten, the whole calculation takes a hundred times longer. Engineers therefore tend to restrict their calculations to problems that take less than quadratic time.

[18] The same argument, of course, applies to functions just as to predicates.

3.6 Comprehending and Constructing Roofs

Now let us turn to the opposite end of the scale: large proofs. We have already noted the question of which functions are feasible to compute. We now turn to the question of what proofs are, in a somewhat different sense, feasible. Parikh in his [55] discussed the question, brought up by Yesenin-Volpin [64] about having an upper bound for the length of human activities, in particular the construction of proofs. Work by Appel and Haken [1, 2] in the late 1970s, which reduced the proof of the four-colour theorem, after a very long conventional proof, to a billion cases which, clearly, could not be done by hand in one lifetime but were done by computer.[19] (In further development the number of cases was very significantly reduced but that does not affect our argument.) Perhaps inevitably, the use of the computer immediately made the status of the proof a subject for debate.

Two other major results have also caused debate about the status of very long proofs. One is the classification of finite simple groups. Even though the classification was claimed to be complete in the 1950s apart from a small number of so-called "sporadic groups", the very length of the proof, amounting to hundreds of printed pages, is still debated (see [6]). In this case no computer was required. A similar case is cited by Devlin [27] (and discussed at many places on the web and in a symposium "Paradise Lost? The changing nature of mathematical proof" at the 2006 AAAS Annual Meeting), namely Hales's claimed solution of Kepler's Sphere Packing Conjecture. These are not questions about the reliability of computers; they are questions about human comprehension and certification.

We have already mentioned the shift from mathematics to metamathematics in Section 3.4 above. How do we check a proof? Devlin [27] roughly distinguishes between what he calls right- and left-wing proofs.

> The right wing ('right-or-wrong', 'rule-of-law') definition is that a proof is a logically correct argument that establishes the truth of a given statement. The left wing answer (fuzzy, democratic, and human centered) is that a proof is an argument that convinces a typical mathematician of the truth of a given statement.

Most logicians aspire to right-wing proofs. Mathematicians on the other hand are usually content with left-wing proofs. Training as a mathematician shows one how to understand or check a proof by breaking it up into pieces. These pieces are then regarded as wholes in themselves and not further analyzed once they are understood. If there is a failure to understand then one of these pieces may be analyzed further. The process will continue until understanding is achieved or the proof rejected. (See [48].) Thus the mathematician's understanding of what a proof is changes over time, as is usually the case with human concepts. Why should logic be any different? Indeed, I believe it is not. Many years ago, Bertrand Russell complained in his autobiography [58] about the amount of effort required to produce all the formal proofs in [63]. Sadly for him, very few people have studied even a

---

[19] Recently a totally formal proof using the proof assistant *Coq* has been used by Gonthier [36] to produce a formal proof, which itself has been formally checked. Of course this changes the kind of guarantee that we have the result from the ordinary formal language level to a meta-level.

few of those proofs. They are routine exercises in a predicate calculus and, once the basic tricks are learnt, they are exceedingly repetitious rather than intellectually demanding.[20] The subsequent development of logic led to the use of metamathematical arguments, and the subsequent development of mathematics usually led to proofs that were accepted by mathematicians, though an increasing number of theorems have now been given proofs in formal logic and verified–by computers (cf. Gonthier's work [36] mentioned above).

So logicians having developed the machinery for right-wing proofs, have now moved to giving left-wing ones. Future formalization, perhaps following the layered approach sketched in [43], may lead to some of those becoming right-wing proofs. It is hard to see that left-wing proofs will ever cease to be found and cherished. The layering, that is the use of metamathematics at higher and higher levels, reduces the scale of (some proofs) to dimensions that are accessible to practising human mathematicians.

3.7 Conclusion

We have traced a little of the history of the development of formal proofs in logic and, to a much lesser extent, mathematics. Until the 20th century the development was at a manageable level, but the advent of computing machines brought with it both the capacity to perform large calculations and the practical need to complete those computations in a "reasonable" time. Large machine computations or proofs are very hard to comprehend, though we have begun to develop a layered approach, which treats them in a coarser way. Moreover, absolute limitations on what can be calculated in a given time means that certain proofs and calculations can not be performed on a machine in a human lifetime.[21]

The mathematicians' view that we submit proofs to our peers for appraisal has worked reasonably well, even though it is possible to find examples, as Lakatos [44] did, where proofs and definitions have had to be changed over time. In this case there was increasing precision in both as mathematicians became more knowledgeable and sophisticated. It seems reasonable to expect a similar history unrolling for formal logic.

I believe it is important to bear in mind that our expectations of proofs, and our production of them, and systems for them, emerge in response to our (human) environment and must therefore change as that environment changes.

---

[20] The whole work contains little deep mathematics, being principally a very formal exposition of others' work.

[21] Although the speed and capabilities of machines continue to increase, there is always a limit to what a given machine can achieve in a specified time.

References

1. Kenneth I. Appel and Wolfgang Haken. Every planar map is four colorable. I. Discharging, *Illinois J. Math.*, 21: 429–490, 1977.
2. Kenneth I. Appel and Wolfgang Haken. Every planar map is four colorable. II. Reducibility, *Illinois J. Math.*, 21: 491–567, 1977.
3. Aristotle. *On Interpretation*. In [51], pages 40–64.
4. Aristotle. *Posterior Analyics*. In [51], pages 110–186.
5. Aristotle. *Prior Analyics*. In [51], pages 65–106.
6. Aschbacher M. The status of the classification of finite simple groups, *Notices of the Amer. Math. Soc.*, 51(7): 736–740, 2004.
7. Barendregt P. H. *The Lambda Calculus, Its Syntax and Semantics*. North Holland Publishing Company, Amsterdam, 1995.
8. Beeson M. J. *Foundations of Constructive Mathematics*. Number 6 in Ergebnisse der Mathematik und ihren Grenzgebiete; 3. Folge. Springer, Berlin, 1985.
9. Bertalanffy L. v. *General Systems Theory: Foundations, Development, Applications*. George Braziller, New York, NY, 1968.
10. Bolyai J. *The Science of Absolute of Space: Independent of the Truth or Falsity of Euclid's Axiom XI*, translated by G. Halsted 4th ed. The University of Texas, Austin, 1896.
11. Burnside W. *Theory of Groups of Finite Order*, 2nd ed. Dover Phoenix Editions, New York, NY, 2004. This edition was originally published in 1911.
12. Buss S. R. *BoundedArithmetic*. Bibliopolis, Naples, 1986. A revised version of his PhD thesis.
13. Buss S. R. Bounded arithmetic, proof complexity and two papers of Parikh, *Ann. Pure Appl. Logic*, 96: 43–55, 1999.
14. Church A. An unsolvable problem of elementary number theory, *Proc. Nat. Acad. Sci. USA*, 20: 584–590, 1934.
15. Couturat L. *La logique de Leibniz d'apr'es des documents inédits*. Feliz Alcan, Paris, 1901. Reprinted Georg Olms, Hildesheim, 1961.
16. Couturat L. *Opuscules et fragments inédits de Leibniz*. Feliz Alcan, Paris, 1903. Reprinted Georg Olms, Hildesheim, 1961.
17. Crossley J. N. What is a logic? Presented at the International Conference on Logic, Navya-Nyāya and Applications: A Homage to Bimal Krishna Matilal, January 2007.
18. Crossley J. N. What is the difference between proofs and programs? Lecture at the First International Conference on Logic and and its application to other disciplines, IIT Bombay, 2005, submitted for publication.
19. Crossley J. N. A note on Cantor's theorem and Russell's paradox, *Austral. J. Phil.*, 51: 70–71, 1973.
20. Crossley J. N. Fifty years of computability, *Southeast Asian Bull. Math.*, 11: 81–99, 1988.
21. Crossley J. N. Samsara. In D. Gabbay, S. S. Goncharov, and M. Zakharyaschev, editors, *Mathematical Problems from Applied Logic. New Logics for XXI Century (International Mathematical Series)*, pages 234–276. Springer, Berlin, 2005.
22. Crossley J. N., Mathai G. L., and Seely R. A. G. A logical calculus for polynomial-time realizability, *Methods Logic Comput. Sci.*, 1: 259–278, 1994.
23. Crossley J. N., and Remmel J. B. Proofs, programs and run times, *Methods Logic Comput. Sci.*, 1: 183–215, 1994.
24. Curry H. B. Functionality in combinatory logic, *Am. J. Math.*, 58: 345–363, 1936.
25. Curry H. B., and Feys R. *Combinatory Logic*, vol. 1. North-Holland, Amsterdam, 1958.
26. Davies A. J. *The Finite Element Method*. Oxford University Press, Oxford, 1980.
27. Devlin K. Devlin's Angle: When is a proof? In *MAA Online*. Mathematical Association of America, June 2003. http://www.maa.org/devlin/devlin_06_03.html, accessed 8.iii.05
28. Dipert R. R. Logic, history of [electronic version]. In *Encyclopdia Britannica Online*, 2006. http://search.eb.com/eb/article-65939, accessed 1.vii.06

29. Dutens L., editor. *Gothofredi Guillelmi Leibnitii Opera Omnia: Nunc prima collecta, in classes distributa, prfationibus & indicibus exornata, studio Ludovici Dutens ....* Apud Fratres de Tournes, Geneva, 1768. 6 vols., reprinted Georg Olms, Hildesheim, 1989.
30. Gerhardt C. I., editor. *Die philosophische Schriften von Gottfried Wilhelm Leibniz*, 7 vols. Georg Olms, Hildesheim, 1978.
31. Girard J. Y. Linear logic, *Theor. Comput. Sci.*, 50: 1–102, 1987.
32. Girard J. Y. *Proof theory and Logical Complexity*. Bibliopolis, Naples, 1987.
33. Girard J. Y., Lafont Y., and Taylor P. *Proofs and Types*. Cambridge University Press, Cambridge, 1989.
34. Girard J. Y., Scott P. J., and Seely R. A. G. Bounded linear logic: a modular approach to polynomial-time computability, *Theor. Comput. Sci.*, 97: 1–66, 1992.
35. Gödel K. Über formal unentscheidbare Sätze der Principia Mathematica und verwandter Systeme, *Monatshefte für Mathematik und Physik*, 38: 173–198, 1931.
36. Gonthier G. A computer-checked proof of the Four Colour Theorem. http://research. microsoft.com/\%7Egonthier/4colproof.pdf, accessed 13.xii.06
37. Gow J. *A Short History of Greek Mathematics*. Chelsea Pub. Co., New York, NY, 1968. Revised edn.
38. Heath T. L., editor. *The Thirteen Books of Euclid's Elements*, 2nd ed. Cambridge University Press, Cambridge, 1925. Reprinted, Dover Books, New York, 1956.
39. Hilbert D. On the foundations of logic and arithmetic, 1904. In [15], pages 129–138.
40. Hilbert D. The foundations of mathematics, 1927. In [15], pages 464–479.
41. Hilbert D. *Foundations of Geometry*, 2nd ed., Open Court, La Salle, IL, 1971. Taken from the 10th German edn.
42. Howard W. The formulae-as-types notion of construction. In J. R. Hindley and J. Seldin, editors, *To H.B. Curry: Essays on Combinatory Logic, Lambda Calculus, and Formalism*, pages 479–490. Academic Press, New York, NY, 1969.
43. Jeavons J. S., Poernomo I., Basit B., and Crossley J. N. A layered approach to extracting programs from proofs with an application in Graph Theory. In R. Downey, D. Decheng, T. S. Ping, Q. Y. Hui, and M. Yasugi, editors, *Proceedings of the 7th and 8th Asian Logic Conferences, Chongqing, China, 29 August – 2 September 2002*, pages 193–222, Singapore, 2003. Singapore University Press and World Scientific.
44. Lakatos I. In J. Worrall and E. Zahar, editors, *Proofs and Refutations: The Logic of Mathematical Discovery*. Cambridge University Press, Cambridge, New York, 1976.
45. Lauchli H. An abstract notion of realizability for which intuitionistic predicate calculus is complete. In A. Kino, J. Myhill, and R. E. Vesley, editors, *Intuitionism and proof theory; Proceedings of the Summer Conference at Buffalo, NY, 1968*, pages 227–234. North Holland Publishing Company, Amsterdam, 1970.
46. Leibniz G. W. XIX. [Non inelegans specimen demonstrandi in abstractis]. In C. I. Gerhardt, editor, *Die philosophische Schriften von Gottfried Wilhelm Leibniz*, vol. VII, pages 228–235. Georg Olms, Hildesheim, 1961. Reprint of the 1890 Berlin edition.
47. Lobachevsky N. I. Pangeometry, 1855 translated by A. Papadopoulos. *Heritage of European Mathematics*. European Mathematical Society (EMS), Zürich, 2010.
48. Lucas J. R. Mathematical tennis, *Proceedings of the Aristotelian Society*, n.s. LXXXIV: 63–72, 1983/4.
49. Matilal B. K. *Perception: An Essay on Classical Indian Theories of Knowledge*. Clarendon Press, Oxford, 1986.
50. Maurolico F. Arithmeticorum libri duo. In *Opuscula Mathematica*. Francisci, Venice, 1575.
51. McKeon R. P., editor. *The Basic Works of Aristotle*. Random House, New York, NY, 1941.
52. Mendelson, E. *Introduction to Mathematical Logic*, 4th ed. Chapman & Hall, Boca Raton, FL, 1997.
53. Nerode A., Remmel J. B., and Scedrov A. Polynomially graded logic I: A graded version of system T. In *Proceedings, Fourth Annual Symposium on Logic in Computer Science*, pages 375–395. IEEE Computer Society Press, 1989.
54. Parikh R. J. Existence and feasibility in arithmetic, *J. Symb. Logic*, 36: 494–508, 1971.

55. Parikh R. J. Some results on the lengths of proofs, *Trans. Amer. Math. Soc.*, 177: 29–36, 1973.
56. Prior A. N. *Formal Logic*. Clarendon Press, Oxford, 1962.
57. Proclus. *A commentary on the first book of Euclid's Elements: Proclus, translated with introduction and notes by Glenn R. Morrow; [with a new foreword by Ian Mueller]*. Princeton University Press, Princeton, NJ, 1970.
58. Russell B. A. W. *The Autobiography of Bertrand Russell*, 3 vols. Allen and Unwin, London, 1967–1969.
59. Schönfinkel M. Ueber die Bausteine der mathematischen Logik. In J. van Heijenoort, editor, *From Frege to Gödel*, pages 495–515. Harvard University Press, Cambridge, MA, 1967. Originally appeared in Mathematische Annalen, vol. 92, 305–316, 1924.
60. Schütte K. *Beweistheorie*, 1st ed. Springer, Berlin, 1960.
61. Szabo M. E., editor. *The Collected Papers of Gerhard Gentzen*. North Holland Publishing Company, Amsterdam, 1969.
62. Turing A. M. Computability and lambda-definability, *J. Symb. Logic*, 2: 153–163, 1937.
63. Whitehead A. N., and Russell B. A. W. *Principia Mathematica*, 3 vols. Cambridge University Press, Cambridge, 1925–27. first published in 1910–13.
64. Ésénine-Volpine A. S. [Yesenin-Volpin]. Le programme ultra-intuitionniste des fondements des mathématiques. In *Infinitistic methods, Proceedings of the Symposium on Foundations of Mathematics, Warsaw, 2–9 September 1959*, pages 201–223. Pergammon Press, Oxford; Państwowe Wydawnictwo Naukowe, Warszawa, 1960.

Appendix: A Brief Guide to the Symbols

Symbol	*Read*	*Symbol*	*Read*
$\forall$	for all	$\exists$	there exists
$\wedge$	and	$\vee$	or
$\rightarrow$	implies		
$\Box$	necessarily	$\Diamond$	possibly
$\begin{matrix}\vdots \\ A\end{matrix}$	a proof of A	$\begin{matrix}A \\ \vdots \\ B\end{matrix}$	a proof of B from A
$\dfrac{A \quad (A \rightarrow B)}{B}$	from A and $(A \rightarrow B)$ infer B		

$\neg$ not (Sometimes a swung dash, $\sim$, is used for 'not'.)

Chapter 4
A Visit to Tarski's Seminar on Elimination of Quantifiers

Wilfrid Hodges

In Spring 1928, 13 years before I was born, I paid an imaginary visit to Warsaw in Poland and attended Alfred Tarski's seminar on the methodology of the deductive sciences. The sanserif text below is my imaginary record of what was said in the seminar. But the serif text, interspersed and at the end, is factual.

Though the seminar proceedings are imaginary, they are based on published texts, chiefly the paper of Presburger [16] and reports by Tarski himself on the achievements of his seminar.

Seminar, 10 April 1928.

Leader: Mr Mojżesz Presburger.

Today we are going to study the science of integers with the concept of addition, using logic without quantification over classes or relations. Our method will be the method of elimination of quantifiers, which Dr Tarski showed us how to apply to the calculus of classes and to the concepts '*b* lies between *a* and *c*' and 'the distance from *a* to *b* equals the distance from *c* to *d*'. Dr Tarski has explained to us that this method was first used by Thoralf Skolem and has recently been applied to the theory of dense order by C. H. Langford in America.

In the 1920s it was common to think of mathematics as consisting of various "sciences", each with its own subject matter. These sciences were said to be "deductive", meaning that they proceeded by making logical deductions from stated axioms. In 1919 Bertrand Russell ([17] p. 5) had referred to "the whole series of sciences that have been deduced from the theory of the natural numbers"; I presume he meant at least the sciences of integers, rational numbers, real numbers and complex numbers, and possibly some kinds of geometric science too. Tarski sometimes referred to deductive sciences as "deductive theories"; this notion evolved into the 'theories' of model theory.

First we must establish a domain of individuals and a set of concepts relating to this domain. Our domain will be the set of integers. For concepts we choose addition, the number zero and the number one. This domain and these concepts are intuitively clear. We choose the symbols $+$, 0 and 1 to represent addition, zero and one. We call these symbols the *primitives*.

Wilfrid Hodges
Queen Mary, University of London, London, UK, e-mail: w.hodges@qmw.ac.uk

J. van Benthem et al. (eds.), *Proof, Computation and Agency*, Synthese Library 352,
DOI 10.1007/978-94-007-0080-2_4,

A student: Don't we also need a symbol to represent the concept 'integer'? Langford used a symbol **K** for his domain of individuals.

Dr Tarski: If we were looking at 'models' of the formal language, i.e. re-interpretations, then it would be convenient to have a symbol N for the domain. It would allow us to describe a model by re-interpreting N, $+$, 0 and 1. But our present task is to study the integers, and for this task the symbol N is not needed.

In fact Tarski did use a symbol for the domain of individuals when he was discussing general theory (for example the illustration in [22] §34, or §37 in the English translation), but generally not when he was studying particular theories in depth. In 1948 Tarski's ex-student Andrzej Mostowski ([15], chapter ix) systematically used a symbol for the domain in his examples of formal theories. On the other hand symbols for the domain are missing from Tarski's address to the 1935 Paris Congress of Scientific Philosophy (xvi "On the concept of logical consequence" in [24]); evidently Tarski regarded their use as a matter of convenience rather than principle. Domain symbols disappeared from general use with the advent of model theory around 1950.

We build up a language for talking about addition of integers. We need a countably infinite set of symbols to serve as variables ranging over individuals; for this we will use the symbols

$$x, x', x'', x''', \dots$$

(though informally we will write x, y, z etc. for variables, following usual mathematical practice). A *term* is either a variable or 0 or 1, or an expression built up from these by applying $+$ any number of times. An *atomic formula* is an expression $(s = t)$ where s and t are terms. A *formula* is either an atomic formula or an expression built up from atomic formulas by applying any of the following three operations any number of times: (1) Given a formula ϕ, form its negation $(\neg\phi)$; (2) given formulas ϕ and ψ, form the formula $(\phi \to \psi)$ expressing 'if ϕ then ψ'; (3) given a formula ϕ and a variable x, form the formula $\exists x\phi$ which expresses 'there is x such that ϕ'. We write L for the set of all formulas.

A *boolean combination* of formulas $\phi_1, \dots, \phi_n$ is a formula that is either one of these formulas, or can be built up from them using $\neg$ and $\to$. Two examples of boolean combinations are $((\neg\phi) \to \psi)$, which we will abbreviate as $(\phi \vee \psi)$, and $(\neg(\phi \to (\neg\psi)))$ which we will abbreviate as $(\phi \wedge \psi)$. We will leave out brackets in the usual ways; for example $(\phi \wedge \psi \wedge \chi)$ is an abbreviation for $((\phi \wedge \psi) \wedge \chi)$. We also write $\forall x\phi$ as an abbreviation for $(\neg\exists x(\neg\phi))$, and $(s \neq t)$ for the inequation $(\neg(s = t))$.

We will assume known the rules of logic for a language of this kind.

My notes have modernised the logic used in the seminar. Presburger's logic in [16] differed from standard first-order logic in the following ways:

- The function and constant symbols had fixed meanings, just as the logical symbols did.
- There were two sets of variables, one for integers and the other for propositions. Both sets were infinite (though this was said by writing "...", not explicitly). In the 1920s the need for infinitely many variables over individuals was presumably understood, but it was often ignored. In 1922 Hilbert [7] proposed to base "the whole of mathematics" on a logical system with exactly eight individual variables, and in 1936 Tarski himself ([23] p. 1) said he would use the lowercase italic letters – a set of size around two dozen – as variables. The first paper

known to me which says precisely what the variables are is Tarski [21]; it defines the individual variables to be the expressions $x'_{\prime}$, $x''_{\prime}$ etc., with any finite number of superscript dashes.

- As the previous point implies, the definition of "formula" was not precise.
- No consistent distinctions were made between formulas, sentences and sentence schemas. An axiom containing free variables was interpreted as if the free variables were universally quantified. So for example an axiom with propositional variables could be used to deduce a formula got by substituting formulas for the variables.
- The axioms of the logic (including the axioms for identity) were listed among the axioms of the theory. Today we make a stricter separation between logical and non-logical axioms.

Presburger used the logical axioms and inference rules proposed by Jan Łukasiewicz in [12] (1929), which Presburger edited for publication.

Every formula of L says something meaningful about addition of integers. The formulas that say something true or false are called *sentences*. In fact sentences are those formulas of L in which every occurrence of a variable x is bound by a quantifier $\exists x$.

If a formula ϕ has some variables with free occurrences, say $x_1, \ldots, x_n$, then by assigning integers $\alpha(x_1), \ldots, \alpha(x_n)$ to these variables we again make ϕ into a true or false statement about the integers. We say that the assignment α *satisfies* ϕ if this statement is true. We say that two formulas with the same free variables are *equivalent* if they are satisfied by exactly the same assignments.

Tarski discovered his definition of truth, and along with it the formal definition of the notion "assignment α satisfies formula ϕ", in 1929 ([24] p. 152). But in [21], the first paper to apply the formal definition of satisfaction, he describes this notion as "intuitive" and "seemingly clear and unambiguous" ([24] p. 117).

Our main work will be to choose three things, namely

- A set of formulas of L that we call *basic*;
- a set of sentences of L that we call *axioms*;
- for each formula ϕ of L, a boolean combination $|\phi|$ of basic formulas of L,

so that the following hold:

(a) The axioms are all evidently true.
(b) Each formula ϕ of L has the same free variables as $|\phi|$, and we can prove from the axioms that ϕ and $|\phi|$ are equivalent.

We can satisfy these requirements in a trivial way by taking all formulas to be basic, the set of axioms to be empty, and $|\phi|$ to be equal to ϕ. The aim of the method of elimination of quantifiers is to do it more helpfully than this.

For example the following would be helpful to have:

(c) There is an effectve procedure for telling, given a basic formula ϕ, whether ϕ is true (if ϕ is a sentence), and which assignments satisfy ϕ (if ϕ is not a sentence).
(d) There is an effective procedure for finding $|\phi|$, given a formula ϕ.
(e) For every basic sentence ϕ, we can prove ϕ from the axioms if ϕ is true, and we can prove $\neg\phi$ from the axioms if ϕ is false.

If we can make (c) and (d) true, then we have a *decision procedure* for addition of integers: in other words we can compute, for each sentence ϕ of L, whether or not ϕ is true. If (e) is true, then by (b) the axioms are sufficient to deduce all true sentences of L, and by (a) they do not allow us to deduce any false sentences of L; we can summarise these two properties by saying that the set of axioms is a *complete axiomatisation* of addition of the integers.

FIRST ATTEMPT. We take the basic formulas to be the atomic formulas. If natural numbers are assigned to all variables in the terms s and t, then we can easily calculate the resulting integer values of s and t, and so check whether the statement $(s = t)$ is true. This assures (c).

For (e) we require a set of axioms that will allow us to prove all true atomic sentences. Our first attempt is

$$\begin{array}{ll} \zeta_{\text{ass}}: & \forall x_0 \forall x_1 \forall x_2 (((x_0 + x_1) + x_2) = (x_0 + (x_1 + x_2))). \\ \zeta_{\text{com}}: & \forall x_0 \forall x_1 ((x_0 + x_1) = (x_1 + x_0)). \\ \zeta_{\text{zero}}: & \forall x_0 ((x_0 + 0) = x_0). \\ \zeta_{\text{canc}}: & \forall x_0 \forall x_1 \forall x_2 ((x_0 + x_2) = (x_1 + x_2) \rightarrow (x_0 = x_1)). \\ \zeta_{\text{diff}}: & \forall x_0 \forall x_1 \exists x_2 ((x_0 + x_2) = x_1). \end{array} \tag{4.1}$$

These axioms tell us that the integers form an abelian group under addition, and so they are clearly true, guaranteeing (a). For each non-negative integer n we write $\bar{n}$ for the term defined as follows: $\bar{0}$ is 0, and for each n, $\overline{n+1}$ is the term $(\bar{n} + 1)$. Also we write nx for $x + \ldots + x$ (where 'x' occurs n times), taking $0x$ to be 0.

We can easily check that for every term s with no variables, the axioms imply $(s = \bar{n})$ for the unique non-negative integer n that is the value of s. If $(s = t)$ is a true atomic sentence, then the terms s and t have the same value n, and so the axioms entail $(s = \bar{n})$ and $(t = \bar{n})$, and hence they entail $(s = t)$ by the rules for equality.

So half of (e) holds. But we must also show that if ϕ is a false atomic sentence then the axioms entail $(\neg\phi)$. For example the axioms should entail $(0 \neq 1)$. We can see no way of deducing this sentence from the five axioms above.

The axioms ζ_{ass}, ζ_{com}, ζ_{zero} and ζ_{diff} include the usual axioms for abelian groups, and hence they imply ζ_{canc}. (Presburger removed the redundant axiom in his supplementary note to [16].) In fact the presence of a redundant axiom does no significant harm to the quantifier elimination procedure. For the same reason, there is no need to *prove* that our axioms don't entail $(0 \neq 1)$ before we set out to repair them. The worst that could happen is that we add a new axiom which we didn't need.

SECOND ATTEMPT. We keep the same basic formulas and axioms as before, but we add axioms to refute all false basic sentences. The new axioms are as follows:

$$\eta_n : \forall x (0 \neq nx + 1)$$

where n is any integer $\geqslant 2$.

These axioms give us $(0 \neq \bar{k})$ for each $k > 2$. But $(0 = \bar{2})$ implies

$$0 = 0 + \bar{2} = \bar{2} + \bar{2} = \bar{4},$$

contradicting $(0 \neq \bar{4})$; so the axioms prove $(0 \neq \bar{2})$. Putting 0 for x in any η_n yields $(0 \neq 1)$ too. Hence our present axioms imply that the additive group of integers is torsion-free, and from this we can deduce the negations of all false atomic sentences. We have ensured (a), (c) and (e). For (b) and (d) we need to define $|\phi|$ for each formula ϕ.

The seminar is in danger of falling into a trap here. The present axioms express that the integers with addition form an abelian group with an element 1 that is not divisible by any integer $n > 1$. This implies that the subgroup generated by 1 is torsion-free; and in fact this subgroup is the whole group of integers. But the axioms don't logically imply that the group is torsion-free; the group $\mathbb{Z} \times (\mathbb{Z}/2\mathbb{Z})$, with 1 interpreted as $(1,0)$, is a counterexample. Hence the axioms can't be complete yet. However, we will see that the method of quantifier elimination itself leads to a correction.

If the seminar had noticed this trap, they might have been tempted to try to find axioms saying that the whole group is generated by 1. The compactness theorem for first-order languages shows that this is impossible with first-order axioms. The compactness theorem (for countable languages) first appeared as Theorem X of Gödel [5] in 1930, but apparently the first person to see how to apply it to get useful information was Mal'tsev [13] around 1940. We should remember that Tarski's seminar belonged to an age before model theory, and admire it all the more for its achievements.

We begin by defining $|\phi|$ in the following cases. First, if ϕ is basic then we take $|\phi|$ to be ϕ. Second, if $|\phi|$ and $|\psi|$ have been defined but $|(\phi \rightarrow \psi)|$ has not, we take $|(\phi \rightarrow \psi)|$ to be

$$(|\phi| \rightarrow |\psi|).$$

Similarly if $|(\neg\phi)|$ is not yet defined, we take it to be $(\neg|\phi|)$. We note that if $|\phi|$ and $|\psi|$ are boolean combinations of basic formulas, then so are $|(\phi \rightarrow \psi)|$ and $|(\neg\phi)|$, as required. Suppose that $|\phi|$ has been defined, and we seek to define $|\exists x\phi|$. Now $\exists x|\phi|$ is not a boolean combination of basic formulas. Instead of enlarging the class of basic formulas, we try to show that $\exists x|\phi|$ is equivalent to a boolean combination of basic formulas, provably from the axioms. This is the stage at which we *eliminate a quantifier*.

We can simplify the problem. Since $|\phi|$ is a boolean combination of basic formulas, we can find an equivalent formula of the form

$$(\phi_1 \vee \ldots \vee \phi_m)$$

where each ϕ_i is a conjunction of one or more formulas that are either basic or negations of basic formulas. The formula $\exists x|\phi|$ is equivalent to the boolean combination

$$(\exists x\phi_1 \vee \ldots \vee \exists x\phi_m).$$

This reduces our problem to the case where $\exists x|\phi|$ has the form

$$\exists x(\chi_1 \wedge \ldots \wedge \chi_k)$$

where each χ_i is either basic or the negation of a basic formula. We note also that if for some j the variable x does not occur in χ_j, then we can move χ_j outside the scope of the quantifier; so we can assume that x occurs in each formula χ_i.
typical case might be the formula

$$\exists x(3x + \bar{2} = y \wedge 1 + z \neq 2x).$$

The coefficients of x are 3 and 2, with least common multiple 6. Since we are in a torsion-free abelian group, each equation or inequation is equivalent to the formula that results from multiplying its terms by a nonzero integer. In particular $3x + \bar{2} = y$ is equivalent to $6x + \bar{4} = 2y$, and $1 + z \neq 2x$ is equivalent to $\bar{3} + 3z \neq 6x$. Also we can add the same number to both sides of an equation, so that $\bar{3} + 3z = 6x$ is equivalent to $\bar{7} + 3z = 6x + \bar{4}$. So the formula that we are analysing is equivalent to

$$\exists x(6x+\bar{4}=2y \wedge \bar{7}+3z \neq 6x+\bar{4}).$$

We can use the equation on the left to rewrite the inequation on the right, and the whole formula becomes

$$\exists x(6x+\bar{4}=2y \wedge \bar{7}+3z \neq 2y),$$

and now since x does not occur in the inequation, this formula is equivalent to

$$(\exists x(6x+\bar{4}=2y) \wedge (\bar{7}+3z \neq 2y)).$$

If we can find a boolean combination of basic formulas equivalent to $\exists x(6x+\bar{4}=2y)$, our analysis succeeds.

Unfortunately we meet difficulties here. We consider a simpler case: $\exists x(2x=1)$. This is clearly false, and in fact the axiom η_2 allows us to refute it. If $2m=1$ then $2(-m)+1=0$, and this contradicts η_2. So there is hope that we can reduce $\exists x(6x+\bar{4}=2y)$ in a similar way.

Dr Tarski closes the seminar after inviting the members to think about this problem.

Seminar reconvenes on 17 April 1928. Again Mr Presburger leads.

It appears that the set of basic formulas must be expanded. The formula $\exists x(2x=y)$ expresses that y is even, and it is intuitively clear that no boolean combination of atomic formulas expresses this. It also appears that we need basic formulas to express that the difference between two terms is divisible by a certain positive integer. We write

$$(n|s-t)$$

as an abbreviation for the formula $\exists x(s=nx+t)$, which expresses that $s-t$ is divisible by n. Here n is any integer >0.

We call a formula $(n|s-t)$ a *congruence*. We abbreviate its negation to $(n \nmid s-t)$, and we call this formula an *incongruence*. We call the number n the *modulus* of $(n|s-t)$, and of $(n \nmid s-t)$.

Habits change. If Presburger had been a model theorist writing in 1960 or later, he would probably not have extended the class of basic formulas within L. Instead he would have expanded the language L to a first-order language with new relation symbols $M_n(x,y)$ that are interpreted so that $M_n(s,t)$ means $(n|s-t)$. The advantages of the modern view are technical: (1) it simplifies the statement of the required recursions (see Comment 3 at the end), and (2) it relates quantifier elimination to model-theoretic preservation results. Keisler [9] is a revealing pointer to the change in habits that set in around 1960.

THIRD ATTEMPT. We keep our axioms as in our Second Attempt, but now we take as basic formulas all atomic formulas and all congruences.

Adding new basic formulas is a costly move, since we have to re-prove (c) and (e). In the present case (c) is straightforward, because for any integers k, m and n we can easily check whether $k-m$ is a multiple of n. But we need to check that our axioms will prove all true sentences $(n|s-t)$ and refute all false ones.

If the sentence $(n|s-t)$ is true then there is an integer c such that $s=nc+t$. If $c \geqslant 0$ then we can write the true sentence $(s=nc+t)$ in L, and we saw in our First Attempt that this equation is provable from our axioms. Hence so is $(n|s-t)$. On the other hand if $c<0$ then we can write $(s+nd=t)$, where $d=-c$, and proceed as before. So all true basic sentences are provable from the axioms.

If the sentence $(n|s-t)$ is false, suppose (transposing s and t if necessary) that $s \geqslant t$. Then by the division algorithm there are integers $q \geqslant 0$ and r with $0 < r < n$, such that $s - t = nq + r$; as a true sentence in L this equation is provable from our axioms. Also this equation and $(n|s-t)$ together entail that n divides r; so it remains only to show that our axioms refute the false basic sentence $(n|\bar{r}-0)$. We can do this as follows.

Let d be the greatest common divisor of n and r, and put $n' = n/d$, $r' = r/d$. Then the sentence $(n'|\overline{r'}-0)$ is also false (whence $n' \geqslant 2$). We refute this sentence as follows. For some positive integers a and b, $ar' = bn' + 1$. Suppose for contradiction that for some x, $n'x = \overline{r'}$. Then

$$(an')x = \overline{ar'} = \overline{bn'} + 1$$

and hence by rearrangement $0 = \overline{n'}y + 1$ for some y. Since $n' \geqslant 2$, this contradicts the axiom $\eta_{n'}$. The false sentence $(n|\bar{r}-0)$ is refuted by first assuming for contradiction that for some z, $nz = r$, and dividing this equation by d.

A student objects: How do our axioms imply that we can divide an equation by a common factor of its coefficients? After some discussion, it is agreed that further axioms are required to prove this when the equation contains free variables.

The seminar has just realised the trap that they nearly fell into at the Second Attempt. They discovered it by trying to carry out a first-order proof required by the method of quantifier elimination.

FOURTH ATTEMPT. We adopt axioms telling us that we can divide an equation by a common factor. They are the sentences

$$\theta_n : \ \forall x \forall y\ (nx = ny \rightarrow x = y)$$

for each positive integer n.

It would have been simpler to write θ_n as

$$\forall x\ (nx = 0 \rightarrow x = 0).$$

But the version used in the seminar is not wrong.

By previous arguments we now have (a), (c) and (e). But now we must revisit (b), since we have expanded the class of basic formulas.

We can still take $|\phi|$ to be ϕ when ϕ is basic, and proceed as before for boolean combinations. But the elimination of quantifiers becomes more complicated. As in the Second Attempt, we can suppose that ϕ is a formula

$$\exists x(\chi_1 \wedge \ldots \wedge \chi_k)$$

where each χ_i is either basic or the negation of a basic formula, i.e. it is either an equation or an inequation or a congruence or an incongruence. We are required to find a boolean combination of basic formulas that is equivalent to this formula ϕ.

We can carry out some normalisations. Each χ_i is of the form either $(mx + s = t)$ or $(mx + s \neq t)$ or $(n|(mx+t) - s)$ or $(n \nmid (mx+t) - s)$, where m is an integer $\geqslant 1$ and s, t are terms in which x never occurs. Here m, s and t need not be the same in different formulas χ_i. If x doesn't occur in a formula χ_i, the formula can be brought outside the scope of the quantifier $\exists x$.

In our Second Attempt we saw how to deal with the case where at least one of the χ_i, say χ_1, is an equation. Our axioms imply that if k is any positive integer, then the equation $(mx+s=t)$ is equivalent to the equation $(kmx+ks=kt)$, and the congruence $(n|(mx+s)-t)$ is equivalent to the congruence $(kn|(kmx+ks)-kt)$. Hence we can assume that the coefficient of x is the same, say m, in all the χ_i. The axioms also imply that for any term r, the equation $(mx+s=t)$ is equivalent to the equation $(mx+s+r=t+r)$, and the congruence $(n|(mx+s)-t)$ is equivalent to the congruence $(n|(mx+s+r)-(t+r))$. So we can assume that if χ_1 is $(mx+s=t)$, then the term $mx+s$ occurs in every formula χ_i. As in our Second Attempt, we can use the equation $(mx+s=t)$ to replace each occurrence of $mx+s$ (except in χ_1) by an occurrence of t. After this replacement, x occurs only in χ_1, and so the other conjuncts can be moved out of the scope of the quantifier. There remains $\exists x(mx+s=t)$, which is equivalent to the basic formula $(m|s-t)$. All the variables that were free in ϕ are still free in the resulting boolean combination of basic formulas.

Henceforth we assume that none of the formulas χ_i are equations. Let k be the least common multiple of the moduli of the congruences and incongruences among the χ_i. If α is an assignment to all the free variables, which satisfies all the formulas χ_i that are congruences or incongruences, consider an assignment β which is the same as α except that the integer assigned to x is increased by some positive multiple of k. Clearly β again satisfies the congruences and incongruences. By choosing a large enough multiple of k we can ensure that all the formulas χ_i that are inequations are also satisfied by β. Our axioms are strong enough to justify this argument and show that the formula ϕ is equivalent to the formula got by deleting all the formulas χ_i that are inequations. (If all of the χ_i are inequations, then ϕ is equivalent to the basic formula $(y=y)\wedge(z=z)\wedge\ldots$, where we list all the variables free in ϕ.)

There remains the case where the formulas χ_i are all either congruences or incongruences. A typical example is

$$\exists x\,((2\nmid x-\bar{0})\wedge(2\nmid x-\bar{1})).$$

This sentence is clearly false, since for every integer x, one of x and $x-1$ is even. Dr Lindenbaum has suggested an efficient way of dealing with such sentences. It requires us to add some new axioms.

Presburger records ([16] p. 98) that Adolf Lindenbaum suggested the following idea for eliminating incongruences.

FIFTH ATTEMPT. We keep the same basic formulas as before, but we add to our previous axioms the following new ones. For each integer $n\geqslant 2$ we adopt the axiom

$$\iota_n:\ \forall x\,((n|x-\bar{0})\vee(n|x-\bar{1})\vee\ldots\vee(n|x-\overline{n-1})).$$

These axioms are clearly true, so that (a) still holds.

We have to deal with formulas ϕ of the form

$$\exists x(\chi_1\wedge\ldots\wedge\chi_k)$$

where each χ_i is either a congruence or an incongruence. We can assume the normalisations that we made in our previous attempt.

Suppose that χ_i is an incongruence $(n\nmid(mx+s)-t)$. Assuming that $n\geqslant 2$ (since otherwise the congruence holds trivially), the axiom ι_n tells us that this incongruence is equivalent to the disjunction

$$((n|(mx+s)-(t+\bar{1}))\vee\ldots\vee(n|(mx+s)-(t+\overline{n-1}))$$

of congruences. By elementary logical manipulations this brings us to the case where each formula χ_i is a congruence.

This is the final case that we have to deal with. By a similar adjustment to those in the Fourth Attempt, we can assume that each congruence χ_i has the form $(n|(m_i x+s_i)-t_i)$, where n is independent of i. Now suppose $m_1 \geqslant m_2$. The congruences χ_1 and χ_2 state that there are y and z such that $m_1x+s_1=ny+t_1$ and $m_2x+s_2=nz+t_2$. This pair of equations is equivalent to

$$((m_1-m_2)x+s_1+t_2=n(y-z)+t_1+s_2)$$
$$\wedge(m_2x+s_2=nz+t_2).$$

It follows that in ϕ we can replace χ_1 by the congruence $(n|((m_1-m_2)x+s_1+t_2)-(t_1+s_2))$; the resulting formula is equivalent to ϕ, and our axioms suffice to prove this. If $m_1=m_2$ then we can move this new congruence outside the scope of $\exists x$. We can repeat this move as long as there are at least two congruences within the scope of $\exists x$. The process must eventually halt, since it lowers the sum of the coefficients of x. So eventually we reach a formula

$$\exists x(n|(mx+s)-t).$$

This congruence is equivalent to $(n'|s-t)$ where n' is the greatest common divisor of n and m; again the axioms suffice to prove this.

We have shown:

LEMMA. If ψ is a conjunction of formulas that are basic or negated basic, then the formula $\exists x\psi$ is equivalent to a conjunction ψ' of formulas that are basic or negated basic, where ψ' has the same free variables as $\exists x\psi$, and the equivalence of $\exists x\psi$ and ψ' is provable from the axioms stated above.

We have already described $|\psi'|$; we define $|\exists x\psi|$ to be $|\psi'|$, and (b) and (d) are established.

This lemma, which appears in Presburger [16], is false. A counterexample is the formula

$$\exists x((3\not| u-x)\wedge(3\not| v-x)\wedge(3\not| w-x))$$

which expresses that at least two of u, v and w are congruent modulo 3. Presburger should have said that ψ' is a boolean combination of basic formulas. Also the definition of $|\psi'|$ is unsound as it stands, since ψ' is not uniquely determined by the Lemma. This gap could be plugged by describing explicitly the algorithm for finding ψ'.

THEOREM. For every sentence ϕ of L there is a sentence ϕ' of L which is a boolean combination of basic sentences, and is such that the equivalence of ϕ and ϕ' is provable from the axioms stated above. Moreover there is an algorithm which, given ϕ, will find ϕ' and the proof of equivalence.

PROOF. Given ϕ, we can find a sentence which is logically equivalent to ϕ and is in prenex form (i.e. has the form

$$Q_1x_1\dots Q_nx_n\psi$$

where each Q_i is either $\exists$ or $\forall$, and ψ has no quantifiers). By the Lemma we can find a formula ψ' which is equivalent to $Q_nx_n\psi$ and is a boolean combination of basic formulas. Then we apply the Lemma again to $Q_{n-1}x_{n-1}\psi'$, and so on through the n quantifiers. QED

Dr Tarski: This makes a good place to close today's proceedings. Those of you who attended my lectures last year should observe that the axioms we have reached today are exactly those which I proposed as an axiomatisation of the science of integers with addition.

A student: Next week, should we ask whether we can prove similar results when multiplication is also a primitive?

Mr Presburger: We will not succeed. In the language of integers with symbols for addition and multiplication, it is possible to write down, for each integer $n > 2$, a sentence stating Fermat's Last Theorem for exponent n. A decision procedure for this language would enable us to calculate the truth or falsehood of each instance of Fermat's conjectured theorem.

4.1 Tarski's Seminar

Our primary sources of information about Tarski's Warsaw seminar are the footnotes on page 205 of Tarski [24] and page 95 of Presburger [16]. Tarski studied several "sciences", and in each case his aim was to give a "structural description of all complete [axiom] systems" for the given science ([24] p. 205). One example was the family of subsets of a given set of cardinality k, with set inclusion as primitive notion; he gave a complete axiom system for each finite k and a complete axiom system for the case where k is infinite. These results are reported on pages 201–208 of [24]. The two footnotes mention some other types of structure to which he applied similar methods. In all cases he used the method of quantifier elimination (or at least he left no trace of any other method). Later Tarski and his students applied the same method to other cases. The best known case is probably the field of real numbers, which he began to study in 1929 [21] and published fully in 1951 [23]. His student Wanda Szmielew [20] generalised Presburger's result by applying quantifier elimination to the theory of abelian groups.

Many years later, in a paper reporting another quantifier elimination done jointly with Mostowski ([2] p. 2), Tarski said "the procedure of eliminating quantifiers in a theory may prove valuable as a heuristic source and deductive base for a many-sided metamathematical study of this theory leading to substantial results not involving decidability". The implication is that we learn useful things about a theory by finding suitable axioms and basic formulas for eliminating quantifiers. So the "method" is at least partly a method of research. I first learned this from Robert Vaught. In writing my imaginary notes I tried to emphasise the heuristic side of quantifier elimination. Tarski's reference to his 'seminar exercises' in Warsaw ([23] p. 205) allows us to imagine that some research took place in the seminar itself.

4.2 Presburger Arithmetic

On Presburger himself see Zygmunt [25]. The set of logical consequences of Presburger's axioms, which is the complete first-order theory of the integers with addition, is known as *Presburger Arithmetic*. Presburger attributes the "logical schema" (does he mean the axiomatisation?) of Presburger Arithmetic to lectures of Tarski in 1927–1928 ([16] p. 95 footnote).

A quantity of good work allegedly on "Presburger Arithmetic" is in fact about the complete first-order theory of the natural numbers $\mathbb{N}$ with a symbol for addition (e.g. [14]), a theory that Presburger never considered. Quantifier elimination for $\mathbb{N}$ with addition is not straightforwardly derivable from Presburger's results. (The theory of the natural numbers is unstable since the linear ordering is definable; Presburger Arithmetic is stable, like the theory of any abelian group.) But it is derivable from an extension of Presburger's results where the primitives are 0, 1, $+$ and $>$. Presburger states this extension in the supplementary note to [16]. A direct quantifier elimination for the natural numbers with addition appears in Hilbert and Bernays [8] pp. 368–377, though with a different set of axioms from Presburger's.

What about the theory of $(\mathbb{N}, \times)$, the natural numbers with multiplication but not addition? If we overlook zero, $(\mathbb{N}, \times)$ is isomorphic to the product of countably many copies of $(\mathbb{N}, +)$, one for each prime. Skolem [19] sketched a quantifier elimination for this structure in 1930.

In 1934 Gödel ([6] p. 367) mentioned these results of Presburger and Skolem, and remarked that a consequence of his own incompleteness theorem [5] is that there is no decision method for the arithmetic of the integers with both addition and multiplication. What would have happened if Tarski's seminar had tried to do a quantifier elimination on the ring of integers? It would have found that it kept needing to add further kinds of basic formula. Also quite soon it would have reached sentences whose truth value is unknown.

4.3 The Algorithm

From p. 374 of Tarski [24] it seems that "method of quantifier elimination" means in the first instance the argument used to prove the Theorem at the end of the seminar notes. Namely, "reducing the sentences to normal form and successively eliminating the quantifiers". On the same page Tarski attributes this method to Skolem [18].

Tarski's emphasis would be surprising today. The reduction to prenex normal form is completely unnecessary; its only function is to simplify the statement of an induction needed in the proof of the Theorem. By contrast neither Tarski nor Presburger considers what seem more pressing questions:

(i) to show formally that the algorithm finding $|\phi|$ from ϕ terminates;
(ii) to show formally that for each ϕ the equivalence of ϕ and $|\phi|$ is provable from the axioms.

(A further question is to show that the function $|.|$ is well-defined. But since we defined it in terms of the algorithm, this question reduces to (i).) Today we control both arguments by assigning ordinal ranks to formulas, so that in case (i) each step of the algorithm takes us to formulas of lower rank, and in case (ii) the result holds for ϕ provided that it holds for all formulas of lower rank than ϕ.

It's helpful to split both proofs into two parts. The first part is for formulas with at most one existential quantifier, which occurs (if at all) at the beginning. The second part is for general formulas, assuming the first part.

The first part is complicated by the fact that in removing an incongruence we may greatly increase the size of the formula. So a simple measure of complexity won't work. Instead we should assign (possibly transfinite) ranks so that ϕ has lower rank than ψ whenever ϕ has fewer congruences in negative position than ψ does; I omit details.

The second part is basically an induction on the number of existential quantifiers, but with a twist. Each time we apply the first part, we eliminate an existential quantifier, but we may introduce many new existential quantifiers inside congruences. The twist is to ignore existential quantifiers inside basic formulas. We can avoid this twist by introducing new atomic formulas where Presburger expanded the class of basic formulas.

Martin Davis implemented a decision procedure for Presburger arithmetic on the Johnniac computer, and reported the outcome to the 1957 Cornell Summer Institute [1] (according to [3] p. 229; I have never seen the Cornell volume).

Around 1970 complexity theory came into being, and one of its earliest results was a lower bound on the complexity of decision procedures for Presburger Arithmetic. In 1973 Michael Fischer and Michael Rabin ([4] Theorem 2) showed that "There exists a constant $c > 0$ such that for every decision procedure [for Presburger Arithmetic], there exists an integer n_0 so that for every $n > n_0$ there exists a sentence F of L of length n for which [the procedure] requires more than $2^{2^{cn}}$ computational steps to decide whether F [is true]". The central idea of their argument is that although exponentiation is not definable in L, if n is a large positive integer then we can write down a formula of L that has length much less than n and serves to define exponentiation for positive integers $< n$. In 1977 Tarski and Doner ([2] p. 2) comment that "recent research into the complexity of algorithms has deprived the pure decidability results of some of their luster".

4.4 The Definition of Truth

In 1929 Tarski discovered his mathematical definition of truth for languages of formal deductive theories ([24] p. 152). Its form is in some ways very like the definition of $|\phi|$ from ϕ; in fact he twice suggests that the truth definition is a generalisation of results of quantifier elimination, got by eliminating the "accidental" features of particular theories ([24] pp. 135 on elementary definability, 208 on truth). This could well be how he discovered the truth definition, as follows.

For a boolean combination ϕ, Presburger's definition of $|\phi|$ is not relative to the particular language. The interesting case is where ϕ begins with a quantifier. In order to avoid reduction to an equivalent formula, which obviously does depend on the particular theory, Tarski's truth definition redefines $|\phi|$ in all cases to be the set of assignments that satisfy ϕ. With this definition the clause for $\exists x$ is just as general as the clauses for $\rightarrow$ and $\neg$.

With this adjustment, the recursive form of the definition of $|\phi|$ becomes technically simpler. In fact $|\phi|$ now depends only on the values $|\psi|$ where ψ are the immediate subformulas of ϕ, together with the syntactic rule for building ϕ from these ψ. A definition of a function $|.|$ of expressions, where $|\phi|$ for compound ϕ depends only on $|\psi|$ for the immediate subexpressions ψ of ϕ and the syntactic operation combining them, is said to be *compositional*. Tarski's truth definition seems to be the first place where a semantic function of phrases of a language is given explicitly and precisely by a compositional definition. (People have found partial anticipations in work of Abu Ḥayyān, Leibniz, Boole, Husserl, Frege.)

References

1. Summer Institute for Symbolic Logic, Cornell University 1957. Summaries of Talks. (Duplicated typescript)
2. Doner J. E., Mostowski A., and Tarski A. The elementary theory of wellordering – a metamathematical study. In A. Macintyre, L. Pacholski and J. Paris, editors, *Logic Colloquium '77*, pages 1–54. North-Holland, Amsterdam 1978.
3. Feferman A. B., and Feferman S. *Alfred Tarski: Life and Logic*. Cambridge University Press, Cambridge 2004.
4. Fischer M. J., and Rabin M. O. Super-exponential complexity of Presburger Arithmetic. In R. M. Karp, editor, *Complexity of Computation*, pages 27–41. American Mathematical Society, Providence, RI, 1974.
5. Gödel K. Die Vollständigkeit der Axiome des logischen Funktionenkalküls, *Monatshefte für Mathematik und Physik*, 37: 349–360, 1930; reprint and translation 'The completeness of the axioms of the functional calculus of logic', in [6] pages 102–123.
6. Gödel K. In S. Feferman et al., editor, *Collected Works I: Publications 1929–1936*, pages 102–123. Oxford University Press, New York, NY, 1986.
7. Hilbert D., Neubegründung der Mathematik, Erste Beteilung, *Abhandlungen aus dem Mathematischen Seminar der Hamburgischen Universität*,1: 157–177, 1922; translated as 'The new grounding of mathematics, first report' in Paolo Mancosu, From Brouwer to Hilbert, Oxford University Press, New York, NY, 1998, pages 198–214.
8. Hilbert D., and Bernays P. *Grundlagen der Mathematik I*. Springer, Berlin, 1968. 66 Wilfrid Hodges
9. Keisler H. J. Theory of models with generalized atomic formulas, *J. Symb. Logic*, 25: 1–26, 1960.
10. Langford C. H. Some theorems on deducibility, *Ann. Math.*, 28: 16–40, 1927.
11. Langford C. H. Theorems on deducibility (Second paper), *Ann. Math.*, 28: 459–471, 1927.
12. Lukasiewicz J. *Elementy Logiki Matematycznej*, mimeographed notes edited by M. Presburger, Warsaw University 1929.
13. Mal'tsev A.A general method for obtaining local theorems in group theory (Russian), *Ucheniye Zapiski Ivanov. Ped. Inst. (Fiz-Mat. Fakul'tet)*, 1(1): 3–9, 1941; translation in Anatoliĭ Ivanovič Mal'cev, The Metamathematics of Algebraic Systems: Collected Papers 1936–1967, North-Holland, Amsterdam (1971) pages 15–21.
14. Michaux C., and Villemaire R. Presburger arithmetic and recognizability of sets of natural numbers by automata: New proofs of Cobham's and Semenov's theorems, *Ann. Pure App. Logic*, 77: 251–277, 1996.
15. Mostowski A. *Logika Matematyczna*. Monografie Matematyczne, Warsaw, 1948.

16. Presburger M. 'Über die Vollständigkeit eines gewissen Systems der Arithmetik ganzer Zahlen, in welchem die Addition als einzige Operation hervortritt, *Comptes Rendus du Premier Congrès des Mathèmaticiens des Pays Slaves, Warszawa 1929*, Warsaw, 92–101, 1930; supplementary note *ibid.* 395.
17. Russell B. *Introduction to Mathematical Philosophy*. George Allen and Unwin, London, 1919.
18. Skolem T., Untersuchungen über die Axiome des Klassenkalküls und über Produktations- und Summationsprobleme, welche gewisse Klassen von Aussagen betreffen, *Videnskapsselskapets Skrifter I. Mat.-nat. klasse 1919*, 3: 1919.
19. Skolem T. Über einige Satzfunktionen in der Arithmetik, *Skrifter Vitenskapsakademieti Oslo*, 1: 1–28, 1930.
20. Szmielew W. Elementary properties of Abelian groups, *Fund. Math.*, 41: 203–271, 1955.
21. Tarski A. Sur les ensembles definissables de nombres réels I, *Fund. Math.*, 17: 210–239, 1931; translation 'On definable sets of real numbers', [24] pages 110–142.
22. Tarski A. *O logice matematycznej i metodzie dedukcyjnej*, Biblioteczka Mat. 3–5, Ksiąznica-Atlas, Lwów, 1936; revised translation *Introduction to Logic and to the Methodology of the Deductive Sciences*. Oxford University Press, New York 1994.
23. Tarski A. *A Decision Method for Elementary Algebra and Geometry*. University of California Press, Berkeley, CA, 1951.
24. Tarski A. In J. Corcoran, editor, *Logic, Semantics, Metamathematics*, 2nd ed. Hackett, Indianapolis, IN, 1983.
25. Zygmunt J. Moj˙zesz Presburger: life and work, *Hist. Philos. Logic*, 12: 211–223, 1991.

Chapter 5
Deductive Systems of Fuzzy Logic

Petr Hájek

5.1 Introduction

Lotfi Zadeh [23] is the author of the theory of fuzzy sets. A fuzzy subset A of a (crisp) set X is characterized by assigning to each element x of X the *degree of membership* of x in A In particular, if X is a set of propositions then its elements may be assigned their *degree of truth,* which may be "absolutely true", "absolutely false" or some *intermediate* truth degree: a proposition may be more true than another proposition. This is obvious in the case of vague (imprecise) propositions like "this person is old" (beautiful, rich, etc.). And this leads to *fuzzy logic*. In the analogy to various definitions of operations on fuzzy sets (intersection, union, complement,...) one may ask how propositions can be combined by *connectives* (conjunction, disjunction, negation,...) and if the truth degree of a composed proposition is determined by the truth degrees of its components, i.e. if the connectives have their corresponding *truth functions* (like truth tables of classical logic). Saying "yes" (which is the mainstream of fuzzy logic) makes fuzzy logic to something principally *different from probability theory* since e.g. the probability of conjunction of two propositions is *not determined* by the probabilities of those propositions[1].

In broad sense, the term "fuzzy logic" is often used as synonymous with theory and applications of fuzzy sets in general. *Fuzzy logic in narrow sense* is symbolic logic with a comparative notion of truth developed fully in the spirit of classical logic (syntax, semantics, axiomatization, truth-preserving deduction, completeness, etc.; both propositional and predicate logic). It is a branch of *many-valued logic* based on the paradigm of *inference under vagueness.* (Observe that this is very different from probability theory, which may be said to study inference under

Petr Hájek
Institute of Computer Science, Academy of Sciences of the Czech Republic, 182 07 Prague, Czech Republic, e-mail: hajek@cs.cas.cz

[1] For example, if the probability of φ is 0.5 and probability of ψ is also 0.5 then clearly the probability of $\varphi\&\psi$ can be anything between 0 and 0.5.

J. van Benthem et al. (eds.), *Proof, Computation and Agency*, Synthese Library 352,
DOI 10.1007/978-94-007-0080-2_5,

randomness)[2]. This fuzzy logic is a relatively young discipline, both serving as a foundation for the fuzzy logic in a broad sense and of independent logical interest, since it turns out that strictly logical investigation of this kind of logical calculi can go rather far. A basic monograph is my [11], further recommended monographs are [20, 22]; also recent monographs dealing with many-valued logic (not specifically oriented to fuzziness), namely [2, 9] are highly relevant.

This paper presents information on fuzzy connectives and a survey of a logical system called the basic fuzzy propositional and predicate logic together with three stronger systems – Łukasiewicz, Gödel and product logic; a short discussion on paradoxes and fuzzy logic; some comments on other formal systems of fuzzy logic, on the relation of fuzziness and probability (the fuzzy notion of being probable) and finally, a few remarks on fuzzy description logic.

5.2 Fuzzy Connectives

The *standard set of truth degrees* is the real interval [0,1] with its natural ordering $\leq$. (1 standing for absolute truth, 0 for absolute falsity); but one can work with diferent domains, finite or infinite, linearly or partially ordered. Truth functions of connectives have to *behave classically on the extremal values* $0, 1$. It is broadly accepted that *t-norms* (triangular norms) are possible truth functions of *conjunction.*

Definition 5.1 A binary operation $*$ on the interval $[0,1]$ is a t-norm if it is commutative, associative, non-decreasing and 1 is its unit element, i.e. the following holds for each $x, y, z \in [0,1]$:

$$x * y = y * x, \qquad x * (y * z) = (x * y) * z,$$

$$x \leq y \text{ implies } x * z \leq y * z,$$

$$1 * x = x$$

(It follows that 0 is a zero element, i.e. $0 * x = 0$ for all x.)

Let us remark that minimum $\min(x, y)$ is the most popular t-norm. The truth function of *negation* has to be non-increasing (and assign 0 to 1 and vice versa); the function $1 - x$ (Łukasiewicz negation) is the most well-known candidate. But a serious analysis leads us to different negations for different t-norms, see below. *Implication* is sometimes disregarded but is of fundamental importance for fuzzy logic in the narrow sense. Logically most interesting are *R-implications:* an R-implication is defined as a residuum of a t-norm. This is well-defined only if the t-norm is left-continuous.

[2] For example, the statement "Tom will die today" is crisp, not vague, but has some probability. The statement "Tom is a tall man" is a vague statement, is true in a certain degree.

Definition 5.2 Let $*$ be a (left-) continuous t-norm. The residuum $\Rightarrow$ of $*$ is defined, for each $x,y \in [0,1]$ as $x \Rightarrow y = \max\{x|x*z \leq y\}$. The corresponding pseudocomplement *neg* is defined as $neg(x) = x \Rightarrow 0$.

The following are three most important continuous t-norms, their residua and the corresponding pseudocomplements:[3]
Łukasiewicz: $x*y = \max(0,x+y-1)$,
Gödel: $x*y = \min(x,y)$
product: $x*y = x \cdot y$.
For all of them, $x \Rightarrow y = 1$ if $x \leq y$, otherwise:
Łukasiewicz: $x \Rightarrow y = 1-x+y$,
Gödel: $x \Rightarrow y = y$,
product: $x \Rightarrow y = y/x$.
This gives the following pseudocomplement:
Łukasiewicz: $neg(x) = 1-x$,
Gödel and product: $neg(0) = 1, neg(x) = 0$ for $x > 0$.

Note that there are infinitely (uncountably many) continuous t-norms and their structure is well-known. Without telling details, let us just mention the theorem of Mostert and Shields saying that each continuous t-norm is an ordered sum of copies of Łukasiewicz, Gödel and product t-norms (in a well defined meaning of "ordered sum").

5.3 Basic Fuzzy Propositional Logic

This is the logic of continuous t-norms (developed in [11]). Formulas are built from propositional variables using connectives & (conjunction), $\rightarrow$ (implication) and truth constant $\bar{0}$ (denoting falsity). Further connectives are defined as follows:
$\varphi \wedge \psi$ is $\varphi \& (\varphi \Rightarrow \psi)$,
$\varphi \vee \psi$ is $((\varphi \rightarrow \psi) \rightarrow \psi) \wedge ((\psi \rightarrow \varphi) \rightarrow \varphi)$,
$\neg\varphi$ is $\varphi \rightarrow \bar{0}$,
$\varphi \equiv \psi$ is $(\varphi \rightarrow \psi) \& (\psi \rightarrow \varphi)$.
An *evaluation of propositional variables* is a mapping e assigning to each propositional variable p its truth value $e(p) \in [0,1]$.
This extends, given a continuous t-norm $*$ and its residuum $\Rightarrow$, to the evaluation $e*(\varphi)$ of each formula φ:
$e_*(\bar{0}) = 0$,
$e_*(\varphi \rightarrow \psi) = (e_*(\varphi) \Rightarrow e_*(\psi))$,
$e_*(\varphi \& \psi) = (e_*(\varphi) * e_*(\psi))$.

[3] The Polish logician Jan Łukasiewicz was the first to publish papers on many-valued logics (since 1920, see [19]). Kurt Gödel, the most famous mathematical logician of the past century, celebrated by his completeness and incompleteness theorems, has just one very short but rather influential paper [8] on many-valued logic published in 1932.

Note that the truth function of $\wedge$ is minimum and that of $\vee$ is maximum, independently from the choice of $*$.

Definition 5.3 A formula φ is a $*$-tautology if $e_*(\varphi) = 1$ for each e. A formula φ is a *t-tautology* or *standard BL-tautology* if $e_*(\varphi) = 1$ for each evaluation e and each continuous t-norm $*$. The following t-tautologies are taken to be *axioms of the logic* BL:

(A1) $(\varphi \to \psi) \to ((\psi \to \chi) \to (\varphi \to \chi))$
(A2) $(\varphi \& \psi) \to \varphi$
(A3) $(\varphi \& \psi) \to (\psi \& \varphi)$
(A4) $(\varphi \& (\varphi \to \psi)) \to (\psi \& (\psi \to \varphi))$
(A5a) $(\varphi \to (\psi \to \chi)) \to ((\varphi \& \psi) \to \chi)$
(A5b) $((\varphi \& \psi) \to \chi) \to (\varphi \to (\psi \to \chi))$
(A6) $((\varphi \to \psi) \to \chi) \to (((\psi \to \varphi) \to \chi) \to \chi)$
(A7) $\bar{0} \to \varphi$

The *deduction rule* of BL is modus ponens (saying that from φ and $\varphi \to \psi$ we may infer ψ, for each φ, ψ). Given this, the notions of a *proof* and of a *provable formula* in BL are defined in the obvious way: a proof in BL is a sequence $\varphi_1, \ldots, \varphi_n$ of formulas such that each φ_i either is an axiom or it follows from some preceding members of the sequence by the inference rule. A formula is provable in BL if it is the last member of a proof.

All axioms of BL are $*$-tautologies for each $*$ and so is each formula provable in BL.

Note that Łukasiewicz logic is the extension of BL by the axiom $\neg\neg\varphi \to \varphi$; Gödel logic is the extension of BL by the axiom $\varphi \to (\varphi \& \varphi)$. Finally, product logic is the extension of BL by the following two axioms:

$$\neg\neg\chi \to (((\varphi \& \chi) \to (\psi \& \chi)) \to (\varphi \to \psi)),$$

$$\varphi \wedge \neg\varphi \to \bar{0}.$$

Now we present a more general semantics of the logic BL.

Definition 5.4 A *BL-algebra* is an algebra

$$\mathbf{L} = (L, \cap, \cup, *, \Rightarrow, 0, 1)$$

with four binary operations and two constants such that

(i) $(L, \cap, \cup, 0, 1)$ is a lattice with the largest element 1 and the least element 0 (with respect to the lattice ordering $\leq$),
(ii) $(L, *, 1)$ is a commutative semigroup with the unit element 1, i.e. $*$ is commutative, associative, $1 * x = x$ for all x (thus $\mathbf{L}$ is a residuated lattice, and
(iii) the following conditions hold for all x, y, z:

(1) $z \leq (x \Rightarrow y)$ iff $x * z \leq y$ (residuation),

(2) $x \cap y = x * (x \Rightarrow y)$ (divisibility),
(3) $(x \Rightarrow y) \cup (y \Rightarrow x) = 1$ (prelinearity).
Define $neg(x) = (x \Rightarrow 0)$.

Definition 5.5 MV-*algebras* are BL-algebras satisfying $(-)(-)x = x$. G-*algebras* are BL-algebras satisfying $x * x = x$. *Product algebras* are BL-algebras satisfying

$$x \cap (-)x = 0$$

$$[(-)(-)z \Rightarrow ((x * z = y * z) \Rightarrow x = y)] = 1.$$

Let **L** be a BL-algebra. An **L**-evaluation of propositional variables is any mapping e assigning to each propositional variable p an element $e(p)$ of **L**. This extends in the obvious way to an evaluation $e_{\mathbf{L}}(\varphi)$ of all formulas φ using the operations on **L** as truth functions. A formula φ is in **L**-tautology if $e_{\mathbf{L}}(\varphi) = 1$ for each **L**-evaluation e.

Theorem 5.1 *BL is complete, i.e. for each formula φ the following three conditions are mutually equivalent:*

(i) φ is provable in BL,
(ii) for each linearly ordered BL-algebra **L**, φ is an **L**-tautology;
(iii) for each BL-algebra **L**, φ is an **L**-tautology (say, φ is a BL-tautology).

Note that we also get *strong completeness* (for provability in theories over BL). A theory is just a set of formulas (axioms of the theory). A proof in a theory T (over BL) is a sequence of formulas whose each member is an axiom of BL or an axiom of T or follows from some preceding formulas by modus ponens.

Theorem 5.2 *For each theory T over BL, T proves φ iff for each [linearly ordered] BL-algebra* **L**, *φ is true in all* **L**-*models of T. (Here e is an* **L** *model of T if $e_{\mathbf{L}}(\alpha) = 1_{\mathbf{L}}$ for each axiom α of T.)*

Definition 5.6 (1) A t-algebra is a BL-algebra $([0,1], \cap, \cup, *, \Rightarrow, 0, 1)$ whose lattice part is the real interval $[0,1]$ with min and max and $*$ is a continuous t-norm (whereas $\Rightarrow$ is its residuum).
(2) A formula φ is a t-*tautology* if it is an **L**-tautology for each t-algebra **L**.

Theorem 5.3 *(Cignoli-Esteva-Godo-Torrens [3]) Each t-tautology is provable in BL, i.e. t-tautologies coincide with BL-tautologies (and hence with formulas provable in BL).*

Remark (Decidability) t-tautologicity (= BL-tautologicity) is decidable (is co-NP complete). Similarly for $*$-tautologicity for any fixed continuous t-norm, like Łukasiewicz, product or Gödel t-norm, is provable in BL. (See [1, 11, 16].) Note that the logics of the three named t-norms are completely axiomatized by the following:

Ł by BL plus the axiom $\neg\neg\varphi \rightarrow \varphi$ of double negation,
G by BL plus the axiom $\varphi \rightarrow (\varphi\&\varphi)$ of idempotence of conjunction,
Π by BL plus the axiom $\neg\neg\varphi \rightarrow ((\varphi \rightarrow (\varphi\&\psi)) \rightarrow (\psi\&\neg\neg\psi))$.

5.4 Basic Fuzzy Predicate Logic

Definition 5.7 A *predicate language* consists of *predicates* $P, Q, \ldots$, each together with its *arity* (number of arguments) and *object constants* $c, d, \ldots$. *Logical symbols* are *object variables* $x, y \ldots$, connectives $\&, \rightarrow$, *truth constants* $\bar{0}, \bar{1}$ and quantifiers $\forall, \exists$. Other connectives $(\wedge, \vee, \neg, \equiv)$ are defined as in propositional calculus. *Terms* are object variables and object constants.

Atomic formulas have the form $P(t_1, \ldots, t_n)$ where P is a predicate of arity n and $t_1, \ldots, t_n$ are terms. If φ, ψ are formulas and x is an object variable then $\varphi \rightarrow \psi, \varphi\&\psi, (\forall x)\psi, (\exists x)\varphi, \bar{0}, \bar{1}$ are formulas; each formula results from atomic formulas by iterated use of this rule.

Let $\mathscr{J}$ be a predicate language and let $\mathbf{L}$ be a BL-algebra. An $\mathbf{L}$-*structure* $\mathbf{M} = \langle M, (r_P)_P, (m_c)_c \rangle$ for $\mathscr{J}$ consists of the following:, a set $M \neq \emptyset$ (the domain), for each n-ary predicate P a $\mathbf{L}$-fuzzy n-ary relation $r_P \colon M^n \rightarrow \mathbf{L}$ on M and for each object constant c, an element m_c of M. An $\mathbf{M}$-*evaluation of object variables* is a mapping v assigning to each variable x an element $v(x)$ of M denoted also $\|x\|_{M,v}$; the $\mathbf{M}$-value of a constant c is $\|c\|_{M,v} = m_c$.

Truth values of formulas (in a given interpretation $\mathbf{M}$, with the evaluation v of variables and with the truth functions given by the algebra $\mathbf{L}$) are defined by induction as follows: for atomic formulas

$\|P(t_1, \ldots, t_n)\|^L_{M,v} = r_P(\|t_1\|_{M,v}, \ldots, \|t_n\|_{M,v})$;

furthermore

$\|\varphi \rightarrow \psi\|^L_{M,v} = \|\varphi\|^L_{M,v} \Rightarrow \|\psi\|^L_{M,v}$;

$\|\varphi\&\psi\|^L_{M,v} = \|\varphi\|^L_{M,v} * \|\psi\|^L_{M,v}$;

$\|\bar{0}\|_{M,v}$; $\quad \|\bar{1}\|_{M,v}$;

$\|(\forall x)\varphi\|^L_{M,v} = \inf\{\|\varphi\|^L_{M,v'} | v \equiv_x v'\}$;

$\|(\exists x)\varphi\|^L_{M,v} = \sup\{\|\varphi\|^L_{M,v'} | v \equiv_x v'\}$;

provided the infimum/supremum exists in the sense of $\mathbf{L}$.

The structure $\mathbf{M}$ is $\mathbf{L}$-*safe* if all the needed infima and suprema exist, i.e. $\|\varphi\|^L_{M,v}$ is defined for all φ, v.

Next define the (global) truth degree of a formula in an interpretation using a BL-algebra of truth functions:

$\|\varphi\|_M = \inf\{\|\varphi\|_{M,v}\ \mathbf{M} - \text{evaluation.}\}$

A formula φ of a language $\mathscr{J}$ is an $\mathbf{L}$-*tautology* if $\|\varphi\|_M = 1_{\mathbf{L}}$ for each safe $\mathbf{L}$-structure $\mathbf{M}$.

Definition 5.8 The following are *logical axioms on quantifiers:*

($\forall$1) $(\forall x)\varphi(x) \to \varphi(t)$ (for t substitutable for x in $\varphi(x)$)

($\exists$1)$\varphi(t) \to (\exists x)\varphi(x)$ (for t substitutable for x in $\varphi(x)$)

($\forall$2)$(\forall x)(\nu \to \varphi) \to (\nu \to (\forall x)\varphi)$ (for x not free in ν)

($\exists$2)$(\forall x)(\varphi \to \nu) \to ((\exists x)\varphi \to \nu)$ (x not free in ν)

$\forall$3)$(\forall x)(\nu \vee \varphi) \to (\nu \vee (\forall x)\varphi)$ (x not free in ν)[4]

The *fuzzy predicate calculus* $\mathscr{C}\forall$ (for a given fuzzy propositional calculus C) has the following axioms:

All formulas resulting from the axioms of $\mathscr{C}$ by substituting arbitrary formulas of $\mathscr{J}$ for propositional variables, and

The axioms ($\forall$1), ($\forall$2), ($\exists$1), ($\exists$2), ($\forall$3) for quantifiers.

Deduction rules are modus ponens (from φ, $\varphi \to \psi$) infer ψ and generalization (from φ infer $(\forall x)\varphi$).

In particular, we have the logic BL$\forall$ and three stronger logics: Ł$\forall$ (Łukasiewicz), G$\forall$ (Gödel), $\Pi\forall$ (product).

Lemma 5.1 *The axioms* ($\forall$1) – ($\exists$2) *are* **L***-tautologies for each BL-algebra* **L***;* ($\forall$3) *for each BL-chain.*

A theory is a set T of formulas (special axioms of the theory). An **L**-model of T is a safe **L**-structure in which all special axioms of T are true (have the **L**-truth value $1_{\mathbf{L}}$. One can prove the following

Completeness theorem Let $\mathscr{C}\forall$ be the predicate calculus given by an extension C of BL by some additional axiom schemas, let T be a theory over $\mathscr{C}\forall$ let φ be a formula of the language of T. T proves φ iff for each linearly ordered C-algebra **L** and each **L**-model of T, $\varphi^{\mathbf{L}}_{\mathbf{M}} = 1_{\mathbf{L}}$.

Let us stress that similarly as in classical logic, provability in BL$\forall$ (and in stronger logic like Łukasiewicz, Gödel and product predicate logic) is not decidable; it is of the same degree of unsolvability as classical logic (Σ_1-complete). By completeness, provability in such a logic is equivalent to tautologicity over all corresponding linearly ordered algebras (BL-chains, MV-chains etc.)

In contradistinction to propositional case, tautologicity over t-norm algebras (standard tautologicity) differs from tautologicity over BL-chains (general tautologicity) and is of different degree of unsolvability: the set of standard tautologies for BL$\forall$ it is outside of arithmetical hierarchy (extremely undecidable). For stronger logics the situation is as follows: the set of standard tautologies of G$\forall$ (tautologies over $[0,1]_G$ is Σ_1-complete, for Ł$\forall$ it is Π_2-complete, for $\Pi\forall$ not arithmetical. (See [10, 18].)

This ends our survey of the basic notions and facts of the fuzzy predicate logic BL$\forall$ and some stronger fuzzy logics; next we show some applications and generalizations.

---

[4] Substitutability of a term and freeness of a variable are natural syntactical restrictions defined in the same way as in the classical predicate logic.

5.5 Probability

Probability on formulas (of classical logic) cannot be understood as an assignment of truth values in the sense of a (truth-functional) fuzzy logic; but still there are bridges between probability and fuzziness. Compare the sentence “Temperature is high.” and “The probability of ... is high” or just “... is probable”. Here evidently “probably” acts as a fuzzy modality.

This can be formalized as follows: Work with propositional language with two kinds of formulas: *non-modal* and *modal* formulas. Non-modal formulas are just formulas of the classical propositional calculus. Atomic modal formulas have the form $P\varphi$ where φ is any non-modal formula ($P\varphi$ is read “φ is probable”) and other modal formulas are built from the atomic modal formulas using connectives of Łukasiewicz logic.

A *model* (interpretation) of such a langauge is a (Kripke) structure $\mathbf{K} = (W, e, \mu)$ where W is a non-empty set of possible worlds, e is a boolean evaluation (for each propositional variable p and each possible world $w, e(p,w) \in \{0,1\}$); μ is a finitely additive probability on subsets of W such that for each propositional variable p the set $\{w | e(p,w) = 1\}$ is measurable. Evidently, e uniquely determines for each possible world w which non-modal formulas are true in w and which are false in w.

The truth value of an atomic modal formula $\|P\varphi\|_{\mathbf{K}}$ is the probability of φ, i.e. $\mu\{w | \varphi \text{ true in } w\})$.

The following axioms are postulated:

(FP1) $P(\neg\varphi) \equiv \neg P\varphi$

(FP2) $P(\varphi \to \psi) \to (P\varphi \to P\psi)$,

(FP3) $P(\varphi \vee \psi) \to ((P\varphi \to P(\varphi \wedge \psi)) \to P\psi)$.

Postulating axioms of classical logic for non-modal formulas, axioms of Łukasiewicz logic plus our (FP1) to (FP3) for modal formulas and taking as deduction rules modus ponens and necessitation (from φ infer $P\varphi$) you get a logic complete with respect to the above semantics.

5.6 Sorites Paradox

One grain does not form a heap. Adding one grain to what is not yet a heap does not make a heap. Consequently, there are no heaps.

So in classical logic. The problem is if we can deal with a notion of a *small* or feasible number (a heap must have a large – non-feasible number of grains). There were various attempts to formalize it (Parikh, Vopěnka). We show a solution in fuzzy logic.

We extend the basic predicate logic $BL\forall$ by a new unary connective *At* (*At*(φ) being read “it is almost true that φ”) and by the following two axiom schemata.

$\varphi \to At(\varphi)$,

$(\varphi \to \psi) \to (At(\varphi) \to At(\psi))$.

Extend the language of the well known Peano arithmetic *PA* by a new unary predicate *Fe* ($Fe(x)$ is read "x is feasible"). Work with the variant of $BL\forall$ with function symbols. Thus the language consists of the binary equality predicate =, unary predicate *Fe*, constant $\bar{0}$ (zero), unary function symbol S (successor), binary function symbols $+, \cdot$ (addition, multiplication).

The axioms of PA_{at} are as follows:
$x = y \vee x \neq y$ (crispness axiom for =),
all axioms of Peano arithmetic (including the definition of $x \leq y$ as $(\exists z)(z+x=y)$),
$x < y \rightarrow (Fe(y) \rightarrow Fe(x))$,
$Fe(x) \rightarrow (At(Fe(S(x)) \wedge At(Fe(x+x)) \wedge At(Fe(x \cdot x))$.

Fact (1) For each formula φ of *PA* (i.e. not containing the predicate *Fe*), PA_{at} proves $\varphi \vee \neg\varphi$ (tertium non datur).
(2) A formula not containing *Fe* is provable in *PA* over classical logic iff it is provable in PA_{at} over $BL\forall_{at}$.
(3) PA_{at} proves

$$Fe(x) \wedge Fe(y) \rightarrow (At(Fe(x+y)) \wedge At(Fe(x \cdot y)).$$

In words: if x and y are feasible then it is almost true that also $x+y, x \cdot y$ are feasible.

Example: let p be a truth constant for a non-extremal truth value (say, 0.99) and define $at(\varphi) \equiv (p \rightarrow \varphi)$. Clearly, this satisfies our axioms; furthermore, the following becomes provable:

$$At(\varphi \rightarrow \psi) \rightarrow (\varphi \rightarrow At(\psi)),$$

$$(At(\varphi) \& At(\varphi \rightarrow \psi)) \rightarrow At(At(\psi)).$$

5.7 The Big Family of Fuzzy Logics

We quickly survey important more general logic of left-continuous t-norms (Esteva-Godo [6]): MTL – the monoidal t-norm-based logic has the language of BL extended by an additional binary connective $\wedge$, axioms (A1), (A2), (A3), (A5a), (A5b), (A6), (A7), plus the following three axioms which replace (A4):

(A4a) $(\varphi \& (\varphi \rightarrow \psi)) \rightarrow (\varphi \wedge \psi)$
(A4b) $(\varphi \wedge \psi) \rightarrow \varphi$
(A4c) $(\varphi \wedge \psi) \rightarrow (\psi \wedge \varphi)$

Recall that in BL, $\wedge$ is defined in terms of & and $\rightarrow$: $\varphi \wedge \psi = \varphi \& (\varphi \rightarrow \psi)$. But this is *not* the case in MTL. $\neg\varphi$ and $\varphi \vee \psi$ are abbreviations for $\varphi \rightarrow \bar{0}$ and $((\varphi \rightarrow \psi) \rightarrow \psi) \wedge ((\psi \rightarrow \varphi) \rightarrow \varphi)$ respectively (as in BL).

An MTL *algebra* is a system $\mathbf{L} = \langle L, *, \Rightarrow, \cap, 0, 1 \rangle$ satisfying all the axioms of a BL algebra except divisibility.

Theorem 5.4 (Completeness) *MTL is complete with respect to the class of standard MTL algebras, linearly ordered MTL algebras as well as general MTL algebras. (Similarly for stronger logic IMTL corresponding to Łukasiewicz (add double negation axiom) and some other logics. Furthermore, MTL is decidable.*

The predicate MTL logic, MTL∀ is defined in the obvious way and has general completeness as BL: provability in MTL∀ equals to tautologicity over MTL-chains.

BUT: MTL∀ has also standard completeness: provability in MTL∀ equals to tautologicity over all algebras $[0,1]_*$ where $*$ is a left-continuous t-norm. (Montagna-Ono).

Complexity: MTL∀ is undecidable (Montagna-Ono).

5.8 Making Fuzzy Description Logic More General

Description logic has become extensively studied in last years (in relation to web mining). It is known to be a fragment of classical predicate calculus and has several variants; we restrict ourselves to that named ALC.

The language: unary predicates $A_1,\ldots$ (atomic concepts), binary predicates $R_1,\ldots$ (roles), variables $x_1,\ldots$ and constants $a_1,\ldots$ Concepts are built from atomic concepts using connectives $\wedge,\vee,\neg$ and quantification constructs denoted $\forall R.C, \exists R.C$. Think of instances of atomic concepts as of formulas $A_i(t)$ (t being a fixed constant or variable); this extends to (instances of) concepts defined by connectives.

Furthermore,

$$(\forall R.C)(t) \text{ is to be read as } (\forall y)(R(t,y) \rightarrow C(y))$$

$$(\exists R.C)(t) \text{ as } (\exists y)(R(t,y) \wedge C(y)).$$

(For each construct always a new variable y is used.) Axioms may have the form $C(a_i)$ (C a concept), $R(a_i,a_j)$ (R a role) and $(\forall x)(C(x) \rightarrow D(x))$ (subsumption of concepts). Typical *problems* are:

satisfiability: does $C(a)$ have a model?
validity: is $(\forall x)C(x)$ true in all models?
in particular: subsumption: is $(\forall x)(C(x) \rightarrow D(x))$ true in all models?

Over classical logic, both problems are *decidable* (and are PSPACE-complete).

It is very natural, due to the intended applications, to make this system (and related systems) fuzzy. Our main contribution is a systematic reduction of problems of satisfiability and validity of concepts to problems of satisfiability and validity of some theories in propositional logic, problems that are known to be decidable, using as much of the richness of fuzzy logic as possible (cf. [15] and the references thereof). What follows is a short summary of the paper [15].

Generalized atoms are instances of quantified concepts, i.e. $(\forall R.C)(t)$, $(\exists R.C)(t)$. Evidently, each instance $C(t)$ of any concept is a propositional combination of some atoms and generalized atoms; the latter will be called *generalized atoms of* $C(t)$.

Assign to each generalized atom $G(a)$ (G atomic concept or a quantified concept, a a constant) a propositional variable $p_{G,a}$. Extend this to all instances $C(a)$ of concepts by defining

$prop(\bar{0}(a)) = \bar{0}$,

$prop((C\&D)(a)) = prop(C(a))\,\&\,prop(D(a))$ and similarly for $(C \to D)(a)$.

If T is a set of formulas (instances of concepts) let $prop(T)$ be the set of all $prop(\alpha), \alpha \in T$.

There is an algorithm assigning to each (closed) instance $C_0(a_0)$ of a concept C_0 with a constant a_0 a finite "witnessing" theory $T(C_0(a_0))$ important in what follows.

From now on, fix a continuous t-norm $*$ (and its residuum); this gives semantics of your fuzzy propositional and predicate calculus. In particular, each evaluation e of propositional atoms by truth values from the real interval $[0,1]$ defines uniquely the truth value $e_*(\varphi)$ of each propositional formula built from these atoms.

Definition 5.9 For any $\mathbf{M}$, a closed formula $(\forall y)\varphi(y)$ is $*$-witnessed if $\|(\forall y)\varphi(y)\|^*_{\mathbf{M}}$ (which is the infimum of values of M-instances of $\varphi(y)$) is in fact equal to $\|\varphi(u)\|^*_{\mathbf{M}}$ for some $u \in M$, thus the infimum is in fact the *minimum.* Similarly for witnessed $\|(\exists y)\varphi(y)\|^*_{\mathbf{M}}$ (*maximum*). More generally, $(\forall y)\varphi(y,x_1,\ldots,x_n)$ is $*$-witnessed in $\mathbf{M}$ if for any choice $v_1,\ldots,,v_n \in M$ of values of $x_1,\ldots,x_n$, the truth value $\|(\forall y)\varphi(y,v_1,\ldots,v_n)\|^*_{\mathbf{M}}$ equals to $\|\varphi(u,v_1,\ldots,v_n)\|^*_{\mathbf{M}}$ for some $u \in M$. Similarly for $(\exists y)\varphi(y,x_1,\ldots,x_n)$ and maximum. $\mathbf{M}$ is $*$-*witnessed* if all quantified formulas are $*$-witnessed in $\mathbf{M}$.

Theorem 5.5 *(1) Over Łukasiewicz logic, a concept C is satisfiable iff it is satisfiable by a finite model iff the associated finite set* $prop(C(a)) \cup prop(T(C(a)))$ *is propositionally satisfiable.*
(2) Over Łukasiewicz logic, a concept C is valid iff it is valid in all finite models iff the propositional theory $prop(T(C(a))$ *entails* $prop(C(a))$*, i.e. each evaluation e of propositional variables which is an Ł-model of* $prop(T(C(a)))$ *gives* $prop(C(a))$ *the Ł-value 1.*

Corollary 5.1 *Over Łukasiewicz logic, the (standard) satisfiability and (standard) validity of a concept are decidable problems.*

Theorem 5.6 *For any continuous t-norm $*$ an any concept C, the following are equivalent:*

1. (1) C is valid in all witnessed $*$-models
2. (2) C is valid in all finite $*$-models
3. (3) $prop(T(C(a)))$ $*$-entails $prop(C(a))$.

Corollary 5.2 *For an arbitrary continuous t-norm $*$, witnessed $*$-satisfiability and witnessed $*$-validity of a concept is decidable.*

This is because propositional $*$-satisfiability and propositional $*$-entailment are decidable.

Acknowledgements The work the author was partly supported by grant A100300503 of the Grant Agency of the Academy of Sciences of the Czech Republic and partly by the Institutional Research Plan AV0Z10300504.

References

1. Baaz M., Hájek P., Montagna F., and Veith H. Complexity of t-tautologies, *Ann. Pure App. Logic*, 113: 3–11, 2002.
2. Cignoli R., D'Ottaviano I., and Mundici D. *Algebraic Foundations of Many-Valued Reasoning. Trends in Logic*. Kluwer, Dordrecht, 2000.
3. Cignoli R., Esteva F., Godo L., and Torrens A. Basic logic is the logic of continuous t-norms and their residua, *Soft Comp.*, 4: 106–112, 2000.
4. Cintula P. About axiomatic systems of product fuzzy logic, *Soft Comp.*, 5: 243–244, 2001.
5. Esteva F., and Godo L. Putting together Łukasiewicz and product logic, *Mathware and soft comput.*, 6: 219–234, 1999.
6. Esteva F., and Godo L. Monoidal t-norm based logic, *Fuzzy Sets Syst.*, 124: 271–288, 2001.
7. Esteva F., Godo L., Hájek P., and Navara M. Residuated fuzzy logics with an involutive negation, *Archive for Math. Log.*, 39: 103–124, 2000.
8. Gödel K. Zum intuitionischen Aussagenkalkül. Anzeiger Akademie der Wissenschaften Wien, *Math.-Naturw. Klasse*, 69: 65–66, 1932.
9. Goguen J. A. The logic of inexact concepts, *Synthese*, 19: 325–373, 1968–1969.
10. Gottwald S. *A Treatise on Many-Valued Logic*. Research Studies Press Ltd., Baldock, 2001.
11. Hájek P. Fuzzy logic and arithmetical hierarchy III, *Studia Logica*, 68: 129–142, 2001.
12. Hájek P. *Metamathematics of Fuzzy Logic*. Kluwer, Dordrecht, 1998.
13. Hájek P. On very true, *Fuzzy Sets Syst.*, 124: 329–334, 2001.
14. Hájek P., Paris J., and Shepherdson J. The liar paradox and fuzzy logic, *J. symb. Logic*, 65: 339–346, 2000.
15. Hájek P. Fuzzy predicate calculus and fuzzy rules. In D. Ruan and Kerre, editors, *Fuzzy IF-THEN Rules in Computational Intelligence*, pages 27–36. Kluwer, Dordrecht, 2000.
16. Hájek P. Making fuzzy description logic more general, *Fuzzy Sets Syst.*, 154: 1–15, 2005
17. Haniková Z. A note on the complexity of propositional tautologies of individual algebras, *Neural Networks World*, 12: 453–460, 2002.
18. Klement E. P., Mesiar R., and Pap E. *Triangular Norms*. Kluwer, Dordrecht, 2000.
19. Łukasiewicz J. O logice trójwartosciowej (on the three-valued logic), *Ruch Filozoficzny*, 5: 170–171, 1920.
20. Montagna F. Three complexity problems in quantified fuzzy logic, *Studia Logica*, 68: 143–152, 2001.
21. Novák V. *Fuzzy Sets and thier Applications*. Adam Hilger, Bristol, 1989.
22. Novák V., Perfilieva I., and Močř J. *Mathematical Principles of Fuzzy Logic*. Kluwer, Dordrecht, 2000.
23. Pavelka J. On fuzzy logic I, II, III. Zeitschrift für Math, *Logik und Grundlagen der Math*, 25: 45–52, 119–134, 447–464, 1979.
24. Turunen E. *Mathematics behind fuzzy logic*. Physica Verlag, Heidelberg, 1999.
25. Zadeh L. Fuzzy sets, *Inform. Contr.*, 8: 338–353, 1965.

Part III
Logic and Computation

Chapter 6
What Is the Difference Between Proofs and Programs?

John N. Crossley

6.1 Introduction

About the year 1900 there was just "one true logic": classical logic. In such a logic one would expect that everything was clear. Certainly, in that logic, any statement was either true or false: there was the law of the excluded middle, $(A \vee \neg A)$. But how do we check an infinite number of instances? What does it mean to say that there is no largest pair of twin primes, that is to say that there is an end to such pairs such as 5 and 7; 11 and 13 or even 202 289 and 202 291?

On the other hand, saying there are infinitely many pairs of twin primes has a clear meaning if we can show that, for every pair, there is a larger pair. In this context compare the way that Euclid established that there were infinitely many prime numbers (although he did not phrase it like that). He gave a method for constructing a larger prime from a given (finite) set of prime numbers.

It was because of questioning by Brouwer, a Dutch mathematician, indeed a topologist, of the "one true logic", that "constructive logic" or "intuitionist logic" arose.

When you look at this logic it is not evident, at first glance, that the logic actually gives you the way of performing the necessary construction. However, that is perhaps the wrong way to look at it. Brouwer was concerned only with constructing mathematical objects that were claimed to exist. He did not like mathematical logic and did not consider it relevant. However, when his approach is formalized, the details are buried inside the proof. Are they perhaps buried in the way that algorithms are buried inside computer programs?

Thus arose "constructive logic" or "intuitionist(ic) logic". In the 1960s I went to a course on Intuitionism taught by Michael Dummett. The lectures have since become a book [8]. I was a student then and I must admit that I found it very odd. Odd, but

John N. Crossley
School of Information Technology, Monash University, Clayton, VIC 3800, Australia, e-mail: John.Crossley@infotech.monash.edu.au

J. van Benthem et al. (eds.), *Proof, Computation and Agency*, Synthese Library 352,
DOI 10.1007/978-94-007-0080-2_6,

also very interesting, indeed, fascinating. The subject matter seemed strange to us in the audience because it did not use the law of the excluded middle. Nowadays intuitionist logic has become a very standard subject as you can see from Michael Dummett's book.

The most important thing about constructive proofs is that they contain the information that allows one to construct the objects considered. For example if we prove $\forall x \exists y A(x,y)$ constructively then, given an x, we can actually construct a y such that $A(x,y)$ is true, indeed, is provable. The information required for the construction is embedded in the proof.

For further details of the actual logical system that I use please see the tutorial at this conference [5]. The system is outlined in Fig. 6.1. Here $\Delta, A \vdash_{\mathsf{Int}} B$ means "B can be inferred from A and the formulae in Δ". The restriction to Harrop formulae is a technical one.[1]

6.2 The Lambda Calculus and the Curry-Howard Correspondence

How do we extract the information from a proof in mathematical logic? Curry started, and Bill Howard [13] developed, the basic idea. To show this we need some notation. We begin with the lambda calculus, but this calculus will slowly get extended. (For the present state of the art see the work of Barendregt, in particular [1, 2].) Here is the formal definition of lambda terms.

Definition 6.1 (Lambda terms) The alphabet comprises variables $x_1, y_1, \ldots$, together with λ and ".", and the brackets (and).

The *lambda terms*, Λ, are formed as follows (in Backus-Naur notation):

$$T = x | \lambda x.T | (T_1 T_2)$$

The lambda calculus has two major constructions: *abstraction* and *application*.

Where does the lambda calculus come from? Consider the following:

What is the function denoted by x^y? We have several choices: as a function of two variables, as a function of x only with y held constant and as a function of y only with x held constant. These are usually denoted by

$$\lambda x \lambda y.x^y (\text{often written as } \lambda xy.x^y),\ \lambda x.x^y \text{ and } \lambda y.x^y.$$

This is *abstraction*.

[1] A formula F is a *Harrop formula* if it is (1) an atomic formula, (2) of the form $(A \wedge B)$ where A and B are Harrop formulae, (3) of the form $(A \to B)$ where B (but not necessarily A) is a Harrop formula, or (4) of the form $\forall x.A$ where A is a Harrop formula. Harrop formulae, in a sense, contribute no information for the program. However, the rule ($\perp$-E) easily extends, through a proof by induction, to provide a proof of any formula A from the false formula $\perp$ since atomic formulae are Harrop and so are their negations $(A \to \perp)$.

Application is written in a familiar way: thus $(T_1 T_2)$ denotes the application of the lambda term T_1 to the lambda term T_2. In particular fa is the application of the lambda term f to the lambda term a. (We omit brackets where there is no ambiguity.)

These notations have the obvious interpretations. (Try them on x^y and specific values of x and y.)

In ordinary mathematics if we apply the function $\lambda x.f$ to a then we get $f[a/x]$, which is read "f with a for x". In the lambda calculus however this is *not* the same as the (application) term $(\lambda x.f)a$, i.e. $\lambda x.f$ applied to a. That is to say they are *syntactically* different. We therefore have to introduce the notion of *β-reduction*[2]

$$(\lambda x.f)a \rhd f[a/x]$$

(Here $\rhd$ is read "reduces to".)

Now note the similarities between $\rightarrow$-introduction, the rule ($\rightarrow$-I), and $\rightarrow$-elimination ($\rightarrow$-E) (in Fig. 6.1) on the one hand, and λ-introduction and λ-elimination on the other, the β-rule.

Next consider a proof of B from A from which we get a proof[3] of $A \rightarrow B$ (by the rule ($\rightarrow$-I):

$$\frac{\begin{matrix}[A]\\ \vdots \\ B\end{matrix}}{(A \rightarrow B)}$$

and lambda abstraction (which abstracts a function from the process where $a \in A$ gives us $f(a) \in B$): that is $\lambda x.f$. Consider the figure[4]:

$$\frac{\begin{matrix}a\\ \vdots \\ f(a)\end{matrix}}{\lambda x.f}$$

What is the connexion?

The most obvious thing, I hope, is that the *shapes* are the same!

The *typed* lambda calculus that we shall consider, that is to say lambda calculus with each term having a *type* assigned to it, can be regarded as the amalgam of the two systems: logic, or more precisely, systems of predicate calculus, and the lambda calculus.

A special kind of typed lambda calculus involves taking formulae of logic as the types. Now this is a strange idea to accept but it is easier to work with it if you just think of a type (formula) as the set of proofs of that formula. Instead, therefore, of variables, we use typed variables of the form $a : A$.

[2] α-reduction refers to the simple renaming of one variable by another (without clashes).

[3] The square brackets indicate that A can be discharged, i.e. is not needed for the proof of B, though it is for the proof of B, of course.

[4] Here we have written $f(a)$ to show that we think of a as being involved in f.

Assume that x, y are individual variables, and that t and t' are individual terms.

$$\frac{}{A \vdash_{\mathsf{Int}} A}\ (\text{Ass-I})$$

$$\frac{\Delta, A \vdash_{\mathsf{Int}} B}{\Delta \vdash_{\mathsf{Int}} (A \rightarrow B)}\ (\rightarrow\text{-I}) \qquad \frac{\Delta \vdash_{\mathsf{Int}} A \quad \Delta' \vdash_{\mathsf{Int}} (A \rightarrow B)}{\Delta, \Delta' \vdash_{\mathsf{Int}} B}\ (\rightarrow\text{-E})$$

$$\frac{\Delta \vdash_{\mathsf{Int}} A}{\Delta \vdash_{\mathsf{Int}} \forall x.A}\ (\forall\text{-I}) \qquad \frac{\Delta \vdash_{\mathsf{Int}} \forall x.A}{\Delta \vdash_{\mathsf{Int}} A[t/x]}\ (\forall\text{-E})$$

x is free in A, not free in Δ

$$\frac{\Delta \vdash_{\mathsf{Int}} P[t'/y]}{\Delta \vdash_{\mathsf{Int}} \exists y.P}\ (\exists\text{-I}) \qquad \frac{\Delta_1 \vdash_{\mathsf{Int}} \exists y.P \quad \Delta_2, P[x/y] \vdash_{\mathsf{Int}} C}{\Delta_1, \Delta_2 \vdash_{\mathsf{Int}} C}\ (\exists\text{-E})$$

where x is not free in C

$$\frac{\Delta \vdash_{\mathsf{Int}} A \quad \Delta' \vdash_{\mathsf{Int}} B}{\Delta, \Delta' \vdash_{\mathsf{Int}} (A \wedge B)}\ (\wedge\text{-I})$$

$$\frac{\Delta \vdash_{\mathsf{Int}} (A_1 \wedge A_2)}{\Delta \vdash_{\mathsf{Int}} A_1}\ (\wedge\text{-E}_1) \qquad \frac{\Delta \vdash_{\mathsf{Int}} (A_1 \wedge A_2)}{\Delta \vdash_{\mathsf{Int}} A_2}\ (\wedge\text{-E}_2)$$

$$\frac{\Delta \vdash_{\mathsf{Int}} A_1}{\Delta \vdash_{\mathsf{Int}} (A_1 \vee A_2)}\ (\vee\text{-I}_1) \qquad \frac{\Delta \vdash_{\mathsf{Int}} A_2}{\Delta \vdash_{\mathsf{Int}} (A_1 \vee A_2)}\ (\vee\text{-I}_2)$$

for any Harrop formula A

$$\frac{\Delta \vdash_{\mathsf{Int}} A \vee B \quad \Delta_1, A \vdash_{\mathsf{Int}} C \quad \Delta_2, B \vdash_{\mathsf{Int}} C}{\Delta_1, \Delta_2, \Delta \vdash_{\mathsf{Int}} C}\ (\vee\text{-E})$$

$$\frac{\Delta \vdash_{\mathsf{Int}} \bot}{\Delta \vdash_{\mathsf{Int}} A}\ (\bot\text{-E})$$

provided A is Harrop

$A[a/x]$ is read "A with a for x" aand denotes the formula A with a substituted for the variable x.

Fig. 6.1 The basic rules of intuitionistic logic, Int.

The rule of *modus ponens* then becomes:

$$\frac{a : A \quad g : (A \rightarrow B)}{(ga) : B} \tag{6.1}$$

where we have changed f to g to avoid confusion in what follows.

If we had a proof of B from A then we would get an expression $\lambda x : A.f : B$ by the rule of ($\rightarrow$-I) which has type $(A \rightarrow B)$. If the g in the expression (6.1) is actually of the form $(\lambda x : A.f : B) : (A \rightarrow B)$, then we get

$$\frac{a : A \qquad (\lambda x : a.f : B) : (A \rightarrow B)}{((\lambda x : A.f : B) : (A \rightarrow B))a : A) : B}$$

which is somewhat hard to read. However the bottom line has the formula B as its type, and the expression reduces to

$$f : B[a : A/x : A] \tag{6.2}$$

where the substitution of $a : A$ for $x : A$ takes place throughout the term $f : B$.

If we translate this back into proofs it means that the corresponding proofs look as follows. On the one hand we have the complicated proof:

$$\dfrac{\begin{array}{c}\vdots\\ A\end{array} \qquad \dfrac{\begin{array}{c}[A]\\ \vdots\\ B\end{array}}{(A \to B)}}{B} \tag{6.3}$$

and on the other hand, by putting the proof of A from the left on top of the proof of B, and *not* introducing the $\to$, we no longer need the hypothesis $[A]$ in the proof on the right in order to get a proof of B.

That is to say, we *reduce* the proof in (6.3) to a simple proof of B of the form

$$\begin{array}{c}\vdots\\ A\\ \vdots\\ B\end{array}$$

This corresponds in the lambda calculus to the reduction[5] that resulted in (6.2). So we have a direct correspondence between proofs and terms of our typed lambda calculus. This is called the *Curry-Howard correspondence*.[6]

6.3 Strong Normalization and Program Extraction

Now it is obvious that a long and complicated formal proof has an even longer typed lambda calculus expression associated with it. If, however, all the possible reductions are carried out it may become considerably simpler. Indeed, in the cases with which we are concerned, we can usually omit all the types. (They will have served their purpose of ensuring that we get a result of the correct type when the proof is complete. This is related to the use of types in computer programming languages.)

The maximum benefit is when we have a *Strong Normalization Theorem* for the system. Such a theorem says that, whatever the order of the reductions (and

[5] This process of reduction is also called *cut elimination*.

[6] Some people use the term *isomorphism* but there are technical difficulties involved in making the correspondence one to one, so I prefer the weaker terminology.

there may be many possible different reductions for a long lambda term) the process always stops. (One reason the process might be expected *not* to stop is clear when you look at substituting $x+x$ for x: the number of xs goes up each time and the expression gets longer!)

The Curry-Howard correspondence can be extended to the other logical connectives by modifying the lambda calculus. Surprisingly, in addition to the above operations involving lambdas, we only need the formation of ordered pairs and the projections onto the first and second elements of those pairs in order to capture all first order logic.[7] We give only a few examples; the full details can be found in [7]. The Curry-Howard term for a conjunction $(A \wedge B)$ obtained by the rule $(\wedge)$-I) is the ordered pair $(p : A, q : B)$ where $p : A$ is the Curry-Howard term for the proof of A, and similarly for B. Conversely we use the projections fst and snd for the rules $(\wedge)$-E$_1$) and $(\wedge)$-E$_2$). For the rule $(\exists)$-I we get the term $(t, p : A(t))$ where the premise has the Curry-Howard term $p : A(t)$. Thus the Curry-Howard term contains the term t that was proved to exist.

The major consequence of the Strong Normalization theorem is then that, if we prove a formula of the form $\exists x A(x)$, we can actually extract, from the normalized proof (i.e. the lambda, or Curry-Howard, term in which no more reductions are possible), an x such that $A(x)$. Further, if we can prove $\forall x \exists y A(x,y)$ then we can actually get a program such that, given an x, it will compute a corresponding y. Moreover, we have a proof of $A(x,y)$ for this x and y so the program is "correct" in the sense that it meets its specification.[8]

Curry-Howard terms are, in general, a generalization of the idea known variously as formulae-as-types or, better, as proofs-as-types: the terms code up a whole proof by successively encoding the applications of the logical rules in a proof.

Not surprisingly, not all rules of logic allow us to prove a strong normalization theorem. One major obstacle is the law of double negation: From $\neg\neg A$ infer A. If we had a rule that would allow us to prove $\exists x A(x)$ from $\neg\neg\exists x A(x)$, how do we obtain such an x? So we generally restrict ourselves to constructive logic and all is well.

Changing to other systems, e.g. arithmetic, may bring in other axioms. Here the most dramatic is the rule of induction. Fortunately the induction axiom

$$\frac{A(0) \qquad \forall x(A(x) \rightarrow A(x+1))}{\forall x A(x)}$$

gives rise to a reduction *exactly* corresponding to the recursion

$$f(\overline{a},0) \;=\; g(\overline{a})\;;\; f(\overline{a},x+1) \;=\; h(\overline{a},x,f(\overline{a},x))$$

Happily we can prove a strong normalization theorem for arithmetic (see [7]).

---

[7] The process can also be extended to higher order logic.

[8] Intuitively speaking, the specification is the statement about the result of the program. See also below Section 6.4.1.

6.4 Beyond Traditional Logic

6.4.1 Algebraic Specifications

We now turn to an application of the above ideas to software engineering. Producing programs that satisfy their specifications is a primary goal of software engineering.

What is an algebraic specification? It is a description in formal logic of a structure, for example, the natural numbers.

We use the *Common Algebraic Specification Language* (*CASL*, see [4]) but the technique could be employed in other specification languages, indeed originally we ourselves used a different language.

Structured specifications in *CASL* are built from *basic (or flat)* specifications by means of *translation (or renaming)*, written **with**, taking *unions* of specifications, written **and**, *hiding* signatures, and the *extension of specifications*. A typical example of a flat specification, this one is for natural numbers, is given in Fig. 6.2.

When we change a specification, then what is true changes – even if simply because we use new names, e.g. "car" instead of "auto", "boot" instead of "trunk", etc. but we may also add new predicates (relations).

We have developed logical systems to reflect the interaction between such changes and the logic statements.

Originally Martin Wirsing studied a logical calculus for structured specifications (see [17]). This was subsequently extended by Wirsing and his student Peterreins (see [15]). Next Wirsing and the present author extended the idea to algebraic specifications, and then we went even further with Iman Poernomo to include even parametrized specifications in the language *CASL*.

Abstractly speaking we have an annotated or labelled deductive system.[9] The basic form of a rule in such a logic can be written in the form

spec NAT_0 = **isorts**
Nat **iops***0* : *Nat*; *s* : *Nat* → *Nat*; + : *Nat* × *Nat* → *Nat* **ipreds**
≥: *Nat* × *Nat* **iaxioms**

$\forall x : Nat \bullet x + 0 = x$	%(Nat_0.1)%	
$\forall x; y : Nat \bullet x + s(y) = s(x + y)$	%(Nat_0.2)%	
$\forall x : Nat \bullet x \geq 0$	%(Nat_0.3)%	**iend**
$\forall x; y : Nat \bullet x + y = y + x$	%(Nat_0.4)%	
$\forall x : Nat \bullet s(x) \geq x$	%(Nat_0.5)%	
$\forall x; y; v; w : Nat \bullet x \geq v \wedge y \geq w \rightarrow x + y \geq v + w$	%(Nat_0.6)%	

Fig. 6.2 The specification NAT_0.

[9] The logical system that we then have is therefore related to the *labelled deduction systems* of Gabbay [9].

$$\frac{p:A \quad q:B}{s(p,q):\sigma(A,B)}$$

It is convenient to use "contexts" also. That is to say, the actual hypotheses with which we are working. These will be written in the standard logical style using the "turnstile" symbol *vdash*. We shall use Γ, possibly with subscripts, to denote a set of logical formulae. Thus we write $\Gamma \vdash A$ to indicate that A is provable in the context Γ (or equivalently, from the hypotheses Γ).

When we wish to extract programs from proofs from algebraic specifications the Curry-Howard terms that we use are now more complicated for two reasons. In addition to the information from, for example, the logical rule being used, the Curry-Howard term also has to "remember" the specification. We have a similar situation for the structural rules. However, the message is as before: the Curry-Howard term carries *all* the information as to how we have constructed the proof so far.

The annotations we shall use will also involve Curry-Howard terms, specification names and the logical connectives. We have two kinds of rules: those for the logical connectives, *logical rules*; and those for the structural changes in the specifications, *structural rules*. Even with the purely logical rules, the specification of the conclusion depends on those in the premises. Thus the full rule for *modus ponens* we have is

$$\frac{\Gamma_1 \vdash a.\textsc{Sp}:A \quad \Gamma_2 \vdash d.\textsc{Sp}:(A \to B)}{\Gamma_1 \cup \Gamma_2 \vdash (da).\textsc{Sp}:B}\ (\to \mathrm{E})$$

Let me explain what is going on in the rule for implication elimination ($\to$ $-$E). We are working within a single specification $\textsc{Sp}$. We have Curry-Howard terms d for the proof of $(A \to B)$ and a for the proof of A. Therefore we have, just as in (6.1), (da) as the Curry-Howard term for the resulting proof of B. As usual the contexts Γ_1 and Γ_2 are added together. The specification $\textsc{Sp}$ is unchanged throughout.

Now here is the rule for implication introduction:

$$\frac{\Gamma, x:A \vdash d.\textsc{Sp}::B}{\Gamma \vdash \lambda x:A.d.\textsc{Sp}:(A \to B)}\ (\to\text{-I})$$

For the structural rules, the change in the structure is reflected in the specification of the conclusion. The full rule for translations, including the lambda calculus elements and the specifications, is as follows:

$$\frac{\Gamma \vdash d.\textsc{Sp} \bullet A}{\rho\,\text{`}(\Gamma) \vdash \rho \bullet d.\textsc{Sp}\mathbf{with}\rho \bullet \rho \bullet A}\ (\mathrm{trans})$$

In the structural rule for translation the formula A is unchanged in meaning but the language is changed, therefore A has to be changed into its translation $\rho \bullet A$. Similarly the context Γ has to be translated. This is written (by us) as $\rho\text{`}(\Gamma)$. In addition, the Curry-Howard term for A has to be changed (translated) into the new language. So the new Curry-Howard term is $\rho \bullet d$. Finally the new specification is the translated one: $\textsc{Sp}\mathbf{with}\rho$.

$$\frac{}{A \vdash \text{SP} \bullet A} \text{(Ass-I)} \quad \text{where } \text{Sig}A \subseteq \text{Sig}(\text{SP})$$

$$\frac{}{\emptyset \vdash \langle \Sigma, Ax \rangle \bullet A} \text{(Ax-I)} \quad \text{where } Ax \text{ are the axioms}$$

Fig. 6.3 Two new logical rules for structured specification logic.
SP is any structured specification expression. Sig indicates taking the signature.

$$\frac{\Gamma \vdash \text{SP} \bullet A}{\rho`(\Gamma) \vdash \text{SP } \textbf{with } \rho \bullet \rho \bullet (A)} \text{(trans)}$$

$$\frac{\Gamma \vdash \text{SP} \bullet A}{\Gamma \vdash \text{SP } \textbf{hide } SL \bullet A} \text{(hide)}$$
where $\Gamma \cup \{A\} \subseteq WFF(Sig(\text{SP})/SL, Var)$

$$\frac{\Gamma \vdash \text{SP\_1} \bullet A}{\Gamma \vdash \text{SP\_1 } \textbf{and } \text{SP\_2} \bullet A} \text{(union}_1\text{)}$$

$$\frac{\Gamma \vdash \text{SP\_2} \bullet A}{\Gamma \vdash \text{SP\_1 } \textbf{and } \text{SP\_2} \bullet A} \text{(union}_2\text{)}$$

$$\frac{\Gamma \vdash \text{SP\_1} \bullet A}{\Gamma \vdash \text{SP\_1 } \textbf{then } \text{SP\_EXT} \bullet A} \text{(ext}_1\text{)}$$

$$\frac{\Gamma \vdash \text{SP\_EXT} \bullet A}{\Gamma \vdash \text{SP\_1 } \textbf{then } \text{SP\_EXT} \bullet A} \text{(ext}_2\text{)}$$

Fig. 6.4 The structural rules of structured specification logic.
Here the condition $\Gamma \cup \{A\} \subseteq WFF(\text{Sig}(\text{SP})/SL, Var)$ means that none of the well-formed formulae in Γ and A contains any letter from the hidden signature SL.
We have omitted the lambda calculus parts of the Curry-Howard terms for clarity.

The logical rules for our system *Structured Specification Logic* are very similar to those in Fig. 6.1 with two exceptions that we give in Fig. 6.3. In order to make the figures less complicated the rules are presented without their Curry-Howard terms. Likewise the structural rules may be found in Fig. 6.4. The complete set of rules we have for *CASL*, with their Curry-Howard terms, may be found in [6] or [16].

In this situation we are again able to prove strong normalization.

From this strong normalization theorem we are then able to give an *extraction map*, that is to say, we give a formal process that, given a Curry-Howard term for a proof of $\forall x \exists y A(x,y)$ for a given specification, the extraction map returns a suitable y for a given x. Indeed it gives a program in the programming language *ML*. The extraction map works recursively and, in particular, the cases for $\rightarrow$-introduction and elimination correspond directly to the procedure we have outlined above.

6.4.2 Imperative Programming

My recent PhD student, Iman Poernomo, now at King's College, London, has developed a protocol for integrating ordinary computer programs into the kind of deductive system we have been discussing. This protocol he calls the *Curry-Howard* protocol. The logical system for such a situation includes the state of the system (i.e. the contents of registers in the machine) and accounts for the changes that take place when a program is run. Despite the complications this produces it is still possible to produce a constructive version of a Hoare logic (cf. [12]) for reasoning about imperative programs to which the Curry-Howard isomorphism may be adapted.

Note that a theorem in the Hoare logic consists of an imperative program and a truth about the program. Because of this the logic can be used to synthesize programs.

However we are also concerned to use programs already in the programming language that we regard as "reliable". We do not use the word "correct" here, reserving that word for programs that have been formally proved to meet their specifications. Here we simply mean that we have programs that we are satisfied will give the correct answers. Such programs include very simple ones such as programs for the multiplication of natural numbers. This achieves a significant saving in the length of the programs extracted. Otherwise we would have to prove a formula in formal arithmetic that allows us to extract a program, for example for the multiplication function. The proof would be inordinately long, involving several applications of induction and its corresponding program would then involve the same number of recursions. This is obviously very uneconomical, because we know it is possible to write a relatively simple program for multiplication (if one is not built into the computer already).

Imperative computer programs have *side-effects*: they change the *state* of the machine and, in particular, the values in various registers. The presence of side-effects is what distinguishes the imperative programming paradigm from the functional one. However, side-effect-free functions are also important in imperative programs because they enable access to data, obtaining views of state and producing return values. Imperative programs involve both side-effects and side-effect-free return values. Consider, for instance, a program that triples the number in the register s and returns the value twice the value in s. In *Standard ML* the program is

$$\mathsf{s} := !\mathsf{s} * 3; !\mathsf{s} * 2$$

It has a side-effect producing assignment statement, $\mathsf{s} := !\mathsf{s} * 3$, followed by the return value $!\mathsf{s} * 2$. In many popular imperative languages such return values are potentially complex, involving higher-order functional aspects that are difficult to program correctly.

Our goal is to specify, reason about and synthesize both aspects of imperative programs – side-effects and functional return values.

Our approach is as follows. We use a version of Hoare logic to synthesize the side-effect producing aspect of a program, specified in terms of pre- and post-conditions. Hoare logic involves considering triples of the form

$$\{\textit{pre-condition}\}\texttt{program step}\{\text{post-condition}\}$$

The *pre-condition* is true before the program step commences and the *post-condition* is true after the step.
The formula

$$s_f > s_i$$

specifies a side-effect where the final value of state s, denoted by s_f, is greater than the initial value, denoted by s_i. We can use Hoare logic to synthesize a *Standard ML* program that satisfies this specification, by producing, for example, a theorem of the form

$$\vdash \texttt{s :=!s} * 3 \bullet s_f > s_i$$

where the left-hand-side of the • symbol is the required *Standard ML* program (written in teletype font), and the right-hand-side is a true statement about the program.

The structural rules may be found in Fig. 6.5. The operation $\mathsf{tologic_i}$ indicates the operation of the Curry-Howard protocol taking the name of the variable in the programming language into the logical language. Thus $\mathsf{tologic_i}(\texttt{b})$ yields the variable coming from the register b. The remaining items in teletype font are standard imperative programming constructs. Thus p; q means the program p is pipelined into the program q and while b do q; done is the usual while-loop that repeats the program q while the assertion q is satisfied.

The logical rules are the same as we had much earlier in Fig. 6.1, but with the state information added. This is because the intuitionistic rules are concerned with

$$\frac{}{\vdash_{\mathsf{K}} \texttt{s := v} \bullet s_f = \mathsf{tologic_i}(\texttt{v})}\ (\text{assign})$$

where s is a state reference.

$$\frac{\vdash_{\mathsf{K}} \texttt{p} \bullet (\mathsf{tologic_i}(\texttt{b}) = \textit{true} \to C) \quad \vdash_{\mathsf{K}} \texttt{q} \bullet (\mathsf{tologic_i}(\texttt{b}) = \textit{false} \to C)}{\vdash_{\mathsf{K}} \texttt{if b then p else q} \bullet C}\ (\text{ite})$$

$$\frac{\vdash_{\mathsf{K}} \texttt{p} \bullet (A[\bar{s}_i/\bar{v}] \to B[\bar{s}_f/\bar{v}]) \quad \vdash_{\mathsf{K}} \texttt{q} \bullet (B[\bar{s}_i/\bar{v}] \to C[\bar{s}_f/\bar{v}])}{\vdash_{\mathsf{K}} \texttt{p;q} \bullet (A[\bar{s}_i/\bar{v}] \to C[\bar{s}_f/\bar{v}])}\ (\text{seq})$$

where A and B are free of state identifiers.

$$\frac{\vdash_{\mathsf{K}} \texttt{q} \bullet (\mathsf{tologic_i}(\texttt{b}) = \textit{true} \wedge A[\bar{s}_i/\bar{v}]) \to A[\bar{s}_f/\bar{v}]}{\vdash_{\mathsf{K}} \texttt{while}\ b\ \texttt{do q; done} \bullet A[\bar{s}_i/\bar{v}] \to (A[\bar{s}_f/\bar{v}] \wedge \mathsf{tologic_f}(\texttt{b}) = \textit{false})}\ (\text{loop})$$

where A is free of state identifiers.

$$\frac{\vdash_{\mathsf{K}} \texttt{p} \bullet P \quad \vdash_{\mathsf{Int}} (P \to A)}{\vdash_{\mathsf{K}} \texttt{p} \bullet A}\ (\text{cons})$$

Fig. 6.5 The structural rules for our logic for imperative programming.
Intuitionistic deduction $\vdash_{\mathsf{Int}}$ is given in Fig. 6.1.

truths that are universal to all programs. That is to say, they can be used to infer properties that hold over any side-effect.

Example 6.1 For instance, an application of the logical ($\wedge$-I) rule

$$\frac{s_f = s_i * 2 \vdash_{\mathsf{Int}} s_f \geq s_i \quad s_f = s_i * 2 \vdash_{\mathsf{Int}} Even(s_f)}{s_f = s_i * 2 \vdash_{\mathsf{Int}} s_f \geq s_i \wedge Even(s_f)} \ (\wedge\text{-I})$$

tells us that, any program that makes $s_f = s_i * 2$ true, makes $s_f \geq s_i$ and $Even(s_f)$ true, and therefore the statement: $s_f \geq s_i \wedge Even(s_f)$ must also be true of the program.

To specify and synthesize return values of a program we adapt *realizability* and the extraction of programs from proofs. We have already treated the latter,so now we consider realizability.

When we extract a program we wish to demonstrate that it is "correct". This requires the notion of *realizing*. This is a different way of verifying proofs in intuitionistic logic by means of computable functions. It was first developed by Kleene, see the last chapter of [14]. The basic idea is that we produce a program for a (partial) recursive function that is a *witness* to the proof of an assertion. Such witnesses can be produced recursively by going down through the proof. Such a program can be regarded as a number (for example, the binary string that encodes the program). For example if we have partial recursive functions with programs p, q realizing A, B, then we take (p, q) as the realizer of $(A \wedge B)$. The full details, which may be found in Kleene [14] for the basic system of intuitionist logic and in our book [16] for the systems we discuss here.

Here is an example. Given the theorem

$$\mathtt{s} := \mathtt{s} * 3 \bullet s_f > s_i \wedge (\exists x : int.Even(x) \wedge x > s_i)$$

we can synthesize a program of the form

$$\mathtt{s} := \mathtt{s} * 3; \mathtt{f}$$

where the function f is a side-effect-free function (such as $!\mathtt{s} * 2$) that realizes the existential statement of the post-condition $(\exists x : int.Even(x) \wedge x > s_i)$, by providing a witness for the x.

With our program extraction users will have no need to manually code the return value, instead they can work within the Hoare logic. There they prove a theorem from which the return value is then synthesized.

Here is an outline of a specific example about an electronic banking system. In the logic here we are using a *many-sorted* system, that is to say, individual variables have their own sorts or varieties. This is a small and natural modification of the logic.[10]

[10] It can be avoided by adding extra predicates, one for each sort. In this case instead of $x : s$ meaning "x is of sort s" we have a predicate s and we write $s(x) \rightarrow \ldots$, which can be read as "If x is of sort s, then ...".

Consider an Automatic Bank Teller machine (ATM) example with the following domain conditions:

1. The ATM permits the user to enter a Personal Identification Number (PIN) and to withdraw money. In order to withdraw money, the user must enter their PIN and a database connection to the bank's server must be made. The machine has a screen on which it displays messages to the user.
2. The integer state reference `pin` stores the PIN number entered by the user, the boolean state reference `canWithdraw` stores a flag to determine whether or not the user may withdraw money from the machine, and the boolean state reference `isConnected` stores a flag to determine whether or not there is a connection to the bank's server.
3. We use the predicate $appMessage(m)$ to assert that a string m is an appropriate message to display on the screen for the user, given that the ATM is in some particular state.
4. There is a program p satisfying the following property. Given the user has entered their Personal Identification Number (PIN) correctly, the program allows the user to withdraw money. This property is formally given by an axiom

$$\vdash_{\mathsf{K}} \mathtt{p} \bullet PINCorrect(pin_i) \rightarrow canWithdraw_f = true$$

5. There is a program q such that, if the user is permitted to withdraw money, then a database connection is established, and also it is the case that there is an appropriate message that can be displayed. These properties are formally given by the axiom

$$\begin{aligned}\vdash_{\mathsf{K}} \mathtt{q} \bullet canWithdraw_i = true \rightarrow \\ (isConnected_f = true \wedge \exists x : string \bullet appMessage(x))\end{aligned}$$

For the sake of argument, we simplify our domain with the following assumptions:

1. We assume two *Standard ML* record datatypes have been defined, `user` and `account`. Instances of the former contain information to represent a user in the system, while instances of the latter represent bank accounts. We do not detail the full definition of these types.
   However, we assume that an `account` record type contains a `user` element in the `owner` field to represent the owner of the account. So the owner of the account element `myAccount` : `account` is accessed by `myAccount.owner`.
   We also assume that `user` is an equivalence type in *Standard ML*, so that its elements may be compared using the boolean valued comparison function $=$.
   We assume a constant `currentUser` : `user` that represents the current user who is the subject of the account search.
2. The database is represented in *Standard ML* as an array of accounts,

$$\mathtt{db} : \mathtt{account\ array}$$

Following the *Standard ML* API, the array is 0-indexed, with the *i*th element accessed as

$$\texttt{sub(db, i)}$$

and the size of the array given as

$$\texttt{length db}$$

Assume we have an array of size `Size`, called `accounts`. Although *Standard ML* arrays are mutable, for the purposes of this example, we will consider `db` to be an immutable value. Consequently, it will be represented in our logic as a constant.

3. We assume a state reference `counter : int ref`, to be used as a counter in searches through the database.

   We take a predicate

$$allAccountsAt(u : user, x : acount\ list, y : int)$$

whose meaning is that x is a list of all accounts found to be owned by the user u, up to the point y in the database `db`. The predicate is defined by the following axioms in $\mathscr{A}\mathscr{X}$

$$\forall u : user \bullet \forall x : (acount\ list) \bullet \forall y : int \bullet (allAccountsAt(u,x,y) \rightarrow (\forall z : int \bullet z \leq y \rightarrow sub(db,z).owner = u)) \quad (6.4)$$

$$\forall u : user \bullet \forall x : (acount\ list) \bullet \forall y : int \bullet ((y < (length\ db) - 1) \wedge sub(db,y+1).user = u \wedge allAccountsAt(u,x,y)) \rightarrow allAccountsAt(u, sub(db,y+1) :: x, y+1) \quad (6.5)$$

$$\forall u : user \bullet \forall x : (acount\ list) \bullet \forall y : int \bullet (y < (length\ db) - 1 \wedge \neg sub(l,y+1).user = u \wedge allAccountsAt(u,x,y)) \rightarrow allAccountsAt(u,x,y+1) \quad (6.6)$$

$$\forall u : user \bullet \forall y : int \bullet y = 0 \rightarrow allAccountsAt(u, [], y) \quad (6.7)$$

Observe that these are intuitionistic axioms that will be used in Hoare logic.

We develop a program that satisfies the following property: Given a user's details, it is *possible* to obtain a list of all accounts held at the bank by the user, by searching through the database. This is formally stated as the following
Specification:

$$\exists y : (acount\ list) \bullet listAllAccounts(currentUser, y, counter_f) \wedge (counter_f < (length\ db) - 1) = false \quad (6.8)$$

The post-condition requirement of $counter_f$ signifies that a complete search of the database should be completed by the program.

We also have an axiom that is needed because, since we are working in intuitionist logic, we do not automatically have the law of the excluded middle.

$$y < (length\,db) - 1 \rightarrow sub(l, y+1).owner = u \vee \neg sub(l, y+1).owner = u \quad (6.9)$$

From these we can eventually prove the theorem

$$\begin{aligned}&\vdash \texttt{counter} := !\texttt{counter} + 1;\\ &\quad \texttt{while}\ !\texttt{counter} < (\texttt{length db}) - 1\ \texttt{do counter} := !\texttt{counter} + 1 \bullet\\ &\qquad \exists y : (acount\ list) \bullet allAccountsAt(currentUser, y, counter_f) \wedge\\ &\qquad\qquad (counter_f < (length\,db) - 1) = false \quad (6.10)\end{aligned}$$

The program that we extract by a special *extraction function* satisfies the specification (6.8). We omit details of the extraction function. Suffice it to say that it works recursively in the same spirit as we extracted our programs in Section 6.4.1.

Essentially, when viewed as a specification of a return value, the specification (6.8) requires a program that, given a user's details, will search through a database to obtain all accounts held at the bank by the user, and then return this list.

6.5 Programs and Proofs

So far we have seen how to obtain programs from proofs in constructive systems of logic. Therefore we could conclude that all proofs are already programs, or at least, that every proof in (constructive) logic contains a program.[11]

What if we were to write the program first? Would we automatically have a proof? The answer is obviously "No!" if we simply write computer programs as many people do. However, a thoughtful computer programmer would wish to know that the program written would do what it was expected to do, that is to say, would meet its specification.[12] Therefore, as part of the task of writing the program, a proof should be produced at the same time.

The approach that we have presented shows how to accomplish both of these tasks at the same time. It does not require a separate investigation to produce a proof that the program will be correct.

From a practical point of view it is sometimes obvious how to write the proof. I studied a program for quicksort.[13] Then I wrote a proof corresponding to the program and extracted a program from it. The resulting program was essentially the

[11] The restriction to constructive systems of logic is essential for us.

[12] This is a very serious issue when it comes to the control of powerful systems, in particular, the control of nuclear weapons.

[13] This was inspired by looking at work of Helmut Schwichtenberg on program extraction in [3].

quicksort program from which I had started. However I have not yet been able to formalize the procedure that I used in producing the proof from the program. It would appear that one needs to know the *algorithm* rather than the program in order to construct the proof. This in itself indicates that one also needs to know that the program is a *correct* implementation of the algorithm. But this is work for the future.

6.6 Conclusion

The techniques we have presented here are based on a variant of Gabbay's labelled deductive systems [9]. Our logical rules are of the form

$$\frac{\text{Logical context, State, Curry-Howard term} \vdash \text{Formula}}{\text{New Logical context, New State, New Curry-Howard term} \vdash \text{New Formula}}$$

although the actual order may vary. Further, each of the items on the lower line may depend on, that is to say, be functions of, any or all of those on the top line, and of course there may be two or more sequences on the top line.

The semantics of these rules will depend on the structures that we are using. Also the interpretation of the informal terms: Logical context, State, etc. will also vary.

What seems to be most important is that we have extended the notion of logic in two ways. First of all we now have programs or other constructions (for example, specifications) interacting with the standard logical connectives. Secondly, the context of the logic may change in the course of a proof. This certainly happens in the context of algebraic specifications. Thirdly, we are now discussing logics (plural) and we arrive at such logics by an analysis of a technical setting. This seems to me to be following Aristotle's approach of looking at the real world, or a small part of it, and then abstracting the logical principles that work in that arena. But we have come a long way from the "one true logic" that I mentioned at the beginning of the introduction!

Acknowledgements Many years ago Georg Kreisel said that I should not work in proof theory, which is the setting of the work described here. I have not taken that advice but am grateful to him for his enthusiasm and stimulation over many decades. I was originally introduced to the area of program extraction when visiting my old friend Anil Nerode in Cornell in the 1990s where we studied Girard's thesis [10], essentially published as [11]. Later John Shepherdson (Bristol) and I extended the Curry-Howard terms to cover all the standard logical connectives directly, that is to say, without going through what I regard as the tortuous translations of Girard into higher order logic. Martin Wirsing of Ludwig-Maximilians Universität, Munich, and I, inspired by work of Martin and his student Hannes Peterreins, began to produce elegant rules for the context of algebraic specifications. From then on, with my former student Iman Poernomo, now at King's College, London we have extended the logical systems even further to structured specifications and then to imperative programming. The idea of the Curry-Howard protocol is due to Iman Poernomo. His thesis, part of which is the basis for Section 6.4.2, has recently been published in a revised version as [16]. I am extremely grateful to all of these colleagues for their friendship, ideas and stimulation.

References

1. Barendregt P. H. *The Lambda Calculus, Its Syntax and Semantics*. North Holland Publishing Company, Amsterdam, 1995.
2. Barendregt P. H. Lambda calculi with types. In S. Abramsky and D. Gabbay and T. Maibaum, editors, *Handbook of Logic in Computer Science*, vol. 2, pages 117–309 (Background: Computational Structures). Clarendon Press, Oxford, 1992.
3. Berger U., and Schwichtenberg H. Program extraction from classical proofs. In D. Leivant, editor, *Logic and Computational Complexity, International Workshop LCC '94*, Indiapolis, IN, USA, October 1994, pages 77–97, 1995.
4. CoFI Language Design Task Group on Language Design. *CASL, The Common Algebraic Specification Language, Summary, 25 March 2001*, March 2001. Available at http://www.brics.dk/Projects/CoFI/Documents/CASL/Summary/ (accessed 3.i.05)
5. Crossley J. N. What is mathematical logic? A survey. Tutorial at the First International Conference on Logic and and its application to other disciplines, IIT Bombay, 2005, submitted for publication.
6. Crossley J. N. Iman Poernomo, and Martin Wirsing. Extraction of structured programs from specification proofs. In D. Bert, C. Choppy, and P. Mosses, editors, *Workshop on Algebraic Development Techniques*, vol. 1827 of LNCS, pages 419–437, 1999.
7. Crossley, J. N., and Shepherdson J. C. Extracting programs from proofs by an extension of the Curry-Howard process. In J. N. Crossley, J. B. Remmel, R. A. Shore, and M. E. Sweedler, editors, *Logical Methods: In honor of AnilNerode's Sixtieth Birthday*, pages 222–288. Birkhäuser, Boston, MA, 1993.
8. Dummett M. *Elements of Intuitionism*. Oxford University Press, Oxford, 1977.
9. Gabbay D. *Labelled Deductive Systems*. Oxford University Press, Oxford, 1996.
10. Girard J. Y. *Interprétation functionelle et élimination des coupures dans l'arithmétiqued'ordre supérieure*. PhD thesis, Université Paris VII, Paris, 1972.
11. Girard J. Y., Lafont Y., and Taylor P. *Proofs and Types*. Cambridge University Press, Cambridge, 1989.
12. Hoare C. A. R. An axiomatic basis for computer programming, *Commun. Assoc Comput. Machinery*, 12(10): 576–80, 1969.
13. Howard W. The formulae-as-types notion of construction. In J. R. Hindley and J. Seldin, editors, *To H.B. Curry: Essays on Combinatory Logic, Lambda Calculus, and Formalism*, pages 479–490. Academic Press, New York, NY, 1969.
14. Kleene S. C. *Introduction to Metamathematics*. North-Holland, Amsterdam, 1952.
15. Peterreins H. *A natural-deduction-like calculus for structured specifications*. PhD thesis, Ludwig-Maximilians-Universität, München, 1996.
16. Poernomo I. H., Crossley J. N., and Wirsing M. *Adapting Proofs-Asprograms*. Springer, New York, NY, 2005.
17. Wirsing M. Structured specifications: Syntax, semantics and proof calculus. In M. Broy, editor, *Informatik und Mathematik, Festschrift für F. L. Bauer*, pages 269–283. Springer, Berlin, 1991.

Chapter 7
Zero-One Laws: Thesauri and Parametric Conditions

Andreas Blass and Yuri Gurevich

Quisani: I've been thinking about zero-one laws for first-order logic. I know it's a rather old topic, but I noticed something in the literature that I'd like to understand better. The first proof [6] established that, for any first-order sentence σ in a finite relational vocabulary Υ, the proportion of models of σ among all Υ-structures with base set $\{1, 2, \ldots, n\}$ approaches 0 or 1 as n tends to infinity. Fagin [5] rediscovered the result (with a simpler proof) and added, near the end of his paper, some remarks about what happens if, instead of considering all Υ-structures, we consider only those satisfying some specified sentence τ. He pointed out that for some but not all choices of τ, there is still a 0-1 law: The proportion of models of σ among models of τ with base set $\{1, 2, \ldots, n\}$ approaches 0 or 1 as n tends to infinity. He gave two examples of such τ, both in the language with just a single binary relation symbol E, the language of digraphs. One example was the sentence saying that τ is symmetric and irreflexive, so the models are undirected loopless graphs. The other example defined the class of tournaments. The case of undirected graphs was rediscovered in [3], where another example was added, pure d-dimensional simplicial complexes, formulated using a completely symmetric and completely irreflexive $(d+1)$-ary relation.

I'd think there should be some general result explaining these variants of the 0-1 law.

Authors: There is such a result in Oberschelp's paper [7], but it seems he never published the proof. His result covers the variants that you mentioned as well as others, for example involving graphs with several colors of edges. It is based on the

Andreas Blass
Mathematics Department, University of Michigan, Ann Arbor, MI 48109–1043, USA, e-mail: ablass@umich.edu

Yuri Gurevich
Microsoft Research, One Microsoft Way, Redmond, WA 98052, USA, e-mail: gurevich@microsoft.com
Originally published in the Bulletin of European Association for Theoretical Computer Science, Number 91, February 2007.

DOI 10.1007/978-94-007-0080-2_7, © Springer Science+Business Media B.V. 2011

same approach, via extension axioms, as the work in [5] and many later works. So it doesn't cover the 0-1 laws obtained by other methods, for example by Compton (see [4] and the references given there) for slowly-growing classes of structures.

Later, not knowing of Oberschelp's work, we introduced in [1] the notion of a thesaurus as a suitable context for Shelah's proof of the 0-1 law for choiceless polynomial time. It also provides a suitable context for the 0-1 law in the more restrictive context of first-order logic, once that logic is appropriately defined for thesauri.

Q: Does the thesaurus approach also depend on the extension axioms? And should it be combined with parametric conditions to produce a common generalization?

A: Both of your questions are answered – the first affirmatively and the second negatively – by the fact that the two approaches, parametric conditions and thesauri, are essentially equivalent, at least when applied to the case of uniform probability distributions on structures of any given size.

Q: That leaves me with a lot of questions: What are parametric conditions? What are thesauri? What exactly does "essentially equivalent" mean in this context? And what happens when the probability distributions aren't uniform?

A: Let's start with Oberschelp's parametric conditions. These are conjunctions of first-order universal formulas, for a relational vocabulary, having the special form

$$\forall x_1 \dots \forall x_k \, (D(\mathbf{x}) \rightarrow C(\mathbf{x}))$$

where $\mathbf{x}$ stands for the n-tuple of variables $x_1, \dots, x_k$, where $D(\mathbf{x})$ is the formula $\bigwedge_{1 \leq i < j \leq k} x_i \neq x_j$ saying that the values of these variables are distinct, and where $C(\mathbf{x})$ is a propositional combination of atomic formulas such that, in each atomic subformula of $C(\mathbf{x})$, all k of the variables occur.

Q: I assume that k is allowed to vary from one conjunct to another in a parametric condition, and that when $k \leq 1$ the empty conjunction D is interpreted as true. So, for example, irreflexivity of a binary relation is expressed by the parametric condition $\forall x(\mathtt{true} \rightarrow \neg R(x,x))$.

A: That's right, and it's easy to express the other conditions you mentioned earlier – symmetry for undirected graphs, asymmetry for tournaments, and complete irreflexivity and symmetry for simplicial complexes – as parametric conditions.

Q: I see that, but I don't yet see the significance of the requirement that all atomic subformulas of C use all the variables.

A: The simplest explanation is that if you drop this requirement then the extension axioms need not have asymptotic probability 1. For example, for almost all finite partially ordered sets, the longest chain has length three (see the proof of [4, Theorem 5.4]). So there are configurations, like a four-element chain, that can arise in partial orders but are absent with asymptotic probability 1. Traditional extension axioms, in contrast, imply that any configuration permitted by the underlying assumption τ must occur. The trouble comes from the transitivity clause in the definition of partial orders; it involves three variables but each atomic subformula uses only two of them.

Q: The example shows that some requirement is needed to eliminate the case of partial orders, but how does the "use all the variables" requirement connect with extension axioms? I guess what I'm really asking for is a sketch of Oberschelp's proof.

A: The crucial contribution of parametricity is that it permits a reformulation of the uniform probability measure on structures of a fixed size n in terms of independent choices of the truth values of instances of the relations.

Recall that, when we consider the class of all structures (of a given relational vocabulary) with universe $\{1,2,\dots,n\}$, the uniform probability measure on these structures can be described by saying that each instance $R(a_1,\dots,a_r)$ (where R ranges over the relations of the structure and $\mathbf{a}$ over tuples of appropriate length from $\{1,2,\dots,n\}$) is independently assigned truth value true or false, with equal probability. When we deal with, say, loopless undirected graphs, this description must be modified, since $R(a,a)$ must be false and since $R(a_1,a_2)$ must have the same truth value as $R(a_2,a_1)$. Nevertheless, the uniform distribution can still be described in terms of independent flips of a fair coin: flip a coin for each 2-element subset $\{a_1,a_2\}$ of $\{1,2,\dots,n\}$ to determine both $R(a_1,a_2)$ and $R(a_2,a_1)$. Similarly in the case of tournaments, a single flip of a fair coin decides which one of $R(a_1,a_2)$ and $R(a_2,a_1)$ shall hold. And similarly in the other examples.

Something similar happens for arbitrary parametric conditions τ. To describe it, we need the notion of a k-type relative to τ. Temporarily fix a positive integer k, less than or equal to the maximum arity of the relation symbols in the vocabulary of τ. Consider all the atomic formulas that use exactly the variables $x_1,\dots,x_k$, possibly more than once. A *k-type* is an assignment of truth values to these atomic formulas that makes $C(\mathbf{x})$ true whenever $\forall\mathbf{x}\,(D(\mathbf{x})\to C(\mathbf{x}))$ is a conjunct of τ (up to renaming bound variables, so that $\mathbf{x}$ is $x_1,\dots,x_k$). In other words, a k-type is an assignment of truth values that can be realized by a k-tuple of distinct elements in a model of τ.

Now the uniform distribution on models of τ with base set $\{1,2,\dots,n\}$ admits the following equivalent description: For each k and each k-element subset $\{a_1 < a_2 < \cdots < a_k\} \subseteq \{1,2,\dots,n\}$, choose, uniformly at random, a k-type to be realized by the k-tuple $(a_1,\dots,a_k)$. This works because each of these types is realized by $(a_1,\dots,a_k)$ in equally many models of τ with base set $\{1,2,\dots,n\}$ and because different increasing tuples $(a_1,\dots,a_k)$ behave independently. Furthermore, once the k-types of increasing tuples $\mathbf{a}$ are chosen, they determine all the relations of the structure.

Q: What about instances of the relations where the arguments are not in increasing order?

A: They're included, because the atomic formulas to which a k-type assigns truth values include those in which the variables $\mathbf{x}$ occur out of order.

Once one has this alternative description of the uniform distribution on models, one can easily imitate the traditional proof of 0-1 laws. There is an extension axiom for each k-type with $k > 0$; it says that, for any distinct $x_1,\dots,x_{k-1}$, there is an x_k, distinct from all of them, such that the tuple $(x_1,\dots,x_{k-1},x_k)$ realizes the given k-type. It is easy to check that each extension axiom has asymptotic probability 1 and that the theory axiomatized by the extension axioms is complete. (To prove completeness, one can proceed as in [5] because the theory is $\aleph_0$-categorical, or one can eliminate quantifiers as in [3].)

The role of parametricity in this argument is to ensure that all the information about any k-tuple of distinct elements can be isolated in its k-type and the k'-types of its subtuples, a finite amount of information, whose size is independent of the size n of the base set. That allows us to formulate extension axioms and verify their asymptotic validity. Contrast this with the situation for, say, partial orders. Here the requirement of transitivity imposes correlations between a truth value $R(a,b)$ and many other truth values $R(a,c)$ and $R(b,c)$, for all c in the base set. The number of relation instances correlated with a single $R(a,b)$ thus grows with the structure and the proof described above breaks down. Oberschelp [7] summarizes this (in the case of a vocabulary with only one relation symbol) by saying "A parametric property defines a class of relations which can be determined by the independent choice of values (parameters) in fixed regions of the adjacency array."

Q: OK. I see that parametricity seems to be just what's needed to carry out the traditional proof of the 0-1 law via extension axioms. Now what are thesauri?

A: A *thesaurus* is a finite set of signa, so of course we have to say what a signum is, but let's first deal with a simplified notion of signum, which turns out to have the same generality as parametric conditions. A *signum* in this simplified sense consists of

- a symbol R (assumed to be different for the different signa in a thesaurus),
- a natural number j called the arity,
- a finite set V called the value set[1],
- a group G of permutations of $\{1,2,\dots,j\}$, and
- a homomorphism from G to the group of permutations of V.

For notational convenience, one often writes simply the symbol R when one really means the whole signum.

Q: That sounds pretty complicated; what's really going on here?

A: The symbol R and the arity j are analogous to what you have in a relational vocabulary of ordinary first-order logic – a symbol and the number of its argument places. Our R's, however, will not necessarily be 2-valued as in first-order logic, but v-valued, where v is the cardinality of the value set V. So we could, for example, treat a graph with colored edges by having a single binary signum where V is the set of colors plus one additional value to indicate the absence of an edge.

The other two constituents of the signum, the group G and the homomorphism h, describe the symmetry properties that we intend R to satisfy. The idea is that permuting the j arguments of R by a permutation π in G results in a change of the value given by $h(\pi)$. More precisely, a structure $\mathfrak{A}$ for a thesaurus consists of a base set A together with, for each signum $\langle R,j,v,G,h\rangle$ (often abbreviated as just R), an interpretation $R^{\mathfrak{A}}$ assigning to each j-tuple of distinct elements $a_1,\dots,a_j$ in A a value $R^{\mathfrak{A}}(\mathbf{a})$, subject to the symmetry requirement

[1] In [1], the value set was always of the form $\{1,2,\dots,v\}$ for some positive integer v. Allowing arbitrary finite sets of values makes no essential difference but is technically convenient.

$$R^{\mathfrak{A}}(a_1,\dots,a_j) = h(\pi)(R^{\mathfrak{A}}(a_{\pi(1)},\dots,a_{\pi(j)})).$$

Q: The following variation seems more natural to me:

$$R^{\mathfrak{A}}(a_{\pi(1)},\dots,a_{\pi(j)}) = h(\pi)(R^{\mathfrak{A}}(a_1,\dots,a_j))$$

It explains how to obtain $R^{\mathfrak{A}}(a_{\pi(1)},\dots,a_{\pi(j)})$ from $R^{\mathfrak{A}}(a_1,\dots,a_j)$.
A: This doesn't work unless you either put π^{-1} on one side of the equation or make h an anti-homomorphism. Here's the calculation, using your proposed variation. Let π and σ be two permutations in the group, let $\mathbf{a}$ be a j-tuple of elements of A, and let $\mathbf{b}$ be the j-tuple defined by $b_i = a_{\sigma i}$.

$$\begin{aligned} h(\pi)h(\sigma)R^{\mathfrak{A}}(a_1,\dots,a_j) &= h(\pi)R^{\mathfrak{A}}(a_{\sigma 1},\dots,a_{\sigma j}) \\ &= h(\pi)R^{\mathfrak{A}}(b_1,\dots,b_j) \\ &= R^{\mathfrak{A}}(b_{\pi 1},\dots,b_{\pi j}) \\ &= R^{\mathfrak{A}}(a_{\sigma\pi 1},\dots,a_{\sigma\pi j}) \\ &= h(\sigma\pi)R^{\mathfrak{A}}(a_1,\dots,a_j), \end{aligned}$$

where I've applied your variation three times, once with σ, once with π, and once with $\sigma\pi$. So for this to work, we'd need $h(\sigma\pi) = h(\pi)h(\sigma)$, i.e., h should be an anti-homomorphism.
Q: I suppose using an anti-homomorphism wouldn't be a disaster, but it would defeat the purpose of my suggestion, increased naturality.

Now why does $R^{\mathfrak{A}}$ apply only to j-tuples of distinct elements?
A: Distinctness is technically convenient. For example, in tournaments, one wants the truth value of $R(a,b)$ to be negated if a and b are interchanged, except when $a = b$. So we think of a binary relation R as being given by two signa, one binary signum for distinct arguments and one unary signum for equal arguments. Similarly, a relation of higher arity would be represented by several signa, one for each way of partitioning the argument places into blocks with equal arguments.
Q: I see that, just as with parametric conditions, you can represent the uniform probability distributions on structures with base set $\{1,2,\dots,n\}$ in terms of independent random choices for some instances $R(\mathbf{a})$. For each signum R, choose a representative from each G-orbit of j-tuples of distinct elements, and assign R random values at these representatives. Then propagate these assignments through the whole orbits by means of the symmetry requirement.
A: That's right. To be precise about these G-orbits, one should say that G acts naturally on the set of j-tuples of elements from any set by

$$\pi(a_1,\dots,a_j) = (a_{\pi^{-1}(1)},\dots,a_{\pi^{-1}(j)}).$$

Q: With this formulation in terms of independent random choices, it should be possible to prove something analogous to extension axioms for the thesaurus context. I'd expect almost all Υ-structures to have the following property, for each n: Given any n distinct points $a_1,\dots,a_n$, there is a point b, distinct from all the a_i, and giving prescribed values for all signum instances $R^{\mathfrak{A}}(c_1,\dots,c_j)$ where one of the c_i is b and the others are distinct elements of $\{a_1,\dots,a_n\}$.

A: That's right, provided the prescribed values obey the symmetry requirement for thesaurus models. We're pleased that you remembered that the arguments of a signum are supposed to be distinct, so that b should occur only once in $R^{\mathfrak{A}}(c_1,\dots,c_j)$ and each a_i should occur at most once.

Q: This result should yield 0-1 laws for thesauri, except that you haven't yet defined first-order logic in the context of thesauri.

A: Indeed, we have not introduced a syntax to go with these semantical notions in [1], but it is not difficult to do so. Take atomic formulas to be $R(\mathbf{x}) = c$ where R is a signum (or the symbol part of it), $\mathbf{x}$ is a sequence of variables of length equal to the arity of R, and $c \in V$. Also allow equality as usual in first-order logic. Then form compound formulas using propositional connectives and quantifiers, just as in ordinary first-order logic. The semantics is obvious. (If the values of the variables in $\mathbf{x}$ are not all distinct, then $R(\mathbf{x}) = c$ is naturally taken to be false.)

If one is willing to stretch the notion of syntax a bit, then it would be appropriate to identify the atomic formulas $R(x_1,\dots,x_j) = c$ and $R(x_{\pi^{-1}(1)},\dots,x_{\pi^{-1}(j)}) = h(\pi)(c)$ for any π in the group of the signum R, since the symmetry requirement for structures implies that these will always have the same truth value.

Once these definitions are in place, it is, as you said, not difficult to show, via extension axioms, that first-order sentences have asymptotic probabilities 0 or 1 over the class of all structures of a thesaurus.

Q: This should also be clear for another reason, once you explain how thesauri and parametric conditions are essentially equivalent. Having the 0-1 law for parametric conditions, we should be able to use the essential equivalence to deduce the 0-1 law for thesauri. But what exactly did you mean by essential equivalence?

A: Essential equivalence has several components. First, for each thesaurus Υ, there is a parametrtic condition τ (in some vocabulary) such that the Υ-structures with any particular base set (for example $\{1,2,\dots,n\}$) are in (natural) one-to-one correspondence with models of τ on the same base set. Second, for each first-order sentence of the thesaurus, there is a first-order sentence of the vocabulary of τ such that the models of these sentences match up under the correspondence above. Third, conversely, for each parametric condition τ there is a thesaurus Υ with a one-to-one correspondence as before. And fourth, for every first-order sentence of the vocabulary of τ, there is a first-order sentence of Υ with the corresponding models.

Q: That seems to be exactly what's needed in order to convert 0-1 laws from either of the two contexts to the other. So how do these correspondences work?

A: One direction is implicit in the syntax for thesauri described above. Given a thesaurus Υ, form a first-order vocabulary Υ' with the same atomic formulas. That is, for each j-ary signum R of Υ and each value c, let R_c be a j-ary relation symbol in Υ'. The intended interpretation is that $R_c(\mathbf{a})$ should mean $R(\mathbf{a}) = c$. Every Υ-structure gives, in this way, a structure (in the ordinary, first-order sense) for Υ'. The converse is in general false, but the collection of Υ'-structures arising from Υ-structures in this way can be described by a parametric condition τ. The conjuncts in τ express the symmetry requirements of the thesaurus, i.e.,

$$\forall \mathbf{x}\,(D(\mathbf{x}) \to (R_c(\mathbf{x}) \to R_{h(\pi)(c)}(x_{\pi^{-1}(1)},\ldots,x_{\pi^{-1}(j)})))$$

for each signum R and each π in its group. There are also conjuncts saying that every j-tuple of distinct elements satisfies R_c for exactly one $c \in V$ and that R_c is false whenever two of its arguments are equal. We trust this makes the first two parts of "essentially equivalent" clear.

Q: Yes; "essentially equivalent" is now half clear. But I suspect that this was the easier half. How do you handle the reverse direction?

A: Here we have to convert a parametric condition τ into a thesaurus Υ. Let Υ consist of one j-ary signum T_j for each j up to the maximum arity of the relation symbols in the vocabulary of τ.

Q: Just one signum per arity, no matter how rich the vocabulary of τ is?

A: That's right. We compensate by using a rich set of values. Take the values of the j-ary signum T_j to be the j-types relative to τ. (This is one place where it's convenient to allow a signum to have any finite set of values, rather than only an initial segment of the positive integers as in [1].)

The group associated to the j-ary signum T_j is the symmetric group of all permutations of $\{1,2,\ldots,j\}$. To describe its action on the set of values, i.e., on the set of j-types, just let it act on the j variables occurring in the types. That is, the truth value assigned to an atomic formula θ by the type $h(\pi)(c)$ is the same as the truth value assigned by c to the formula obtained from θ by substituting $x_{\pi^{-1}(i)}$ for x_i for all i.

A structure for this thesaurus provides, for each j-tuple $\mathbf{a}$ of elements of the base set, a j-type to be realized by this j-tuple. This specifies which atomic formulas are to be true of this $\mathbf{a}$ and of all permutations of $\mathbf{a}$. The symmetry requirement on thesaurus structures is exactly what is needed to ensure that these specified truth values for the various permutations of $\mathbf{a}$ are consistent and thus describe a structure for the vocabulary of τ. Furthermore, since we use only j-types relative to τ, the resulting structures will be models of τ. And it is easy to check that every model of τ arises from exactly one Υ-structure.

Q: That takes care of the third part of essential equivalence. For the fourth part, you have to translate formulas in the vocabulary of τ into Υ-formulas. Since the syntax and semantics of thesauri treats connectives and quantifiers the same way as first-order logic does, it suffices to consider atomic formulas.

A: Right, and handling these is mainly just bookkeeping. Given an atomic formula θ in the vocabulary of τ, let $\{x_1,\ldots,x_k\}$ be the set of distinct free variables in it. For each equivalence relation E on this set of variables, we can write a quantifier-free formula θ'_E, in the syntax associated to the thesaurus Υ, with variables $x_1,\ldots,x_k$, saying that

- the equality pattern of the x_i is given by E, i.e.,

$$\bigwedge_{(i.j)\in E} (x_i = x_j) \wedge \bigwedge_{(i,j)\notin E} \neg(x_i = x_j),$$

and

- the type realized by distinct values of x_i's gives θ the value true, i.e.,

$$\bigvee_c (T_r(x_{i_1},\dots,x_{i_r}) = c),$$

where $x_{i_1},\dots,x_{i_r}$ are chosen representatives of the equivalence classes of E and where c ranges over those r-types that assign true to the formula obtained from θ by replacing each variable by the chosen representative of its equivalence class.

Then the disjunction of these formulas θ'_E, over all equivalence relations E, is satisfied by a tuple of elements $\mathbf{a}$ in exactly those Υ-structures that correspond to models of τ in which θ is satisfied by $\mathbf{a}$.

Q: This bookkeeping sounds complicated, but I think I get it. Your Υ-translation of θ just says that the values of the x_i's satisfy some pattern of equations and negated equations (an equality-type) and that the distinct ones among those values give T_r a value that corresponds to θ being true.

A: Right. So this finishes the explanation of how parametric conditions and thesauri are essentially equivalent.

Q: Yes, except you said earlier that you were using a simplified notion of signum. What's the full-scale notion?

A: In [1], our definition of a signum included, in addition to R, j, V, G, h as above[2], a probability distribution p on the set $\{1,\dots,v\}$ of values, subject to the requirements that each value has non-zero probability and that the distribution is invariant under the group $h(G)$.

The purpose of p is to modify the probability distribution on the structures with a fixed base set $\{1,2,\dots,n\}$. Previously, we chose a representative j-tuple from each G-orbit in $\{1,2,\dots,n\}^j$, and we chose the values of R at these representatives uniformly at random. Now, we choose these values according to the probability distribution p. And then, as before, we propagate the chosen values to all the other j-tuples by means of the symmetry requirement for thesaurus-structures.

Q: You require that p is invariant under $h(G)$ to ensure that the probability distribution on structures is independent of the choice of representative tuples.

A: Exactly. And of course the requirement that each value have non-zero probability is just a normalization. Any value whose probability is zero could be omitted, since it is unused in almost all structures.

Q: Is this more general notion of thesaurus equivalent to anything in the parametric condition world?

A: As far as we know, parametric conditions have been used only in connection with the uniform probability distribution on the structures that satisfy the conditions. But there is nothing to prevent one from introducing non-uniform distributions in this context, in a way that is equivalent to general thesauri, via the essential equivalence described above.

---

[2] Actually, as noted in an earlier footnote, it had a number v rather than a set V, but the difference is irrelevant.

Q: Apart from non-uniform probabilities, are the other advantages to thinking in terms of thesauri rather than in terms of parametric conditions?

A: Yes, we see a few.

First, certain structures are naturally thought of as having multi-valued relations. For example, colored graphs, where the values of the edge relation would be the colors and "false" (the latter for where there is no edge). Similarly, a tournament in which the outcome of a game can be a tie or a win for either player seems to be naturally viewed as a three-valued binary relation on the set of players.

Second, thesauri suggest a generalization that may be worth exploring. Instead of a fixed set of values for each signum, we could let the number of values grow slowly with the size of the structure.

Q: Won't that mess up the extension axioms?

A: Not if "slowly" is taken seriously. The usual computation showing that extension axioms have asymptotic probability 1 still works if the number of values of any relation grows more slowly than n^{ε} for each positive real number ε, as n, the number of elements in the structure, tends to infinity. So for example, $\log n$ values would be OK.

Q: What does it mean that the computation still works?

A: It means that, using the same ideas as in the proof of the usual first-order extension axioms, one finds that almost all finite models for a generalized thesaurus of this sort have the following property for each fixed natural number k. Take k variables, say $x_1,\dots,x_k$ and specify possible values for all the (finitely many) atomic formulas that use only these variables, at most once each, and really do use x_k. Here "posisble values" means that

- if an atomic formula begins with the signum R then the value must be in the value set of that signum, and
- if two atomic formulas begin with R, and their variables differ only by a permutation π of argument places, and π is in the group G of the signum R, then the values must correspond via $h(\pi)$, as required in the definition of thesaurus structures.

Then, if one interprets $x_1,\dots,x_{k-1}$ as any $k-1$ distinct elements of the structure, there will be an interpretation for x_k, distinct from these, and realizing all the given values for atomic formulas.

Q: OK, so it's really analogous to traditional extension axioms. How does the "usual computation" work in this situation.

A: Well, fix k and fix an assignment of values to atomic formulas as above. We

want to show that, with asymptotic probability 1, given any distinct $x_1,\dots,x_{k-1}$, there is some x_k realizing the given values. Consider a structure of size n, and let v be the largest of the cardinalities of the value sets, in this structure, for all the signa. Notice that the number of atomic formulas to which values are assigned is a constant f, because k and the thesaurus are fixed. So there are at most v^f possible assignments of values to these atomic formulas.

Temporarily, fix the interpretations of $x_1,\dots,x_{k-1}$. There are $n-k+1$ possible interpretations, distinct from these, for x_k. Each of these will, with the given interpretations of $x_1, \ldots, x_{k-1}$, realize one of the $\leq v^f$ possible assignments of values to atomic formulas, so it has probability $\geq 1/v^f$ of realizing the assinment we want. Therefore, the probability that no interpretation of x_k realizes our desired assignment is at most

$$\left(1 - \frac{1}{v^f}\right)^{n-k+1}.$$

Since $1 - t \leq e^- t$ for all t, this is at most $e^{-n/(2v^f)}$, where the factor 2 (more than) compensates for omitting $-k+1$ once n is larger than $2k$. Our assumption that v grows slowly compared with n means, in particular, that $2v^f < \sqrt{n}$ once n is large enough. For such large n, we conclude that the probability that no x_k realizes the desired values is $< e^{-\sqrt{n}}$.

Now un-fix the interpretation of $x_1, \ldots, x_{k-1}$. There are at most n^{k-1} such interpretations, so the probability that at least one of them has no suitable x_k is $< n^k e^{-\sqrt{n}}$. That's the probability that our analog of the k^{th}extension axiom fails, and the upper bound $n^k e^{-\sqrt{n}}$ approaches 0 as $n \to \infty$.

We have not studied possible applications of this idea, or even the appropriate extensions of the syntax and semantics of first-order logic. Once the number of values isn't constant, relations seem intuitively to behave more like functions than like relations in the traditional first-order world. One could also equip the set V of values with some relations or functions so that it becomes a structure in its own right.

Q: You'll need some structure on V, at least implicitly, in order to formulate first-order sentences over such a generalized thesaurus. For ordinary thesauri, your first-order language had, in effect, names for all the values $c \in V$. But with V now allowed to grow, that would make the language vary with the structure – not a good idea if you want to talk about the asymptotic probability of fixed sentences as the structure grows.

A: The idea could still work if, as the structure grows, the language also grows, so that any fixed sentence would make sense in all sufficiently large structures. But in fact, in at least one situation, V is naturally a relational structure.

Q: What situation is that?

A: Consider a tournament in which the result of each game (i.e., the relationship between a pair of players) is not merely a win for one or the other (as in traditional tournaments) or a tie (as in a generalization mentioned above) but can be any one of several "degrees of victory", where these degrees come from a small, linearly ordered set V. As before, "small" should mean of cardinality $< n^\varepsilon$ for each fixed $\varepsilon > 0$ and all large n. The outcomes of the games are modeled by a V-valued function R subject to the (anti-)symmetry requirement that $R(x,y)$ and $R(y,x)$ are symmetrically located in V, i.e., each is the image of the other under the unique order-reversing permutation of V.

In this situation, it is natural to admit the linear ordering of V as part of the structure, so that there are atomic formulas like $R(x,y) < R(u,v)$.

Q: Do you also want to allow names for the elements of V? If so, then it seems that your suggestion of a growing language depends on making some arbitrary conventions here. How shall the interpretation of a particular name vary while n and therefore V grow? There would be no problem with names for the first element, the second, the last, etc., but what about names for the element one third of the way up, or the element in position $\lfloor\sqrt{|V|}\rfloor$?

A: For the purposes of the present discussion, we'd want a language for which the 0-1 law holds. Apart from that, the choice of language would be guided by potential applications.

Notice, though, that as long as we have the ordering relation on V available, names for the first element, the second, the last, etc. are not really needed, with asymptotic probability 1, because these elements are definable in terms of the ordering.

Q: Wait a minute. Those definitions use quantifiers. Do you intend to allow quantification over V in your language?

A: With asymptotic probability 1, we have quantification over V automatically, since the elements of V are almost surely the same as the values of R. So, for example, we can almost surely express "$R(x,y)$ is the first element of V" by the formula

$$\forall u \forall v \neg (R(u,v) < R(x,y))$$

Q: I get it. This formula doesn't express the desired property in all structures, but it does in those structures where, as u and v range over the players, $R(u,v)$ ranges over *all* elements of V – and that's almost all structures.

A: Such expressive power requires some caution in the definition of structures, specifically in the choice of how V is allowed to grow. For example, we had better insist that the cardinality of V has the same parity in almost all structures, because that parity is definable by the truth value of the sentence

$$\exists x \exists y (R(x,y) = R(y,x))$$

Fortunately, this parity issue turns out to be the only such problem, in this particular example. That is, if we restrict $|V|$ to have a fixed parity and to grow slowly with n, then there will be a 0-1 law for the first-order properties of tournaments of this sort. (Here the atomic sentences are of the forms $x = u$ and $R(x,y) < R(u,v)$, where x and y are distinct variables and u and v are distinct variables, but other pairs of variables might coincide. It would make no difference if we also allow $R(x,y) = R(u,v)$, since it can be defined in terms of $<$.) We'll put a sketch of the argument into an appendix of this paper.

There is another possibility suggested by thesauri, namely the possibility of playing with the groups in the signa, and perhaps bringing some group theory to bear on these topics.

Q: That reminds me of something that occurred to me while you were proving the equivalence between thesauri and parametric conditions. Although thesauri allow arbitrary groups of permutations of $\{1,2,\ldots,j\}$ for a j-ary signum, you used only the full symmetric group in the thesauri that simulate given parametric conditions. Combining this with the simulation in the other direction, you've shown, in effect, that any thesaurus is equivalent to one in which all the groups involved in the signa are full symmetric groups. I'd think this equivalence, involving only thesauri, should have a proof that involves only thesauri, rather than going from thesauri to parametric conditions and back.

A: Yes, one can obtain the same result by working only with thesauri. Let $\langle R, j, V, G, h\rangle$ be a signum, and let S be the full symmetric group of all permutations of $\{1,2,\dots,j\}$. Then the signum $\langle R, j, V, G, h\rangle$ is essentially equivalent to a signum $\langle R_+, j, V_+, S, h_+\rangle$, where of course we have to define V_+ and h_+ (not R_+ because it's just a label).
Q: Before you start defining V_+ and h_+, please tell me what "essentially equivalent" means here.
A: It means that there is, for each set A, a canonical bijection between the possible interpretations in A for the two signa.
Q: OK. Now what are V_+ and H_+.
A: V_+ is the set of functions $f : S \to V$ that are G-equivariant in the sense that, for each $\pi \in G$ and each $\sigma \in S$,

$$f(\pi\sigma) = h(\pi)(f(\sigma)).$$

The action h_+ of S on V_+ is given by

$$(h_+(\sigma)(f))(\tau) = f(\tau\sigma).$$

It is straightforward to calculate that h_+ is a homomorphism[3] from S into the group of permutations of V.
Q: On the basis of my recent experience, I suppose that, if you had written $\sigma\tau$ instead of $\tau\sigma$, then h_+ would have been an anti-homomorphism.
A: That's right.

The essential equivalence between the two signa is obtained as follows. Given an interpretation $R^{\mathfrak{A}}$ of the original signum, i.e., a G-equivariant map $A^j \to V$, we obtain an interpretation $R_+^{\mathfrak{A}+}$ of the new signum, an S-equivariant map $A^j \to V_+$ by letting $R_+^{\mathfrak{A}+}(\mathbf{a})$ be the element of V_+ defined by

$$R_+^{\mathfrak{A}+}(\mathbf{a})(\sigma) = R^{\mathfrak{A}}(a_{\sigma^{-1}(1)}, \dots, a_{\sigma^{-1}(j)}).$$

The transformation in the other direction takes any S-equivariant map $R_+^{\mathfrak{A}+} : A^j \to V_+$ to the G-equivariant map $R^{\mathfrak{A}} : A^j \to V$ defined by

$$R^{\mathfrak{A}}(\mathbf{a}) = (R_+^{\mathfrak{A}+}(\mathbf{a}))(1),$$

where 1 is the identity permutation. Of course, there's a lot to be checked here: that the two transformations are well-defined and that they're inverse to each other. That should be a good exercise for you.
Q: OK, I'll check it when I have some spare time. But, in view of this equivalence, why did you define thesauri in terms of arbitrary permutation groups G rather than just the full symmetric group?
A: We weren't aware of the equivalence until recently. But it seems worthwhile to retain the generality of arbitrary groups, since some forms of symmetry are more

[3] Category theorists would describe the operation sending any G-set (V,h) to the S-set (V_+, h_+) as the right adjoint of the forgetful functor from S-sets to G-sets.

naturally expressed using groups smaller than the full symmetric group. Notice also that the conversion from a thesaurus using arbitrary groups to one using full symmetric groups makes the value sets V considerably larger and more complicated, especially if the original groups were small.

Appendix

We outline here the proof of the 0-1 law for the generalized tournaments described above. In fact, we prove somewhat more, namely the 0-1 law for first-order sentences in certain two-sorted structures. Here is the precise formulation.

Define (for the purposes of this appendix only) a *generalized tournament* to be a finite two-sorted structure $\mathfrak{A} = (A, V, <, \pi, R)$, where A and V interpret the two sorts, where $<$ is a linear ordering of V, where π is the unique order-reversing permutation of V, and where R is a function $A^2 \to V$, subject to the requirements that

- $R(a,b) = \pi(R(b,a))$ for all $a, b \in A$ and
- $|V|$ is odd.

The main reason for the second of these requirements is that, as we saw above, the parity of $|V|$ is definable and must therefore be fixed (at least almost surely) if we are to have a 0-1 law. We chose "odd" rather than "even" so that the first requirement makes sense even when $a = b$. For even $|V|$, a similar argument would apply, but we would have to either make R a partial function defined only when the arguments are distinct, or modify the first requirement to accommodate some convention for the case of equal arguments.

We say that a generalized tournament $\mathfrak{A}$ as above has *size* (n, v) if $|A| = n$ and $|V| = v$. For a first-order sentence θ in the language of generalized tournaments, we define the (n, v)-probability of θ as the proportion of models of θ among generalized tournaments of size (n, v). We say that θ has *asymptotic probability* 1 if, for each $\varepsilon > 0$, there exist $M \in \mathbb{N}$ and $\delta > 0$ such that, whenever $M < v < n^{\delta}$, then the (n, v)-probability of θ is at least $1 - \varepsilon$. Note that our definition of asymptotic probability incorporates the requirement that v is small compared to n, namely $v < n^{\delta}$; just how small this is (i.e., the choice of δ) depends on θ and ε. This definition of asymptotic probability is intended specifically for use with generalized tournaments; we do not propose it for more general contexts, though we expect that something similar would be appropriate in greater generality.

Theorem 7.1 *Let θ be any sentence in the first-order language of generalized tournaments. Then one of θ and $\neg\theta$ has asymptotic probability 1.*

Proof. Consider the first-order theory $\mathscr{RGT}$ (which stands for "random generalized tournaments"), in the language of generalized tournaments, given by the following axioms.

- R is a binary function from A to V.
- $<$ is a linear ordering of V.
- V has a first element and a last element.
- Each element of V but the first (resp. last) has an immediate predecessor (resp. successor).
- V is infinite (i.e., infinitely many axioms, saying $|V| > k$ for each $k \in \mathbb{N}$).
- π is an order-reversing permutation of V.
- $R(x,y) = \pi(R(y,x))$.
- π has a fixed point.
- The extension axioms

$$(\forall \mathbf{x})(\forall \mathbf{u}) \left[D(\mathbf{x}) \rightarrow (\exists y) \bigwedge_{i=1}^{k} (y \neq x_i \wedge R(x_i,y) = u_i)\right]$$

where the variables $\mathbf{x} = x_1, \ldots, x_k$ and y range over the first sort and $\mathbf{u} = u_1, \ldots, u_k$ over the second.

We computed above that each extension axiom has asymptotic probability 1. Each of the axioms saying $|V| > k$ has asymptotic probability 1; just take $M > k$ in the definition of asymptotic probability. The remaining axioms of $\mathscr{RGT}$ have asymptotic probability 1 for the trivial reason that they are true in all generalized tournaments.

Just as with the usual (one-sorted) notion of asymptotic probability, one sees that any conjunction of finitely many sentences of asymptotic probability 1 also has asymptotic probability 1 (just take the largest of the relevant M's and the smallest of the δ's) and that any logical consequence of a sentence of asymptotic probability 1 also has asymptotic probability 1. Therefore, the theorem will follow if we can show that the theory $\mathscr{RGT}$ is complete.

We prove completeness by presenting a winning strategy for Duplicator in an Ehrenfeucht – Fraïssé game of an arbitrary but specified length (i.e., number of rounds) r, between two models of $\mathscr{RGT}$. The essential part of Duplicator's work takes place in V. Here, Duplicator uses a familiar strategy for discrete linear orders with endpoints. The involution π and the A of the models are handled by suitable bookkeeping devices, described below in terms of imaginary pebbles.

For convenient reference, we first summarize the usual strategy for the Duplicator in the r-round Ehrenfeucht – Fraïssé game on two discrete linear orders *without* endpoints. In order to win, Duplicator must (by definition of the game) ensure that corresponding pebbles are ordered the same way in both models. His winning strategy is to ensure that, in addition, the distances between corresponding pebbles are not too different. Specifically, if, after m moves, two pebbles are at a distance $\leq 2^{r-m}$ on one model, then the corresponding pebbles on the other model are the same distance apart. Distances greater than 2^{r-m} need not be matched exactly, though of course if the distance between two pebbles is $> 2^{r-m}$ on one model then the distance between the corresponding pebbles on the other model will also be $> 2^{r-m}$. The key to the proof is that, because the distance 2^{r-m}, below which matching is required, decreases by a factor 2 from each move to the next, Duplicator can always move so as to maintain the required matching. By doing so, he ensures that he wins the game.

For infinite discrete linear orders *with* endpoints, Duplicator's strategy is the same except that he pretends that, already at the start of the game, there is a pair of corresponding pebbles on the first elements of the models and another pair of corresponding pebbles on the last elements. (In fact, the strategy doesn't really need that the models are infinite; it suffices that they have large enough cardinalities so that the Duplicator's matching requirements are satisfied by the initial imaginary pebbles.)

With these preliminaries, we now present Duplicator's strategy for models of $\mathcal{RGT}$. It involves a more elaborate scheme of imaginary pebbles. At the start of the game, Duplicator should imagine pebbles already placed on the first, last, and middle elements of the V sort in both models. Here "middle element" means the unique element fixed by π. In addition, during the course of the game, whenever a pebble is on an element $q \in V$, Duplicator should imagine an associated pebble at $\pi(q)$. Furthermore, whenever pebbles are on two elements a and b of A, Duplicator should imagine an associated pebble at $R(a,b)$.

As in the proof for plain linear orderings, Duplicator must ensure that corresponding pebbles on the V sorts of the two models are ordered the same way, and he will, in addition, voluntarily ensure that sufficiently small distances between corresponding pebbles match exactly. But now, "sufficiently small" does not mean $\leq 2^{r-m}$ (after m rounds of an r-round game) but rather $\leq 2^{r^2-m^2}$. The reason for this change will become clear in a moment, but for now notice that, like 2^{r-m}, this new boundary of smallness, $2^{r^2-m^2}$, decreases by at least a factor 2 at every move.

It is clear that, if Duplicator can follow this strategy, then doing so guarantees that he wins. It remains to show that Duplicator always has a move that maintains the required matching of distances. The proof of this is in two cases, depending on which sort Spoiler adds a pebble to.

Suppose first that Spoiler puts a new pebble on the V sort of one of the two models. Then, exactly as in the proof for plain linear orders, Duplicator can find a suitable spot for the corresponding pebble on the other model, because the distance below which matching is required has decreased by at least a factor 2. The new pebbles played in this round give rise to a new pair of imaginary pebbles, located at π of the real pebbles. But the required matching of distances between these imaginary pebbles and the played pebbles at earlier moves is automatic because π preserves distances. We must also consider the distance between the new real pebble and the new imaginary pebble, but since these locations are related by π they lie on opposite sides of the middle of V. Their distance is twice the distance from either of them to the (initially imagined) pebble at the middle. Since the distance to the middle pebble is matched (if small), so is the distance between the real and imaginary new pebbles. This completes the proof for the case where Spoiler's new pebble is on the V part.

Now suppose that Spoiler puts a new pebble on the A sort of one of the models. This results in many new imaginary pebbles on the V part. If there were already m moves before the current one, then there could be as many as m pebbles already in the A part, say at elements $a_1,\dots,a_m$. The new pebble, say at b, gives rise to as many as $2m$ new imaginary pebbles, at the elements $R(a_i,b)$ and (their π-images) $R(b,a_i)$. In this situation, Duplicator should proceed as follows. Pretend that, instead of a single move of Spoiler producing all these imaginary pebbles, there were $m+1$

moves, that during the first m of these moves Spoiler put pebbles on the elements $R(a_i,b)$ one at a time (resulting in imaginary pebbles at $R(b,a_i)$), and that at the last of the $m+1$ moves Spoiler put the (real) pebble at b. Let Duplicator pretend to respond to the first of these m moves as described in the preceding paragraph, i.e., let him put imaginary new pebbles in the appropriate places. At each step, the maximum distance below which he can maintain exact matching decreases by a factor 2, so in these m moves, it has decreased by 2^m. Fortunately, this is still large enough, because $2^{r^2-m^2}$ (from before these moves) decreased by a factor 2^m is still larger than $2^{r^2-(m+1)^2}$, the required distance for matching after these moves. (This is of course why we used $2^{r^2-m^2}$ rather than 2^{r-m}. Our choice is not the smallest, since it could accommodate a decrease by a factor 2^{2m+1}, but it seems to be the simplest.) Finally, after all the imaginary pebbles have been placed in V, Duplicator must find a place for his real pebble in A. It together with previous pebbles must produce, as values of R, the elements of V bearing the imaginary pebbles just placed. Fortunately, an extension axiom guarantees the existence of an appropriate element to receive this pebble.

Bibliographic Remark: Two useful survey papers on 0-1 laws, Oberschelp's [7] and Compton's [4], were not treated kindly by the printing process. The title of [4], though given correctly in the table of contents of the book in which it appears, had "0-1" deleted on the first page of the paper itself. As a result, "0-1" is also missing in the MathSciNet entry and perhaps elsewhere. In [7], some of the pages were printed out of order, and numbered in the printed order rather than the intended order. Fortunately, all the pages are present and can be found in correct order by a local search.

References

1. Blass A., and Gurevich Y. Strong extension axioms and Shelah's zero-one law for choiceless polynomial time, *J. Symbolic Logic*, 68: 65–131, 2003.
2. Blass A., and Gurevich Y. Ordinary interactive small-step algorithms, I, *A. C. M. Trans. Comp. Logic*, 7: 363–419, 2006.
3. Blass A., and Harary F. Properties of almost all graphs and complexes, *J. Graph Theory* 3: 225–240, 1979.
4. Compton K. 0–1 laws in logic and combinatorics. In I. Rival, editor, *Algorithms and Order (Proc. NATO Advanced Study Institute, Ottawa, 1987)*, pages 353–383. Kluwer, Dordrecht, 1989.
5. Fagin R. Probabilities on finite models, *J. Symbolic Logic*, 41: 50–58, 1976.
6. Glebskii Y. V., Kogan D. I., Liogon'kii M. I., and Talanov V. A. Range and degree of realizability of formulas in the restricted predicate calculus, *Kibernetika (Kiev)*, 2: 17–28, 1969; English translation *Cybernetics* 5 (1969) 142–154.
7. Oberschelp W. Asymptotic 0–1 laws in combinatorics. In D. Jungnickel and K. Vedder, editors, *Combinatorial Theory (Proc. Conf. Schloss Rauischholzhausen, 1982)*, Springer-Verlag, Lecture Notes in Mathematics 969: 276–292, 1982.

Chapter 8
Recent Developments of Feedback Coding and Its Relations with Many-Valued Logic

Ferdinando Cicalese* and Daniele Mundici

8.1 Basic Facts in Feedback Coding

8.1.1 Introduction

The basic problem of feedback coding is vividly described by Rényi [23, p. 47] as a problem of fault-tolerant adaptive search with errors, as follows:

> [...] I made up the following version, which I called "Bar-kochba with lies". Assume that the number of questions which can be asked to figure out the "something" being thought of is fixed and the one who answers is allowed to lie a certain number of times. The questioner, of course, doesn't know which answer is true and which is not. Moreover the one answering is not required to lie as many times as is allowed. For example, when only two things can be thought of and only one lie is allowed, then 3 questions are needed [...] If there are four things to choose from and one lie is allowed, then five questions are needed. If two or more lies are allowed, then the calculation of the minimum number of questions is quite complicated [...] It does seem to be a very profound problem [...]

The same problem is posed by Ulam in his book "Adventures of a Mathematician" [25, p. 281], as the problem of searching for an unknown m-bit number by asking a minimum of yes-no questions, in a game of Twenty Questions with up to e lies in the answers.

Ferdinando Cicalese
AG Genominformatik, Technische Fakultät, Universität Bielefeld, D-33594 Bielefeld, Germany,
e-mail: nando@cebitec.uni-bielefeld.de

Daniele Mundici
Department of Mathematics "Ulisse Dini", University of Florence, 50134 Florence, Italy,
e-mail: mundici@math.unifi.it

* The first author was supported by the Sofja Kovaleskvaja Award of the Alexander von Humboldt Foundation. The second author was partially supported by Cofin-2004 Project on Many-valued Logic.

J. van Benthem et al. (eds.), *Proof, Computation and Agency*, Synthese Library 352,
DOI 10.1007/978-94-007-0080-2_8,

Someone thinks of a number between one and one million (which is just less than 2^{20}). Another person is allowed to ask up to twenty questions, to each of which the first person is supposed to answer only yes or no. Obviously the number can be guessed by asking first: Is the number in the first half million? then again reduce the reservoir of numbers in the next question by one-half, and so on. Finally the number is obtained in less than $\log_2(1000000)$. Now suppose one were allowed to lie once or twice, then how many questions would one need to get the right answer? One clearly needs more than n questions for guessing one of the 2^n objects because one does not know when the lie was told. This problem is not solved in general.

The Rényi-Ulam game is an interesting variant of the familiar game of Twenty Questions. Here, the two players, named Carole and Paul, first agree on fixing a finite space S with $|S|$ elements. Then Carole chooses an element $x_{secret} \in S$ and Paul must find it by asking a minimum of yes-no questions. Both Rényi and Ulam were interested in a situation where up to e of Carole's answers may be erroneous/mendacious/inaccurate. From Paul's viewpoint it is immaterial whether wrong answers arise just because Carole is unable to answer correctly, or because she is (moderately) mendacious, or else Carole is always sincere and accurate, but distortion may corrupt up to e of the bits carrying her yes-no answers. Under this latter representation, *Rényi-Ulam games pertain to Berlekamp's communication theory with feedback* [4]. Here one has a noisy channel (for answers) and a noiseless channel (for questions). Carole sends x_{secret} to Paul bitwise, on the noisy channel, each bit b being the correct answer to Paul's question. It may happen, up to e times, that Paul receives a distorted version $b' = 1 - b$ of Carole's bit. In this co-operative model Carole knows Paul's questioning strategy: his $(t+1)$th question Q_{t+1} only depends on the previous bits $b'_1, \ldots, b'_t$. Thus for Carole to know Q_{t+1} and to formulate the correct answer b_{j+1}, it is enough that Paul sends her the sequence $b'_1, \ldots, b'_t$ via the noiseless (feedback) channel. The shortest sequence of bits $b_1, b_2, \ldots, b_q$ constitutes an optimal encoding of x_{secret}, allowing Paul to recover x_{secret} even if $\leq e$ of the received b'_i are different from b_i.

The particular case when all questions are asked non-adaptively corresponds to a situation where no feedback bits are sent by Paul. Strategies in such nonadaptive Rényi-Ulam games boil down to e-error-correcting codes. As we shall see, adaptiveness (=feedback) makes the difference between traditional error-correcting codes and Rényi-Ulam coding with feedback.

8.1.2 Symmetric Coding

Under the assumption that both bits 1 and 0 are equally liable to distortion, one easily sees that x_{secret} cannot be found using less than $N_{\min}(|S|, e)$ questions, where the *sphere-packing bound* $N_{\min}(|S|, e)$ is the smallest integer $q \geq 0$ satisfying the classical inequality

$$2^q \geq |S| \sum_{j=0}^{e} \binom{q}{j}. \tag{8.1}$$

The right hand term is the number of Carole's answering strategies. This is the number of q-bit tuples that Paul may receive when Carole sends her q yes-no answers over a channel that flips up to e of the transmitted bits. Specifically, for any possible choice of $x_{secret} \in S$, and for each $i = 0, 1, \ldots, e$, if exactly i bits are flipped by noise, then Carole's undistorted q-tuple may be received by Paul as one among $\binom{q}{i}$ different corrupted q-tuples. Equation (8.1) amounts to asking that q is large enough for x_{secret} to be uniquely identifiable by Paul.

Definition 8.1 Let us agree to say that the secret number $x_{secret} \in S$ can be guessed by a *perfect* strategy with e errors/lies iff the number of questions coincides with the smallest q satisfying inequality (8.1).

If no feedback is available, the problem of finding a perfect searching strategy for the Rényi-Ulam game with e errors over the search space S, is formulated in the theory of error-correcting codes as follows:

> In the metric space $\{0,1\}^q$ of all q-bit tuples with Hamming distance, find $|S|$ points such that the spheres of radius e centered at any two points do not overlap.

The following negative result [24, 26] (also see [17]) is well known to researchers in error-correcting codes:

Fact 0: Non-existence of perfect error-correcting codes
If no feedback is allowed, a perfect strategy does not exist in general, already in case of 2 *errors.*

The situation is radically different if feedback is allowed. Now Paul can make his ith question dependent on what he already knows from Carole's previous answers.

The following result is the output of extensive research during the last two decades (see the survey papers [9, 21] and references therein):

Fact 1: Existence of perfect error-correcting codes with feedback
If feedback is allowed then for every number e of errors, a number can be guessed in general, using a perfect strategy.

More importantly, it turns out [6, 9, 10] that perfect error-correcting codes also exist under very weak conditions on feedback:

Fact 2: Minimum feedback is sufficient for perfect coding
Perfect strategies that use feedback only once enable Paul to find x_{secret} by asking a first batch of nonadaptive questions, and then, only depending on the answers to these questions, asking a second batch of nonadaptive questions.

In the light of Fact 0, the above result states that perfect error-correcting codes exist provided one allows *minimum* feedback.

We shall sketch a proof of Fact 2. Minimum feedback perfect strategies rely on *perfect* questions and *Gilbert-packing*. To give an operative definition of perfect (= maximum information) question, let us recall that a *question* is a subset Q of the search space S. Answers are most conveniently defined in the non-cooperative model where Carole can lie up to e times. [2] Then an *answer* to Q is just a set $A \in \{Q, S \setminus Q\}$ telling Paul that $x_{secret} \in A$. An element $z \in S$ is said to *satisfy* A iff $z \in A$. Otherwise, z *falsifies* A.

During the game/transmission, Paul's current knowledge is given by the record $\mathbf{A}$ of Carole's answers. Suppose t answers/bits have been received. For each $z \in S$ let $f(z)$ be the number of answers in $\mathbf{A}$ that are falsified by z. Thus, if $x_{secret} = z$ at most $e \dot{-} f(z)$ new errors can affect the remaining $q - t$ answers[3]. Let $w^{\mathbf{A}}(z)$ be the number of Carole's answering strategies for the rest of the game, under the assumption that $x_{secret} = z$. The same counting argument leading to the sphere packing bound (8.1), now yields

$$w^{\mathbf{A}}(z) = \sum_{j=0}^{e \dot{-} f(z)} \binom{q-t}{j}.$$

Therefore, Paul can infallibly identify x_{secret} only if

$$\sum_{z \in S} w^{\mathbf{A}}(z) \leq 2^{q-t}. \tag{8.2}$$

For each $j = 0, 1, \ldots, e$, let $S_j = \{z \in S \mid f(z) = j\}$. The $(e+1)$-tuple $(S_0, S_1, \ldots, S_e)$ is the *state* of the game *determined by* $\mathbf{A}$. Inequality (8.2) is equivalent to

$$\sum_{i=0}^{e} |S_i| \sum_{j=0}^{e-i} \binom{q-t}{j} \leq 2^{q-t}. \tag{8.3}$$

The *weight* of record $\mathbf{A}$ is defined by

$$w(\mathbf{A}) = \sum_{z \in S} w^{\mathbf{A}}(z) = \sum_{i=0}^{e} |S_i| \sum_{j=0}^{e-i} \binom{q-t}{j}.$$

Let Q be the next question asked by Paul, and $\mathbf{A}^{no}$ and $\mathbf{A}^{yes}$ be the new records according as Carole's answer is *no* or *yes*. The following *conservation law* (for the number of answering strategies) is easy to prove:

$$w(\mathbf{A}) = w(\mathbf{A}^{no}) + w(\mathbf{A}^{yes}). \tag{8.4}$$

Then, we say that question Q is *perfect* for the state corresponding to $\mathbf{A}$ iff

$$w(\mathbf{A}^{no}) = w(\mathbf{A}^{yes}). \tag{8.5}$$

Condition (8.5) is related to the familiar balancing property of optimal binary search procedures, e.g., in the game of Twenty Question without lies. Any perfect question Q guarantees that Paul gains from Carole the maximum amount of information: If Paul were able to keep on asking perfect questions for the whole duration

[2] The reader may provide the necessary reformulation of "answer" for the cooperative model where Carole knows Paul's searching strategy.

[3] Here $\dot{-}$ denotes subtraction truncated to 0.

of the game he would identify x_{secret} with the a smallest possible number q questions, as given by the sphere packing bound (8.1).

How many perfect questions can Paul ask? To answer this question, fix integers $e,m \geq 0$. Let $|S| = 2^m$. Let $q = q(m)$ be the smallest integer ≥ 0 satisfying the sphere packing bound (8.1). For each $i = 1,\ldots,m$, the ith question asked by Paul is as follows:

"Is the i-th bit of the number x_{secret} equal to 1?".

It is immediately checked that all these questions are perfect. Thus, no matter Carole's answers, Paul's resulting state of knowledge satisfies the identity

$$|S_i| = \binom{m}{i},$$

for each $i = 0,1,\ldots,e$, independently of the distribution of the i distorted bits. This is so because S_i contains exactly those binary numbers $z \in S = \{0,1\}^m$ differing in exactly i positions from the sequence of bits received by Paul. While the sets S_i's depend on record $\mathbf{A}$ their cardinality is independent of the record.

After receiving the m-bit tuple $\mathbf{b}$ of answers to the above perfect questions, Paul knows that the secret number belongs to $\bigcup_{i=0}^{e} S_i$.[4] Paul also knows that if $x_{secret} \in S_j$ then already j of Carole's answers are wrong, whence at most $e-j$ more erroneous answers can be received ($j = 0,1,\ldots,e$). Only depending on $\mathbf{b}$, Paul will now send a final minibatch of $q-m$ nonadaptive questions enabling him to guess x_{secret}.

To see this we recall the Gilbert packing bound (see, e.g., [10]):

Lemma 8.1 *There exist disjoint sets $\mathscr{C}_1, \mathscr{C}_2$ of binary tuples of length $q-m$ satisfying the following conditions:*

(i) $|\mathscr{C}_1| = \sum_{j=0}^{e-1} |S_j|$;
(ii) $|\mathscr{C}_2| = |S_e|$;
(iii) *For any two distinct tuples $\mathbf{z}_1, \mathbf{z}_2 \in \mathscr{C}_1$ if $\mathbf{z}'_1$ and $\mathbf{z}'_2$ are respectively obtained by flipping up to e bits in $\mathbf{z}_1$ and $\mathbf{z}_2$, then $\mathbf{z}'_1$ and $\mathbf{z}'_2$ are distinct, and neither belongs to $\mathscr{C}_2$.*

Building on this lemma Paul will first of all choose a function φ sending all elements of $\bigcup_{j=0}^{e-1} S_j$ one-one onto $\mathscr{C}_1$, and all elements of S_e one-one onto $\mathscr{C}_2$. Paul's final batch of questions is about the binary number $\varphi(x_{secret})$. More precisely, for each $i = 1,\ldots,q-m$, Paul's $(m+i)$th question is (the subset of $\{0,1\}^{q-m}$ corresponding to)

"Is the ith bit of $\varphi(x_{secret})$ equal to 1?".

Let $\mathbf{a} \in \{0,1\}^{q-m}$ be the uncorrupted tuple of Carole's answers – before distortion. Let $\tilde{\mathbf{a}} \in \{0,1\}^{q-m}$ be the tuple of answers to these questions, as received by Paul. The properties of the sets $\mathscr{C}_1, \mathscr{C}_2$ ensure that from the corrupted tuple $\tilde{\mathbf{a}}$, Paul

[4] The dependence of the S_i's on $\mathbf{b}$ is understood.

will infallibly x_{secret}. As a matter of fact, by Lemma 8.1, **a** coincides with the element of $\mathscr{C}_1 \cup \mathscr{C}_2$ having the shortest Hamming distance from $\tilde{\mathbf{a}}$ – and this is easily computable by Paul who knows $\mathscr{C}_1 \cup \mathscr{C}_2 = \varphi(\bigcup S_i)$. Finally, by computing $\varphi^{-1}(\mathbf{a})$ he will recover x_{secret}, as required.

8.1.3 Asymmetric Coding: Basic Facts

Feedback coding for the asymmetric channel is a different matter. As usual, the channel carrying Carole's answers is said to be *asymmetric* if the two bits 0 and 1 are not equally liable to distortion. Consider, e.g., an optical channel [22] where one transmits the bit 1 by sending a photon, and transmits 0 by not sending a photon. Since photons can be lost but spurious photons cannot be received when none has been sent, it follows that the only possible type of error in this channel occurs when answer 1 is received as 0.

In the classical theory of error-correcting codes, this type of channel is known as the *Z-channel.* Optimization problems for coding in asymmetric channels are especially challenging, because the usual basic facts for symmetric channels, e.g., the sphere packing inequality (8.1), do not seem to have an immediate asymmetric counterpart. Still, the paper [7] gives a complete solution of the case $e = 1$, and a characterization of "perfect" strategies. The techniques introduced in [7] were subsequently generalized in [3, 12].[5]

For the Z-channel with e errors we have the following result: For all sufficiently large q and for all sets S, there exists an encoding of length q for S if, and only if,

$$|S| \leq \frac{2^{q+e}}{\binom{q}{e}}. \tag{8.6}$$

While this inequality is reminiscent of (8.1), from (8.6) no weight function for states, nor any conservation law can be directly obtained from it. This makes it difficult to derive optimal strategies and bounds for the asymmetric channel. In fact, only for the case $e = 1$ an exact (non-asymptotic) analysis is available [7]. The more recent papers [3, 12], provide more comprehensive and general results, but only give asymptotic estimates. For our purposes in this paper we shall limit ourselves to sketch the main ideas leading to the characterization of optimal strategies. We shall closely follow [7, 12].

We shall first argue that inequality (8.6) provides an a priori lower bound on the size of any search strategy for the Z-channel with e errors.

By a *strategy with q questions* we understand a binary tree of depth q, where each node ν is mapped into a question Q_ν, and the two edges $\eta_{\text{left}}, \eta_{\text{right}}$ generated by ν are respectively labeled *yes* and *no*. Let $\eta = \eta_1, \ldots, \eta_q$ be a path in $\mathscr{S}$, from the root to a leaf, with respective labels $b_1, \ldots, b_q$, generating nodes $\nu_1, \ldots, \nu_q$ and associated questions $Q_{\nu_1}, \ldots, Q_{\nu_q}$. Then the state associated to the leaf of η_q is the

[5] This latter paper gives the state of the art on the asymmetric Rényi-Ulam game.

tuple $(S_0, S_1, \ldots, S_e)$ where, for $i = 1, \ldots, e$, S_i is the set of elements in S that satisfy all negative answers along the path η and falsify exactly i of the positive answers. A strategy is said to be *winning* iff for every path η in it, the state associated to the leaf η_q contains at most one element. A strategy is said to be *non-adaptive* iff all nodes at the same depth of the tree are mapped into the same question.

Now, let Σ be a strategy of minimum length q for the search space S. By a path in Σ we shall understand a path from the root to a leaf of Σ. For each $x \in S$ there exists precisely one path μ_x in Σ leading to the final state $(\{x\}, \emptyset, \ldots, \emptyset)$. This final state is obtained if Carole chooses x as the secret number and all her answers are received without distortion. Let yes_x be the number of branches whose label is "yes" in this path. The yes_x many branches of μ_x are a record of Carole's "yes" answers, once she decides to choose x as the secret number. As an effect of noise, the sequence of answers received by Paul may deviate from this path in yes_x many ways, replacing a "yes" true answer by a mendacious "no" answer and entering a new path.[6] Since Σ is a winning strategy, there exist in Σ precisely yes_x many paths leading to a final state $(\emptyset, x, \emptyset, \ldots, \emptyset)$. On the other hand, over each of these paths noise can affect a *yes* edge entering a new path that will lead to a final state of the type $(\emptyset, \emptyset, x, \emptyset, \ldots, \emptyset)$, and so on, to the effect that the strategy will also contain paths leading to a final state $(\emptyset, \ldots, \emptyset, x)$.

Roughly speaking, for all large q and almost all binary trees of height q, almost all paths contain an almost equal number of left and right branches. It follows that, for each $x \in S$ there are $\binom{q/2}{e}$ possible paths in Σ leading to a state of the form $(\emptyset, \ldots, \emptyset, \{x\})$. The assumption that Σ is a winning strategy implies that Σ must contain at least $|S|\binom{q/2}{e} \sim \frac{|S|}{2^e}\binom{q}{e}$ different paths, whence, we have the desired bound (8.6).

Conversely we shall explain why, for any fixed e and all large q, and for all sets S satisfying (8.6), Paul has an e-error correcting encoding for S of length q on the Z-channel.[7] Again, Paul's strategy is based on perfect questions and Gilbert-packing. However, in the absence of a conservation law for information in the asymmetric channel, perfect questions have the following definition:

Definition 8.2 We say that a question is *perfect* for a state $(S_0, S_1, \ldots, S_e)$ iff the two resulting states $(S_0^{yes}, S_1^{yes}, \ldots, S_e^{yes})$ and $(S_0^{no}, S_1^{no}, \ldots, S_e^{no})$, respectively arising from a *yes* or a *no* answer, satisfy the identity

$$|S_i^{yes}| = |S_i^{no}|$$

for each $i = 0, 1, \ldots, e$.[8]

Let $(S_0, S_1, \ldots, S_e)$ be an arbitrary state of the game. If for each $i = 0, 1, \ldots, e$, we have the inequality $|S_i| \geq \Sigma_{j=0}^{i} |S_j|/2^{i-j}$ then each question Q with

---

[6] Note the asymmetric effect of noise.

[7] For a rigorous treatment the interested reader is referred to [3, 7, 12].

[8] Definition 8.1 is a generalization of the present one, but only applies to the symmetric channel.

$$|Q \cap S_i| = \sum_{j=0}^{i} |S_j|/2^{i-j+1} \tag{8.7}$$

is perfect.

To avoid unnecessary complications, let us consider the case when the size of the search space has $2^{m'}$ elements, whence the initial state $(S_0^{(0)}, S_1^{(0)}, \ldots, S_e^{(0)})$ satisfies $|S_0^{(0)}| = 2^{m'}$, and $|S_i^{(0)}| = 0$, for $i = 1, \ldots, e$.

A moment's reflection shows that, for each $j = 0, 1, \ldots, m' - e - 1$, Paul can adaptively choose his $(j+1)$th question $Q^{(j+1)}$ according to the following rule, which is a minor variant of (8.7):

$$|Q^{(j+1)} \cap S_i^{(j)}| = \min\{\sum_{j=0}^{i} |S_j^{(j)}|/2^{i-j+1}, |S_i^{(j)}|\}, \tag{8.8}$$

where $(S_0^{(j)}, S_1^{(j)}, \ldots, S_e^{(j)})$ is Paul's state resulting from the first j answers.

Let $m = m' - e$. Then [3, Lemma 4] (see also [12, Lemma 3.7]) there are positive constants $c_0, c_1, \ldots, c_e$ such that, for all sufficiently large q, Paul's state $(S_0^{(m)}, S_1^{(m)}, \ldots, S_e^{(m)})$ satisfies

$$|S_i^{(m)}| \le c_i \binom{m}{i}$$

for each $i = 0, \ldots, e$.

It turns out that Lemma 8.1 holds for this state, too (the papers [3, 10, 12] provide different proofs of this fact). Thus Paul can use it to finish his search within the allowed q questions: it is sufficient for him to proceed as if the channel were symmetric. [9]

Let $N_{\min}^{asym}(|S|, e)$ be the smallest integer $q \ge 0$ satisfying the following variant of (8.1):

$$2^{q+e} \ge |S| \binom{q}{e}. \tag{8.9}$$

We then say that the secret number can be guessed by a *perfect* strategy in the asymmetric Rényi-Ulam game iff the number of questions coincides with $N_{\min}^{asym}(|S|, e)$. Direct inspection shows that, for any fixed $e \ge 1$ and all sufficiently large cardinalities $|S|$, $N_{\min}^{asym}(|S|, e) < N_{\min}(|S|, e)$.

Our analysis has shown that perfect questions allow strategies for the asymmetric variant of the Rényi-Ulam game that are asymptotically perfect. *Therefore, optimal strategies for the asymmetric variant of the Rényi-Ulam game violate the sphere-packing bound* (8.1).

[9] Since search on a symmetric channel is harder than on the Z-channel, a fortiori Paul can successfully use Gilbert packing for this part of his strategy on the asymmetric channel.

Note that the strategies yielding the above results make extensive use of the feedback channel. Indeed, the appropriate questions satisfying conditions (8.7) and (8.8) crucially depend on Paul's detailed knowledge of the S_i's.

8.2 Part Two: The Logic of Feedback Coding

In [18] it was proved that Paul's states of knowledge in Rényi-Ulam games obey the infinite-valued Łukasiewicz calculus. Also see [19, 20], or [11, Chapter 5]. Generalizing the analysis of [18] we shall prove that Hájek Basic Logic is the logic of *multichannel* Rényi-Ulam games.

To avoid unnecessary syntactic complications we shall work in the context of BL-algebras [14]. We assume the reader of this part of the paper has some acquaintance with the elementary properties and constructions of BL-algebras [14]. For the proof of the main theorem we shall also need some background material from the theory of MV-algebras [11], and Wajsberg hoops [5].

8.2.1 Asymmetry and Multichannels

The prototypical example of asymmetric coding is provided by the Z-channel. The latter can be equivalently represented by equipping Carole with *two* channels, denoted 1 and 2: the more expensive Channel 1 is noiseless; channel 2 is e-noisy. Carole and Paul stipulate that, whenever the correct answer to a question is "no" she sends bit 0 only on channel 1. On the other hand, all 1 bits are to be sent on channel 2 only. Assuming that Paul does not know whether the received bit b has traveled via channel 1 or 2, he can still conclude that any received bit $b = 1$ is undistorted: for, the distortion pattern $0 \mapsto 1$ is impossible. Since the converse pattern $1 \mapsto 0$ may occur up to e times, Paul concludes that up to e of the 0's received by him may be wrong, but all 1's are to be taken for granted – precisely as is the case for the Z-channel with e errors.

As a simple variant of this two-channel game, let us suppose Carole is equipped with m increasingly noisy channels $1, 2, \ldots, m$: when asking a question Paul also tells Carole which channel should be used for her answer. Let us finally assume that for all $i < j$, any piece of negative information supplied via channel i *pointwise supersedes* all information given by the noisier channel j. In other words, once an element $z \in S$ falsifies an answer given via channel i, then any past and future piece of information *about* z supplied via channel j has no value.[10]

[10] Of course, answers sent on channel j may still contain relevant information about any $w \neq z$. In accordance with our aims in this section, we shall not be interested in defining the various types of optimization problems concerning search in a multichannel game: suffice to say that communication on a noisy channel is cheaper than on a low-noise channel. Thus Paul's minimum cost search of x_{secret} will require a careful analysis of which channels should be used when: generally

8.2.2 Search Space, Questions, Answers

Let us inspect a round of the multichannel game, played by Paul (on behalf of all of us) and Carole: Carole and Paul agree to fix a finite nonempty set S of numbers, called the *search space*. We let $|S|$ denote the cardinality of S. Answers can be sent on m channels $1,2,3,\dots,m$, each channel i allowing an expected maximum number e_i of lies/errors/distortions, $0 \le e_1 < \dots < e_m$. All e_i's are known to both players. Then Carole chooses a number $x_{secret} \in S$. Paul must find x_{secret} by asking yes-no questions. Each question comes together with Paul's specification of the channel j for Carole's answer to be sent. In more detail, a *question* Q is a subset of S: thus for instance, the question "is x_{secret} odd?" is the set of all odd numbers in S. Question Q is sent on the noiseless feedback channel, together with an integer $j \in \{1,2,\dots,m\}$; Carole must send her answer on channel j. An *answer* to Q is a pair (j,T), where j is the chosen channel, and $T \in \{Q, S \setminus Q\}$.[11] An element $z \in S$ *satisfies* answer $A = (j,T)$, in symbols, $z \models A$, iff $z \in T$. Otherwise, z *falsifies* A.

8.2.3 The Truth-Value

In the traditional, error-free game of Twenty Questions, a complete *record* of our knowledge about x_{secret} is the *set* of Carole's answers (two equal answers carrying the same information as one). To see when two records $\mathbf{A}$ and $\mathbf{B}$ are equivalent, let the function $\mathbf{A}^{\#}\colon S \to \{0,1\}$ compute for every $z \in S$ the truth-value of z. This is the quantity

$$1 \dot{-} |\{\text{answers in } \mathbf{A} \text{ falsified by } z\}|, \tag{8.10}$$

where $\dot{-}$ denotes truncated addition. Stated otherwise, for each $z \in S$,
$\mathbf{A}^{\#}(z) = 1$ iff z does not falsify any answer; $\mathbf{A}^{\#}(z) = 0$ iff z falsifies at least one answer $A \in \mathbf{A}$. Then two records $\mathbf{A}$ and $\mathbf{B}$ are equivalent iff $\mathbf{A}^{\#} = \mathbf{B}^{\#}$. For example, $\{x_{secret}$ is odd$, x_{secret}$ is even$\}$ and $\{x_{secret}$ is ≥ 6, x_{secret} is $< 6\}$ are equivalent records. Both records are to the effect that no possible candidate for x_{secret} survives in S.

Also in the m-channel game with $\mathbf{e} = (e_1 < \dots < e_m)$ lies, our knowledge about x_{secret} is given by the finite *record* $\mathbf{A}$ of Carole's answers, each answer being a pair (j,T) with $j \in \{1,\dots,m\}$ and $T \subseteq S$. Note however that $\mathbf{A}$ is now a *multiset* – because repeated equal answers to the same repeated question carry more information than single answers. Thus each answer $A \in \mathbf{A}$ carries a multiplicity, telling how

speaking, an expensive channel is to be sparingly used, after the search space has been greatly reduced by preliminary extensive use of noisy channels.

[11] In the co-operative model where Carole knows Paul's strategy and errors are due to distortion, Carole sends honest bits $b = 0,1$ to signify her negative or positive answers. It is expected that up to e_j of these answers may be erroneous/mendacious.

many times A occurs in $\mathbf{A}$. For any channel c we denote $\mathbf{A}_c$ the sub-multiset of those answers $A = (i,T)$ of $\mathbf{A}$ with $i = c$.

Definition 8.3 Fix an element $z \in S$ and a record $\mathbf{A}$. Then the *truth-value* $\mathbf{A}^{\#}(z)$ of *z in* $\mathbf{A}$ is the pair (j,r) where $(j,r) = (m,1)$ if z satisfies all answers in $\mathbf{A}$, and otherwise

$$\begin{cases} j = \text{first channel } c \in \{1,\ldots,m\} \text{ such that } \exists A \in \mathbf{A}_c \text{ with } z \not\models A \\ r = 1 \dot{-} \frac{|\{\text{answers in } \mathbf{A}_c \text{ falsified by } z\}|}{e_c+1}. \end{cases} \tag{8.11}$$

Intuitively, letting c be the least noisy channel of an answer falsified by z, the truth-value of z decreases proportionally to the number of answers sent on channel c and falsified by z. Like a truth-value in Łukasiewicz logic [18–20], [11, (5.3)], the quantity $\mathbf{A}^{\#}(z)$ measures (in units of $e_c + 1$) the distance of z from the condition of being excluded from the possible candidates for x_{secret} *as an effect of the information sent on channel c.* [12] Needless to say, formula (8.11) is also a generalization of (8.10).

The *juxtaposition* $\mathbf{A} \odot \mathbf{B}$ of two records $\mathbf{A}$ and $\mathbf{B}$ is the record obtained by giving each answer A a multiplicity equal to the sum of its multiplicity in $\mathbf{A}$ plus its multiplicity in $\mathbf{B}$.

Truth-values form a totally ordered set, by stipulating

$$(j,r) < (j',r') \text{ iff } [\text{either } j < j' \text{ or } (j = j' \text{ and } r < r')]. \tag{8.12}$$

This is a precise formulation of the condition, stated in Section 8.2.1 that cheap noisy channels are pointwise superseded by low-noise expensive channels.

8.2.4 The Partially Ordered Monoid of States of Knowledge

Paul's current state of knowledge about x_{secret} is uniquely determined by recording Carole's answers. As in the error-free case, two records $\mathbf{A}$ and $\mathbf{B}$ are said to be *equivalent* iff they assign the same truth-value to each $z \in S$, in symbols,

$$\mathbf{A} \equiv \mathbf{B} \text{ iff } \mathbf{A}^{\#}(z) = \mathbf{B}^{\#}(z) \ \forall z \in S. \tag{8.13}$$

We denote by $[\mathbf{A}]$ the equivalence class of $\mathbf{A}$.

[12] Even if Carole's answers sent on a noisy channel $d > 1$ may suggest that z is not a possible candidate for x_{secret}, Carole's further answers on a more reliable channel $d' < d$ may well have the contrary effect. When this happens we are led to revise the error parameter e_d, or to conclude that an exceptionally large number of errors has affected channel d during this particular session. The essential role of each e_d is to tentatively fix an upper bound for our counting of distorted bits sent through channel d: thus, $e_d + 1, e_d + 2, \ldots$ wrong bits are counted as e_d. While it is true that only the first channel can give a definitive verdict about which x's should be excluded as possible candidates for x_{secret}, also the other (less expensive) channels are useful in the search of x_{secret}: in most concrete applications they can be used to substantially reduce the size of the search space.

Definition 8.4 A *state* (of knowledge) [13] in an m-channel Rényi-Ulam game over search space S with $\mathbf{e} = (e_1 < \ldots < e_m)$ lies, is an equivalence class of records. The *initial* state of knowledge 1 is the equivalence class of the empty record (no answer received). It assigns truth value $(m, 1)$ to each $z \in S$. We let $\mathscr{K}_{S,\mathbf{e}}$ denote the set of states.

Given states $a = [\mathbf{A}]$ and $b = [\mathbf{B}]$ we write $a \leq b$ (read: "a is *more restrictive* than b") iff $\mathbf{A}^{\#}(z) \leq \mathbf{B}^{\#}(z)\ \forall z \in S$.

Direct inspection shows that $\mathscr{K}_{S,e}$ has an interesting algebraic structure:

Proposition 8.1 *Adopt the above notation and terminology. We then have:*

1. *The binary relation $\leq$ is a partial order over the set $K_{S,\mathbf{e}}$ of states.*
2. *Juxtaposition $\odot$ of records canonically equips the set $K_{S,\mathbf{e}}$ with an operation, also denoted $\odot$, given by*
$$[\mathbf{A}] \odot [\mathbf{B}] = [\mathbf{A} \odot \mathbf{B}].$$
The $\odot$ operation is commutative, associative, and has the initial state 1 as its neutral element.
3. *Given two states a and b, among all $t \in K_{S,\mathbf{e}}$ such that $a \odot t \leq b$ there is a least restrictive state, denoted $a \to b$.*
4. *The set $K_{S,\mathbf{e}}$ equipped with the $\odot$ and $\to$ operations and the distinguished constant 1, is a BL-algebra.*
5. *The partial order of $K_{S,\mathbf{e}}$ is definable in $\langle K_{S,\mathbf{e}}, \odot, \to, 1\rangle$ via the stipulation $a \leq b$ iff $a \to b = 1$.*

Since the $(\odot, \to, 1)$-structure of $\mathscr{K}_{S,\mathbf{e}}$ is rich enough to reconstruct the order structure, one is naturally led to study the equational properties of the class of BL-algebras $\mathscr{K}_{S,\mathbf{e}}$ for all possible multichannel Rényi-Ulam games.

8.2.5 Multichannel Games vs. Hájek Basic Logic

In this section we shall provide a natural semantics for propositional Hájek Basic Logic in terms of multichannel games.

Let A be an MV-algebra. Define $x \to y = \neg x \oplus y$ and $x \odot y = \neg(\neg x \oplus \neg y)$. Then the structure A' obtained from A replacing $\neg$ and $\oplus$ by $\odot$ and $\to$ is called a *Wajsberg algebra*. Wajsberg algebras are definitionally equivalent to MV-algebras. By [5, p. 241], a *Wajsberg hoop* can be defined as a subalgebra of a reduct of a Wajsberg algebra obtained by deleting 0 from the language. Every Wajsberg hoop inherits the natural MV-algebraic order, and we have [11, Lemma 1.1.2], $x \leq y$ iff $x \to y = 1$. A Wajsberg hoop with an additional constant interpreted as its bottom element is said to be *bounded*. Thus bounded Wajsberg hoops coincide with Wajsberg algebras. By a *finite Łukasiewicz chain* we mean the Wajsberg algebra L corresponding to a finite

[13] This is a generalization of the notion of state given in the first part of this paper.

simple MV-algebra. Thus [11, Corollary 3.5.4] $L = \{0, \frac{1}{n}, \ldots, \frac{n-1}{n}, 1\}$ for some integer $n \geq 1$, equipped with conjunction $x \odot y = \max(0, x+y-1)$ and implication $x \to y = \min(1, 1-x+y)$.

Let $\langle I, \leq \rangle$ be a linearly ordered set with smallest element i_0. For all $i \in I$ let H_i be a Wajsberg hoop such that for $i \neq j$, $H_i \cap H_j = \{1\}$. Assume H_{i_0} has a bottom element. Then the *ordinal sum* $\bigoplus_{i \in I} H_i$ is the structure whose universe is given by $\bigcup_{i \in I} H_i$, and whose operations are defined by[14]

$$x \to y = \begin{cases} x \overset{H_i}{\to} y & \text{if } x, y \in H_i \\ y & \text{if } \exists j < i (x \in H_i, y \in H_j) \\ 1 & \text{if } \exists i < j (1 \neq x \in H_i, y \in H_j) \end{cases}$$

$$x \odot y = \begin{cases} x \odot^{H_i} y & \text{if } x, y \in H_i \\ y & \text{if } \exists j < i (x \in H_i, 1 \neq y \in H_j) \\ x & \text{if } \exists i < j (1 \neq x \in H_i, y \in H_j) \end{cases}$$

and whose bottom element is the bottom element of H_{i_0}.

Theorem 8.1 *Given BL-terms $\sigma = \sigma(x_1, \ldots, x_n)$ and $\tau = \tau(x_1, \ldots, x_n)$, the following conditions are equivalent for the equation $\sigma = \tau$:*

(i) *For every finite set $S \neq \emptyset$ and increasing m-tuple $\mathbf{e} = (e_1 < \ldots < e_m)$ of integers ≥ 0, the equation is valid in the BL-algebra $\mathscr{K}_{S,\mathbf{e}}$ of states in the m-channel Rényi-Ulam game over search space S with* **e** *errors.*
(ii) *For every increasing m-tuple of integers* **e** *and singleton set $\{s\}$, the equation holds in the BL-algebra $\mathscr{K}_{\{s\},\mathbf{e}}$.*
(iii) *The equation holds in every finite ordinal sum of finite Łukasiewicz chains of increasing cardinalities.*
(iv) *The equation holds in every BL-algebra.*

Proof. (i) $\Rightarrow$ (ii) is trivial. To prove (i) $\Leftarrow$ (ii) let s be an arbitrary element of S. Then direct inspection shows that $\mathscr{K}_{S,\mathbf{e}}$ is the product of $|S|$ many copies of $\mathscr{K}_{\{s\},\mathbf{e}}$.

To prove (ii) $\Leftrightarrow$ (iii) in the light of Proposition 8.1 it is sufficient to note that when $S = \{s\}$ is a singleton, the structure $\mathscr{K}_{\{s\},\mathbf{e}}$ is the most general possible finite ordinal sum of increasingly large finite Łukasiewicz chains.

(iv) $\Rightarrow$ (iii) is trivial. For the proof of (iii) $\Rightarrow$ (iv) we prepare

Claim A. [2, 16] Up to isomorphism, linearly ordered BL-algebras coincide with ordinal sums of indexed families of Wajsberg hoops whose first component is a Wajsberg algebra.[15]

Claim B. [2] Suppose the linearly ordered BL-algebra A can be represented as $A = \bigoplus_{i=0}^{n} W_i$ for finitely many Wajsberg components $W_0, \ldots, W_n$. Then the nontrivial subalgebras of A are precisely the ordinal sums $B = \bigoplus_{i=0}^{n} U_i$, where U_0 is a nontrivial subalgebra of W_0, and for each $i = 1, \ldots, n$, U_i is a possibly trivial (= singleton) subalgebra of W_i.

14 Compare with [5, p. 247].

15 For any linearly ordered BL-algebra A, the Wajsberg hoops W_i such that $A \cong \bigoplus W_i$ will be called the *Wajsberg components* of A.

This result has a straightforward generalization to infinite ordinal sums.

Claim C. [5] Wajsberg hoops have the *finite embeddability property.* In other words, every finite partial subalgebra of a Wajsberg hoop can be embedded into a finite Wajsberg hoop.

As an immediate consequence of the definition of ordinal sum, for any BL-chain A and elements $a_1,\ldots,a_n \in A$, all in the same Wajsberg component W of A, the subalgebra generated by $\{a_1,\ldots,a_n\}$ is contained in W. We then have

Claim D. Every finitely generated BL-chain is the ordinal sum of finitely many Wajsberg hoops, in such a way that every Wajsberg component contains at least one generator.

We are now in a position to prove:

Claim E. [1] The variety of BL-algebras is generated by its *finite* linearly ordered members.

Proof of Claim E. Suppose an identity $\sigma(x_1,\ldots,x_n) = \tau(x_1,\ldots,x_n)$ fails in some BL-algebra. Then by [14, Lemma 2.3.18] it fails in some linearly ordered BL-algebra C. Letting e be an evaluation in C that falsifies $\sigma = \tau$, we may replace C by its subalgebra generated be $e(x_1),\ldots,e(x_n)$. Thus we may safely assume C to be finitely generated. Let C' be the finite partial subalgebra of C consisting of all elements of the form $e(\rho)$, where ρ ranges over all subterms of σ and of τ. By Claim D, C is the ordinal sum of finitely many Wajsberg hoops, each containing at least one generator. Arguing by induction on the cardinality n of C', it is not hard to see that C' embeds into a finite BL-chain. Indeed, if $n = 1$, then C consists of a single Wajsberg component, and the desired conclusion follows from Claim C. For the induction step, in case $n > 1$, let W be the top Wajsberg component of C, and write $C = E \oplus W$, for some BL-chain E. Since W has the finite embeddability property, again by Claim C, $C' \cap W$ embeds into a finite linearly ordered Wajsberg hoop S. W contains at least one generator, and the cardinality of $E \cap C'$ is $< n$. By induction hypothesis $E \cap C'$ embeds into a finite BL-chain B. It follows that C' embeds into the finite BL-chain $B \oplus S$, as promised. Let j be such an embedding. Then the composite map $j \circ e$ is an evaluation into the finite BL-chain $B \oplus S$ that falsifies $\sigma = \tau$. The proof of Claim E is complete.

Conclusion of the proof of (iii) $\Rightarrow$ (iv). By Claim A every finite BL-chain can be identified with a finite ordinal sum of finite Wajsberg hoops. Since every finite Wajsberg hoop W is (the reduct of) a bounded Wajsberg hoop, W coincides with some Łukasiewicz chain. Thus if an equation $\sigma = \tau$ is falsified in some BL-algebra, then by Claim E, the equation is also falsified in a finite ordinal sum $\bigoplus_{t=1}^{m} \{0, \frac{1}{n_t}, \ldots, \frac{n_t-1}{n_t}, 1\}$ of finite Łukasiewicz chains, for suitable integers $1 \leq n_1,\ldots,n_m$. Let n^* be the least common multiple of $n_1,\ldots,n_m$. Direct inspection shows that a Łukasiewicz chain $\{0, \frac{1}{r}, \ldots, \frac{r-1}{r}, 1\}$ is embeddable into $\{0, \frac{1}{s}, \ldots, \frac{s-1}{s}, 1\}$ provided[16] r is a divisor of s. We then have an embedding

[16] In fact also the converse is true.

$$\{0, \frac{1}{n_t}, \ldots, \frac{n_t - 1}{n_t}, 1\} \hookrightarrow \{0, \frac{1}{t \cdot n^*}, \ldots, \frac{t \cdot n^* - 1}{t \cdot n^*}, 1\},$$

for each $t = 1, \ldots, m$. By Claim B, we have an embedding

$$\bigoplus_{t=1}^{m} \{0, \frac{1}{n_t}, \ldots, \frac{n_t - 1}{n_t}, 1\} \hookrightarrow \bigoplus_{t=1}^{m} \{0, \frac{1}{t \cdot n^*}, \ldots, \frac{t \cdot n^* - 1}{t \cdot n^*}, 1\}.$$

We conclude that the equation $\sigma = \tau$ can be falsified in an ordinal sum of finite Łukasiewicz chains of increasing size.

Examples By [14, p. 63] adding to the axioms of BL-algebras the involutive law $(x \to 0) \to 0$ one characterizes MV-algebras. By [18, 19], the latter precisely capture the logic of one-channel games, namely Łukasiewicz infinite-valued calculus. Involutiveness together with the *idempotence* equation $x \odot x = x$, characterizes boolean algebras; boolean logic takes care of (one-channel) Rényi-Ulam games with no errors.

Remark The assumption that the m channels have a strictly increasing number of errors $e_1 < \ldots < e_m$ was made only to stress that for $c < d$, all information supplied on channel d about $z \in S$ is made obsolete by the information sent on (the less noisy) channel c. The above proof of Theorem 8.1 shows that BL-algebras can also be characterized via multichannel games where the error m-tuple $\mathbf{e}$ is *arbitrary*.

Further Reading

One can find in the literature several surveys and technical papers treating the Rényi-Ulam game-theoretic interpretation of infinite-valued logic, and its applications to error-correcting codes, fault-tolerant search, algorithmic learning and logic programming [6, 8, 9, 15, 19–21].

The monograph [14] is the main source for Hájek Basic Logic and BL-algebras. Also see [13]. The monograph [11] is entirely devoted to Łukasiewicz logic and its algebras, Chang MV-algebras: these special BL-algebras take care of states of knowledge in the one-channel game (as originally described by Rényi and Ulam). The book also discusses other applications and relations of MV-algebras with various areas of mathematics.

Acknowledgements We are grateful to Franco Montagna for his valuable assistance in the writing of the proof of (iii) $\Rightarrow$ (iv) in Theorem 8.1. We also thank Manuela Busaniche for her further simplification of this proof.

References

1. Aglianó P., Ferreirim I. M. A., and Montagna F. Basic hoops: an algebraic study of continuous t-norms, *Preprint*.
2. Aglianó P., and Montagna F. Varieties of BL-algebras I: general properties, *J. Pure Appl. Algebra*, 181: 105–129, 2003.
3. Ahlswede R., Cicalese F., and Deppe C. Searching with lies under error cost constraints. *Submitted*, 2004.
4. Berlekamp E. R. Block coding for the binary symmetric channel with noiseless, delayless feedback. In H. B. Mann, editor, *Error-Correcting Codes,* pages 330–335. Wiley, New York, NY, 1968.
5. Blok W. J., and Ferreirim I. M. A. On the structure of hoops, *Algebra Universalis*, 43: 233–257, 2000.
6. Cicalese F., and Mundici D. Perfect two fault-tolerant search with minimum adaptiveness, *Adv. Appl. Math.*, 25: 65–101, 2000.
7. Cicalese F., and Mundici D. Optimal coding with one asymmetric error: below the sphere packing bound, *Lecture Notes in Computer Science*, Proceedings COCOON-2000, Springer, 1858: 159–169, 2000.
8. Cicalese F., and Mundici D. Learning and the art of fault-tolerant guesswork. In I. Stamatescu et al., editors, *Handbook Chapter, In: Perspectives on Adaptivity and Learning*, pages 117–143. Springer, 2003.
9. Cicalese F., Mundici D., and Vaccaro U. Rota-Metropolis cubic logic and Ulam-Rényi games. In H. Crapo, and D. Senato, editors. *Algebraic Combinatorics and Computer Science: A Tribute to Gian-Carlo Rota*, pages 197–244. Springer, Milan, 2001.
10. Cicalese F., Mundici D., and Vaccaro U. Least adaptive optimal search with unreliable tests, *Theor. Comput. Sci.*, 270: 877–893, 2002.
11. Cignoli R., D'Ottaviano I. M. L., and Mundici D. *Algebraic Foundations of Many-Valued Reasoning*. Kluwer, Dordrecht, 2000.
12. Dumitriu I., and Spencer J. The Liar Game over an Arbitrary Channel, *Theor. Comput. Sci.*, 2004, to appear.
13. Gottwald S. *A treatise on Many-Valued logics (Studies in Logic and Computation, vol. 9)*. Research Studies Press, Baldock, 2000.
14. Hájek P. *Metamathematics of Fuzzy logic*. Kluwer, Dordrecht, 1998.
15. Klawonn F., and Kruse R. A Łukasiewicz logic based Prolog, *Math. Soft Comput.*, 1: 5–29, 1994.
16. Laskowski L. C., and Shashoua Y. V. A classification of BL-algebras, *Fuzzy Sets Syst.*, 131: 271–282, 2002.
17. MacWilliams F. J., and Sloane N. J. A. *The Theory of Error-Correcting Codes*. North-Holland, Amsterdam, 1977.
18. Mundici D. The logic of Ulam's game with lies. In C. Bicchieri, and M. L. Dalla Chiara, editors, *Knowledge, Belief and Strategic Interaction*, pages 275–284. Cambridge University Press, Cambridge, 1992. (*Cambridge Studies in Probability, Induction and Decision Theory)*.
19. Mundici D. Ulam's games, Łukasiewicz logic, and AF C_-algebras, *Fund. Inform.*, 18: 151–161, 1993.
20. Mundici D. Ulam game, the logic of Maxsat, and many-valued partitions, In D. Dubois, H. Prade, E.P. Klement, editors, *Logics and Reasoning About Knowledge*, pages 121–137. Kluwer Academic Publishers, Dordrecht, 1999.
21. Pelc A. Searching games with errors: fifty years of coping with liars, *Theor. Comput. Sci.*, 270: 71–109, 2002.
22. Pierce J.R. Optical Channels: Practical limits with photon counting, *IEEE Trans. Comm.*, COM-26: 1819–1821, 1978.
23. Rènyi A., *Napló az információelméletről, Gondolat*, Budapest, 1976. (English translation: *A Diary on Information Theory*, J. Wiley and Sons, New York, 1984).

24. Tietäväinen A. On the nonexistence of perfect codes over finite fields, *SIAM J. Appl. Math.*, 24: 88–96, 1973.
25. Ulam S. *Adventures of a Mathematician*. Scribner's, New York, NY, 1976.
26. Zinoviev V. A., and Leontiev V. K., The non-existence of perfect codes over Galois fields, *Probl. Contr. Inform. Theory*, 2: 123–132, 1973.

Chapter 9
Two Applications of Epistemic Logic in Computer Security

Ron van der Meyden

9.1 Introduction

Epistemic logic has, in the past few decades, grown beyond its origins in philosophy, to be embraced in several other disciplines, including economics, linguistics and computer science. Within computer science, its application can be found in several subdisciplines: artificial intelligence, distributed computing and computer security.

In these tutorial notes, we illustrate some of the ways in which epistemic logic has come to be applied in computer science by reviewing a number of examples from our recent work that demonstrate how epistemic logic can bring a clarifying perspective to issues in computer security. A significant thread in this work has been to identify situations in which questions about the security of systems and communication protocol designs are amenable to *automated analysis*. To address this issue, we focus on the decidability of logical problems, seeking to identify decidable problems that encompass interesting questions computer security.

Our examples involve two distinct types of decidability questions: *model checking* and *synthesis*. Model checking involves the question of whether a given formula is satisfied in a given semantic structure. This methodology enables us to study a communications protocol from the perspective of how information flows to its trusted participants, or to a passive adversary, who may observe, but not interfere with, communications between the trusted parties. We discuss a system we have developed that implements this task, and descibe its application to a protocol designed to enable information to be transmitted anonymously: Chaum's Dining Cryptographers protocol.

Synthesis, the second type of decidability question we consider, concerns the existence of strategies that lead to the satisfaction of a formula in a given context. The relevance of this in computer security is that an active adversary, who not only

Ron van der Meyden
School of Computer Science and Engineering, University of New South Wales, Sydney, Australia, e-mail: meyden@cse.unsw.edu.au

J. van Benthem et al. (eds.), *Proof, Computation and Agency*, Synthese Library 352,
DOI 10.1007/978-94-007-0080-2_9,

observes but also interferes with the system, can be viewed as seeking a strategy that satisfies certain conditions on the adversary's knowledge. The system is secure if no such strategy exists.

Logic itself is broadened and enriched by such encounters with new application areas, as these frequently raise fresh questions for research on logic itself. Applications of epistemic logic in computer security are no exception, and raise some quite challenging question for further study, which we discuss in Section 9.7.

9.2 The Logic of Knowledge and Time

To set the scene for the applications we discuss, we first describe a general framework for the semantics of the logic of knowledge in multi-agent systems that has found application not just in computer security, but also in distributed computing and artificial intelligence. Computation being an inherently dynamic phenomenon, the framework includes not just the logic of knowledge, but also temporal logic. We refer the reader to [9] for a more leisurely exposition.

A parameter of the framework is the number n of agents in the systems to be modelled. We refer to agents by a number $i = 1 \dots n$. Let $Prop$ be a finite set of propositional variables. The language of knowledge and linear time is the propositional multi-modal language $\mathscr{L}$ with formulas defined as follows. Each proposition $p \in Prop$ is a formula, and if φ_1 and φ_2 are formulas, then so are $\neg\varphi_1$, $\varphi_1 \wedge \varphi_2$, $\circ\varphi_1$ (meaning that φ_1 holds at the next instant of time) $\varphi_1 U \varphi_2$ (meaning that φ_1 holds until φ_2 does), and $K_i\varphi_1$, for each agent $i = 1 \dots n$ (meaning that agent i knows that φ_1 holds). We use all the usual boolean abbreviations, and write $\Diamond\varphi$ for $trueU\varphi$ (meaning that φ holds at some time in the future).

This language can be given a semantics in *interpreted systems*, a type of semantic structure that models both temporal and epistemic accessibility relations. We later present some additional semantic constructs that describe how interpreted systems arise from concrete computational systems. We suppose that at each moment of time, the system we are modelling is in a *global state* in the set $S = S_0 \times S_1 \times \dots \times S_n$, where S_0 is the set of possible states of the environment in which the agents operate and, for $i = 1 \dots n$, the set S_i is the set of *local states* of agent i. We work with a discrete model of time, representing moments of time by elements of the natural numbers $\mathbf{N}$. A possible history, or *run*, is a mapping $r : \mathbf{N} \to S$ that assigns a global state to each moment of time. Since the behaviour of systems is typically non-deterministic, we model a *system* by a set $\mathscr{R}$ of runs. A *point* in a system is a pair (r,m) consisting of a run r and a time m. We write $r_i(m)$ for the local state of agent i at time m in the run r. An *interpreted system* is a pair $\mathscr{I} = (\mathscr{R}, \pi)$ where $\mathscr{R}$ is a system and $\pi : \mathscr{R} \times \mathbf{N} \to \mathscr{P}(Prop)$ is a function that assigns to each point in the system the set of propositions holding at that point.

The local states play a key role in the interpretation of knowledge in systems. Intuitively, an agent's local state captures all the information available to an agent at a given moment of time. We say two points (r,m) and (r',m') are *indistinguishable* to agent i, and write $(r,m) \sim_i (r',m')$ if the agent has the same local state at those points, i.e. $r_i(m) = r'_i(m')$.

Given a point (r,m) of an interpreted system $\mathscr{I}$, we define what it means for a formula φ to hold at this point, denoted $\mathscr{I},(r,m) \models \varphi$, by:

- $\mathscr{I},(r,m) \models p$ if $p \in \pi(r,m)$, for $p \in Prop$,
- $\mathscr{I},(r,m) \models \neg\varphi$ if not $\mathscr{I},(r,m) \models \varphi$,
- $\mathscr{I},(r,m) \models \varphi_1 \wedge \varphi_2$ if $\mathscr{I},(r,m) \models \varphi_1$ and $\mathscr{I},(r,m) \models \varphi_2$,
- $\mathscr{I},(r,m) \models \circ\varphi$ if $\mathscr{I},(r,m+1) \models \varphi$,
- $\mathscr{I},(r,m) \models \varphi_1 U \varphi_2$ if there exists $m' \geq m$ such that $\mathscr{I},(r,l) \models \varphi_1$ for all l with $m \leq l < m'$, and $\mathscr{I},(r,m') \models \varphi_2$,
- $\mathscr{I},(r,m) \models K_i\varphi$ if $\mathscr{I},(r',m') \models \varphi$ for all (r',m') with $(r,m) \sim_i (r',m')$.

The final clause captures the following intuition about knowledge: agent i knows fact φ if φ is true at all the points that are consistent with the information available to agent i, i.e., at all the points that agent i is not able to distinguish from the actual situation. We write $\mathscr{I} \models \varphi$ if $\mathscr{I},(r,0) \models \varphi$ for all runs r of $\mathscr{I}$.

While interpreted systems provide a clean abstract model for reasoning about knowledge and time, it is useful in applications to identify some additional types of structure. We now present some definitions that show how interpreted systems arise concretely. In broad brush, the picture is as follows: we first define environments, which model how the actions that can be performed by agents affect the state of the world. Next, we view agents as executing some protocol that describe how agents choose an action to perform at each moment of time. When each agent runs a protocol in the environment, this generates a set of runs of that environment. By further selecting for the agents a local state at each moment of time, based on some degree of recollection of the observations the agent has made to that point of time, we obtain an interpreted system in the form described above.

Suppose that for each agent $i = 0 \ldots n$, ACT_i is a finite, non-empty set of *actions*, corresponding to the actions that may be performed by each of the agents, and, in the case $i = 0$, the actions or nondeterministic events due to the environment. We permit these actions to occur in a simultaneous fashion, and represent this by defining the set of *joint actions* by $ACT = ACT_0 \times ACT_1 \times \cdots \times ACT_n$. When $\mathbf{a}$ denotes a joint action, we write $\mathbf{a}_i$ for the action of agent i in $\mathbf{a}$.

An *environment* over a signature is a tuple $E = (S_0, I, P_0, \tau, O, \pi_0)$ where

- S_0 is a finite set of *states,*
- $I \subseteq S_0$ is the set of *initial states,*
- $P_0 \colon S_0 \to \mathscr{P}(ACT_0)$ is the *protocol of the environment,* which says which actions can be performed by the environment in a given state,
- $\tau \colon ACT \to (S_0 \to S_0)$ is the *transition function,* which, for every joint action $\mathbf{a}$ specifies a transition function $\tau(\mathbf{a})$,
- $O = (O_1, \ldots, O_n)$ is a family of *observation functions* $O_i \colon S \to \mathscr{O}$ for some set $\mathscr{O}$ of *observations,*
- $\pi_0 \colon S_0 \to \mathscr{P}(Prop)$ is an *interpretation function* which assigns to each state the propositions it satisfies.

We require $P_0(s) \neq \emptyset$ for each $s \in S_0$. An *execution* of such an environment is an infinite sequence $s_0, s_1, s_2, \ldots$ such that $s_0 \in I$ and such that for all m there exists $\mathbf{a} \in ACT$ with $\mathbf{a}_0 \in P_0(s_m)$ and $\tau(\mathbf{a})(s_m) = s_{m+1}$. When ε denotes such an execution, and $m \in \mathbf{N}$, we write $\varepsilon(m)$ for s_m.

Executions resemble the runs of systems described above, but they lack the local states of the agents $i = 1 \ldots n$. There are many ways these could be determined, depending on factors including how much the agents are able to remember about their past observations and whether or not the system operates synchronously, i.e., with a global clock. We focus here on the assumptions that the sytem is *synchronous* and agents have *perfect recall.* Given an execution ε, we define a run r^ε by $r_0(m) = \varepsilon(m)$, and, for each agent $i = 1 \ldots n$, defining the local states by $r_i^\varepsilon(m) = \langle O_i(\varepsilon(0)), O_i(\varepsilon(1)), \ldots, O_i(\varepsilon(m)) \rightarrow$. Synchronous perfect recall is particularly significant for security analyses. This is because a frequent objective in security settings is to place restrictions on the information that certain agents are able to acquire (while not inhibiting certain other flows of information.) An adversary in a security setting is likely to attempt to make as much use as it can of the information it is able to accumulate by observing the system under attack. Thus, an agent assumed to have perfect recall corresponds to an adversary that makes maximal use of its inferential capabilities. Ideally, we would like the system to be designed so as to be secure against even this most powerful adversary.

Given an environment E, we obtain the interpreted system $\mathcal{I}^{spr}(E) = (\mathcal{R}, \pi)$ by taking $\mathcal{R}$ to be the set of runs r^ε for ε an execution of E, and defining the interpretation π by $\pi(r^\varepsilon, m) = \pi_0(\varepsilon(m))$.

The system $\mathcal{I}^{spr}(E)$ models what agents know if they could perform any action at any time. In general, it is more interesting to consider situations where the agents choose their actions according to some rules of behaviour. We model such rules as follows. A *protocol* for agent i is a function $P_i\colon \mathcal{O}^+ \rightarrow \mathcal{P}(ACT_i)$. Intuitively, a sequence $\sigma \in \mathcal{O}^+$ represents the maximal information that the agent could have acquired at a given moment of time: its synchronous perfect recall local state. Based on this information, the agent nondeterministically selects one of the local actions $P_i(\sigma)$ as the next action it will perform. (In practice, $P_i(\sigma)$ could depend on only a small part of σ, and it may not be necessary for an agent to maintain its complete history of observations in order to determine its next action.)

A *joint protocol* is a family $\mathbf{P} = \{P_i\}_{i \in [n]}$ where each P_i is a protocol for agent i. Given such a joint protocol and a run r of $\mathcal{I}^{spr}(E)$, we say r is *consistent* with the protocol if for every m there exists a joint action $\mathbf{a}$ such that $r_0(m+1) = \tau(\mathbf{a})(r_0(m))$ where $\mathbf{a}_0 \in P_0(r_0(m))$ and $\mathbf{a}_i \in P_i(r_i(m))$ for $i = 1 \ldots n$. We write $\mathcal{I}^{spr}(E, \mathbf{P})$ for the interpreted system obtained from $\mathcal{I}^{spr}(E)$ by restricting its runs to runs consistent with the joint protocol $\mathbf{P}$.

9.3 Model Checking

Model checking is methodology for verification of hardware and software designs [5] that emerged in the 1980s and is now well entrenched in the computer hardware industry, and finding increasing applications to software. Generally, it is applied to finite state systems designs and temporal logic specifications. Given a semantic structure M, often a Kripke structure of some form, and a formula φ, one asks whether φ is satisfied in a particular state, or set of states of M. A *model checker* is a program that computes the answer to this question (and may also provide some output by way of explanation of why this is the answer, e.g., counter-examples in case the answer is false.) Model checkers have been developed for a variety of modal logics, including linear time temporal logic [13], branching time temporal logic [2].

We have developed a model checker MCK [10] that deals with specifications in the logic of knowledge and time, based on the semantic framework discussed in the previous section. MCK accepts as input a description of an environment E, a joint protocol $\mathbf{P}$ for the agents in E, and a specification φ in the language of knowledge and time, and determines whether $\mathcal{I}^{spr}(E,\mathbf{P}),\models \varphi$ for all runs r of $\mathcal{I}^{spr}(E,\mathbf{P})$. In addition to the synchronous perfect recall view that is our focus in this paper, the system can handle several other ways of defining local states.

9.4 Verifying the Dining Cryptographers Protocol

To illustrate how model checking for the logic of knowledge may be applied to security verification, we now discuss the Dining Cryptographers protocol, a protocol for anonymous broadcast due to [6]. Chaum introduces this protocol by the following story:

> Three cryptographers are sitting down to dinner at their favorite three-star restuarant. Their waiter informs them that arrangements have been made with the maitre d'hotel for the bill to be paid anonymously. One of the cryptographers might be paying for the dinner, or it might have been the NSA (US National Security Agency). The three cryptographers respect each other's right to make an anonymous payment, but they wonder if the NSA is paying.

Chaum shows that the following protocol can be used to solve the dining cryptographers' quandary. The protocol assumes that at most one cryptographer is paying.

1. Each cryptographer flips an unbiased coin behind his menu, between him and the cryptographer to his right, so that only the two of them can see the outcome.
2. Each cryptographer then states aloud whether the two coins that he can see - the one he flipped and the one his left-hand neighbour flipped – fell on the same side or different sides.
3. As an exception to the previous step, if one of the cryptographers is the payer, he states the opposite of what he sees.

It can be seen that after running this protocol, all the cryptographers are able to determine whether it was the NSA or one of the cryptographers who paid for dinner.

(Each announcement can be seen as the exclusive-or of the two coin tosses known to the speaker, together with the bit representing whether the speaker paid. Taking the exclusive or of the announcements, the coins cancel out, leaving the exclusive or of the bits about payment.) More interestingly, and more subtle, is that this is all the information that the protocol reveals about the identity of the payer: if a cryptographer is paying neither of the other two learns anything from the utterances about which cryptographer it is.

This claim may be automatically verified using the model checker MCK. Figure 9.1 shows how the situation is represented as an input to the model checker. We being by declaring some of the variables that make up the state of the environment: The boolean vector `paid` represents whether or not each agent paid. We model sharing of the information between the agents by having each cryptographer fip a coin, and then communicate the outcome to an adjacent cryptographer using the communication channels represented by the boolean vector `chan`. The boolean vector `said` is used to represent the public announcements made by the agents. Next, we identify the initial states to be those in which at most one agent paid. The agents themselves are declared in statements of the form "`agent name protocol (parameters)`". The protocol part gives the name of the protocol executed by the agent being declared, and the parameters are the variables of the environment with which the agent interacts.

The protocol itself is declared separately in a generic template. Each `agent` decaration has the effect of binding an instance of the local variables in the template to the matching environment variables in the declaration. An agent's observation at each moment of time consists of the variables declared using the keyword `observable`, together with their values. The `if` statement represents a guarded non-deterministic choice: to execute it, the agent nondeterministically executes one of the statements for which the guard (to the left of `->`) evaluates to true. Variables `v` bound to variables in the environment may be read or written by statements of the form `x:= v.read()` or `v.write(expression)`. (This form is used to highlight the fact that there may be conflicts caused by simultaneous read and writes, though the issue does not arise in this example.)

Finally, the specification statement `spec_spr_xn` indicates that knowledge is to be determined using the synchronous perfect recall view (`spr`), and (indicated by `xn`) that the algorithm to be used for the model checking is a specific algorithm that has been designed to deal with formulas of the form $\circ^n \varphi$, where φ is a formula of the logic of knowledge. In this case, $n = 4$, since the protocol takes 4 steps to execute, and φ is a formula that states that if cryptographer 1 did not pay, then he knows either that no cryptographer paid, or knows that one of the other two paid, but does not know which. (To verify the protocol, it suffices to consider only the specification for cryptographer 1 above, because of symmetry considerations.) As remarked above, the use of the perfect recall semantics for knowledge is significant. In asking for confirmation that the payer's anonymity is not compromised by the protocol, we wish to know that it is not compromised even by an agent that makes maximal inferential use of what it is able to observe during an execution of the protocol.

When MCK is run on this input, it computes the answer that the specification is satisfied in a matter of a few seconds.

```
paid : Bool[3]
chan : Bool[3]
said : Bool[3]

init_cond = ((neg paid[0]) /\ (neg paid[1]) /\ (neg paid[2]))
         \/ ((paid[0])      /\ (neg paid[1]) /\ (neg paid[2]))
         \/ ((neg paid[0]) /\ (paid[1])      /\ (neg paid[2]))
         \/ ((neg paid[0]) /\ (neg paid[1]) /\ (paid[2]))

agent C0 "dc_agent_protocol" (paid[0], chan[0], chan[1], said)
agent C1 "dc_agent_protocol" (paid[1], chan[1], chan[2], said)
agent C2 "dc_agent_protocol" (paid[2], chan[2], chan[0], said)

protocol "dc_agent_protocol"
( paid : observable Bool,
  chan_left : Bool,
  chan_right : Bool,
  said : observable Bool[] )

coin_left : observable Bool
coin_right : observable Bool
  where all_init

begin
  -- This agent decides the coin toss to the right.
  if True -> coin_right := True
  [] True -> coin_right := False
  fi;
  << chan_right.write(coin_right) >>;
  << coin_left := chan_left.read() >>;
  << said[self].write(coin_left xor coin_right xor paid) >>
end

spec_spr_xn = X 4 (neg paid[0]) => ((Knows C1 (neg paid[0])
                  /\ (neg paid[1])
                  /\ (neg paid[2]))
    \/ ((Knows C1 (paid[1] \/ paid[2]))
     /\ (neg (Knows C1 paid[1]))
     /\ (neg (Knows C1 paid[2]))))
```

Fig. 9.1 The Dining Cryptographers Protocol represented in MCK.

9.5 Synthesis from Epistemic Specifications

The type of security analysis considered in the previous section is based on the premise that each of the agents to the protocol is trusted to act according to the rules of the protocol. In general, this is a stronger assumption than we would wish to make in a security analysis. If any of the agents in a system is untrustworthy, we should reason on the assumption that it may behave in any way that may help it to undermine the security objectives that the system is attempting to enforce. This may include not just interception of communications, but also performing actions that are not in accordance with a protocol that it is supposed to be executing, e.g., tampering with communications between other agents. A careful choice of such malicious behaviour may enable it deduce information that it is not entitled to have. To show that the system is secure, we need to prove that however such an untrustworthy agent may behave, it will not succeed in subverting the security aims of the system.

The model checking problem discussed above is not able to express this type of analysis. Instead, we need to consider a version of the *synthesis* problem: given a specification φ, construct a program P, which, when executed, guarantees that the execution will satisfy φ. The classical literature on program synthesis deals with this problem for sequential programs and *input-output* specifications. The literature on automated verification has considered a version of this problem where P is a concurrent program running in a nondeterministic environment, and φ is a specification expressed in *temporal logic* [7, 18, 19]. It has been shown that, under certain circumstances, automata-theoretic ideas can be applied to automate this synthesis process.

For security analyses, a richer set of specifications expressed in the logic of knowledge and time is appropriate. Whereas for temporal specifications, the definition of the synthesis problem need refer only to the individual runs of a system generated by the synthesized protocol, for epistemic specifications we need to consider the interpreted system generated by the protocol. This leads to the following definition:

Let φ be a formula of $\mathscr{L}$ and E an environment. We say a formula φ is *realizable* in the environment E if there exists a joint protocol $\mathbf{P}$ such that $\mathscr{I}(E,\mathbf{P}) \models \varphi$.

Realizability of specifications in the logic of knowledge and linear time under the assumption of perfect recall was first studied by van der Meyden and Vardi [17].

Theorem 9.1 *[17] Realizability of specifications in the logic of knowledge and linear time for a single agent in single-agent environments is decidable in double exponential time.*

The situation with respect to multi-agent systems is more complex. In fact this is the case even for the much weaker set of temporal specifications:

Theorem 9.2 *[20] Realizability of specifications in linear time temporal logic in multi-agent environments is undecidable.*

However, with some restrictions on the class of environments we consider, there exist some decidable cases. Pnueli and Rosner [20] identified some such cases for temporal specifications, and their characterization has been given a detailed treatment in [15].

Here we consider a class of environments whose structure has intuitive content from the perspective of the logic of knowledge. *Hierarchical* environments [8] are those in which for all states s,t and agents $i = 1 \ldots n-1$, we have that $O_i(s) = O_i(t)$ implies $O_{i+1}(s) = O_{i+1}(t)$. Intuitively, this means that each agent in the sequence observes not more than the preceding agent: if agent i is not able to distinguish states s and t by observation, then neither is agent $i+1$. An example of an hierarchical system is a system of three agents with security clearances to read unclassified, secret and top-secret documents, respectively (where a security clearance implies a capability to read documents at or below the security level.)

In order to obtain a decidable case of synthesis from epistemic specifications, we also need a restriction on the class of formulas. We say a formula φ in the language of linear time and knowledge is *positive* if every occurrence of any knowledge modality K_i is positive, that is, under an even number of negations. More precisely, a formula is positive if it can be built from p and $\neg p$ for $p \in Prop$, using $\vee$, $\wedge$, U, R, $\circ$, and K_i for $i = 1 \ldots n$. Here R is the binary operator defined by $\varphi R \psi$ if $\neg(\neg\varphi U \neg\psi)$.

Theorem 9.3 *[22] The synthesis problem for multi-agent hierarchical environments is decidable for positive specifications in the logic of knowledge and linear time.*

Thus, while synthesis from epistemic specifications in multi-agent systems is in general undecidable, there are restrictions on the class of systems and on the class of formulas that render it decidable. We illustrate in the following sections that these restrictions nevertheless encompass an interesting application.

9.6 A Strategic Notion of Deducibility

As an example of the application of these results on synthesis to security analysis, we consider a type of information flow security policy that originates in the context of multi-level security systems for military applications. Such systems are subject to stringent requirements concerning the release of information from the high security domain (which we shall represent by an agent called H, for High) to the low security domain (represented by agent L, for Low). The simplest type of security policy in this setting specifies that information is permitted to flow from Low to High, but not vice versa. (Policies with exceptions to this rule are also of interest, but we will not consider them here.) One of the concerns that needs to be dealt with is that viruses or Trojan Horse programs may be downloaded into the High part of the system, which, when coordinated with actions taken at the Low end of the system, may lead to secret information being leaked. We show how to formulate a version of this question as a realizability problem.

Let $E = (S_0, I, P_0, \tau, O, \pi_0)$ be an environment with agents H and L, describing the possible states of the system whose security we wish to analyse. We would like to answer the question "can even a single bit of information be communicated by H to

L?"To this end, we begin by adding a bit to the system: let $E^+ = (S_0', I', P_0', \tau', O', \pi_0')$ be the system for the same set of agents and the enriched set of propositions $Prop \cup \{\mathtt{b}\}$ defined as follows:

1. $S_0' = S_0 \times \{0,1\}$, i.e., we add a single bit to each state;
2. $I' = I \times \{0,1\}$, so that the extra bit may initially have any possible value and we do not constrain the original initial states;
3. $P_0'((s,b)) = P_0(s)$ for $s \in S$ and $b \in \{0,1\}$, so that the environment's behaviour is not modified;
4. $\tau'(\mathbf{a})((s,b)) = \tau(\mathbf{a})(s)$ for $s \in S$ and $b \in \{0,1\}$, so that actions do not change the value of the new bit and otherwise behave as in E;
5. $O_L'((s,b) = O_L(s)$ and $O_H'((s,b)) = (O_L(s), O_H(s), b)$, for $s \in S$ and $b \in \{0,1\}$;
6. $\pi'((s,b)) = \pi(s) \cup \{\mathtt{b} \mid b = 1\}$, so that $\mathtt{b}$ is interpreted as the value of the new bit.

The observation functions in E^+ can be understood as follows. First, the new bit is not observed by L, but it is observed by H, so it is initially a secret known only to H. The observation of L has also been added to the observation of H. This captures the intuition that since L models "publicly available information," the security analysis should assume that H has access to any information known to L. Note that this makes the environment hierarchical with respect to the ordering H, L on the agents.

We may now formalise the question "Can H and L collude in environment E to reliably pass information from H to L" as the problem of whether the formula $\Diamond(K_L(\mathtt{b}) \vee K_L(\neg \mathtt{b}))$ is realizable in E^+. If this is not the case, we say that E satisfies *strong nondeducibility on strategies*.

The nomenclature is due to the fact that this notion is closely related to, but somewhat stronger than, the notion of "nondeducibility on strategies" of Wittbold and Johnson [23], which intuitively says that L cannot deduce any information about the protocol being executed by H. Some of the differences between these two notions is that nondeducibility on strategies does not take into account L collusion in the information flow, and it also considers a system insecure if there exists a H protocol that makes it *possible* that information will flow from H to L. By contrast, strong nondeducibility on strategies tests for the existence of a *reliable* channel for information flow.

Noting that $\Diamond(K_L(p) \vee K_L(\neg p))$ is a positive formula and applying Theorem 9.3, we obtain the following result.

Theorem 9.4 *Strong deducibility on strategies is decidable.*

This illustrates that the decidability of synthesis from epistemic specifications has some interesting applications in computer security.

9.7 Conclusion

Our discussion in these notes of the applications of epistemic logic in computer security has been quite parochial, and we have not attempted to survey what is a growing field. However, to conclude we briefly mention some of closely related work in this space.

The dining cryptographers problem can be considered as an instance of the more general problem of specifications that require that certain information be kept secret. Halpern and O'Niell [12] have given a general treatment of the notion of secrecy from the perspective of the logic of knowledge. Whereas our focus has been on the logic of knowledge, they also consider probabilistic concerns, which are very important in security settings.

The dining cryptographers protocol is somewhat unusual amongst security protocols in that it makes use of the properties of the exclusive-or operation rather than shared or public-key cryptosystems. One important class of protocols that do exploit cryptography is *authentication* protocols. A variant of epistemic logic called BAN logic (after the initials of its originators) [1] specifically aimed to build a logic to capture the intuitive reasoning of protocol designers about such protocols. The epistemic operator in this logic was called "belief" rather than knowledge, since the reasoning was intended to capture the conclusions that agents could reach in running the protocol from "trust" assumptions, expressed as an initial set of beliefs. This work has spawned a significant literature, but the approach has had some contentious features and involves some quite difficult problems, so that it still constitutes a topic of current research [3, 4, 21].

One of these difficult problems relates to the question of logical omniscience. The semantics of knowledge given above has the property that if ψ is a logical consequence of φ, and $K_i\varphi$ holds, then so does $K_i\psi$. For many applications of the logic of knowledge in computer science, this is not a concern. However, public key cryptography uses a pair of keys, a public key K and a private key K^{-1}, with the property that the key K^{-1} may be used to decrypt any messages encrypted with K. In most implementations, the key K^{-1} can be *deduced* from K. Thus, according to the semantics for knowledge we have discussed, if agent i knows the value of a message when encrypted with a public key, it also knows the value of the message itself. This is far too strong a conclusion to be useful for security analyses: the security of public key cryptography is based on the fact that the amount of time required to compute the private key from the public key is prohibitively large. How best to capture this *computational* aspect of cryptography in an epistemic logic remains a challenging problem. Current approaches are based on the notions of resource-bounded knowledge [16], or algorithmic knowledge [11, 14].

In addition to these semantic concerns, there are many open questions concerning the computational aspects of modal logics with these rich sets of features. For example, we still lack results on synthesis from specifications that combine knowledge with branching temporal operators. Model checking the combination of the logic of knowledge and probability remains largely unaddressed to date. Moreover, decidability does not imply practicability, and often a great deal of additional research is required before we have useful tools for model checking or synthesis. There is, therefore, wide scope for further research of both a pure and an applied nature in this area.

Acknowledgements Work supported by a grant from the Australian Research Council.

References

1. Burrows M., Abadi M., and Needham R. M. A logic of authentication. *ACM Trans. Comput. Syst.*, 8(1): 18–36, 1990.
2. Cimatti A., Clarke E. M., Giunchiglia E., Pistore M., Roveri M., Sebastiani R., and Tachella A. NuSMV 2: An open source toolkit for symbolic model checking. In *Proc. Int. Conf on Computer Aided Verification (CAV'02)*, pages 27–31, 2002.
3. Cohen M., and Dam M. A completeness result for ban logic. In *Proc. Methods for Modalities 4*, pages 121–132, 2005.
4. Cohen M., and Dam M. Logical omniscience in the semantics of ban logics. In *Proc. Found. Comp. Sci.*, pages 121–132, 2005.
5. Clarke E. M., Grumberg O., and Peled D. A. *Model Checking*. MIT Press, 2000.
6. Chaum D. The dining cryptographers problem: Unconditional sender and recipient untraceability, *J. Cryptology*, 1(1): 65–75, 1988.
7. Emerson E. A., and Clarke E. M. Using branching time logic to synthesize synchronization skeletons, *Sci. Comput. Program*, 2: 241–266, 1982.
8. Engelhardt K., van der Meyden R., and Su K. Modal logics with a hierarchy of local propositional quantifiers. In P. Balbiani, N. Suzuki, F. Wolter, and M. Zakharyaschev, editors, *Advances in Modal Logic*, vol. 4, pages 9–30. World Scientific, 2003.
9. Fagin R., Halpern J. Y., Moses Y., and Vardi M. Y. *Reasoning about Knowledge*. MIT Press, Cambridge, MA, 1995.
10. Gammie P., and van der Meyden R. Mck: Model checking the logic of knowledge. In *Proc. Comp. Aided Verification, CAV'04*, pages 479–483, 2004.
11. Halpern J. Y., Moses Y., and Vardi M. Y. Algorithmic knowledge. In R. Fagin, editor, *Theoretical Aspects of Reasoning about Knowledge: Proc. Fifth Conf.*, pages 255–266, 1994.
12. Halpern J. Y., and O'Neill K. Anonymity and information hiding in multiagent systems. In *Proc. of the 16th IEEE Computer Security Foundations Workshop*, pages 75–88, 2003.
13. Holzmann G. J. *The SPIN Model Checker: Primer and Reference*. Addisson-Wesley, 2003.
14. Halpern J. Y., and Pucella R. Modelling adversaries in a logic for security protocol analysis. In *Proc. Formal Aspects of Security*, vol. 2629 of Springer LNCS, pages 115–132, 2003.
15. Madhusudan P. *Control and Synthesis of Open Reactive Systems*. PhD thesis, University of Madras, November 2001.
16. Moses Y. Resource-bounded knowledge. In *Proc. Conf. on Theoretical Aspects of Reasoning about Knowledge*, pages 261–275, 1988.
17. van der Meyden R. and Vardi M. Y. Synthesis from knowledge-based specifications. In *CONCUR'98, 9th International Conf. on Concurrency Theory, Springer LNCS No. 1466*, pages 34–49, September 1998.
18. Manna Z., and Wolper P. Synthesis of communicating processes from temporal logic specifications, *ACM Trans. Program. Lang. Syst*, 6(1): 68–93, January 1984.
19. Pnueli A., and Rosner R. On the synthesis of a reactive module. In *Proc. 16th ACM Symposium on Principles of Programming Languages*, Austin, January 1989.
20. Pnueli A., and Rosner R. Distributed reactive systems are hard to synthesize. In *Proc. 31st IEEE Symposium on Foundation of Computer Science*, pages 746–757, 1990.
21. Ramanujam R., and Suresh S. P. Deciding knowledge properties of security protocols In *Proc. Conf. on Theoretical Aspects of Rationality and Knowledge*, pages 219–235. ACM Digital Library, 2005.
22. van der Meyden R., and Wilke T. Synthesis of distributed systems from knowledgebased specifications. In *Proc. Concurrency Theory, 16th Int. Conf., CONCUR 2005*, pages 562–576, 2005.
23. Wittbold J. T., and Johnson D. M. Information flow in nondeterministic systems. In *Proc. IEEE Symposium on Research in Security and Privacy*, pages 144–161, 1990.

Chapter 10
An Introduction to Quantum Computing

Noson S. Yanofsky

10.1 Intuition

Quantum Computing is a fascinating new field at the intersection of computer science, mathematics and physics. This field studies how to harness some of the strange aspects of quantum physics for use in computer science. Many of the texts to this field require knowledge of a large corpus of advanced mathematics or physics. We try to remedy this situation by presenting the basic ideas of quantum computing understandable to anyone who has had a course in pre-calculus or discrete structures. (A good course in linear algebra would help, but, the reader is reminded of many definitions in the footnotes.)

The reason why we are able to ignore the higher mathematics and physics is that we do not aim to teach the reader all of quantum mechanics and all of quantum computing. Rather, we lower our aim to simply present that part necessary to offer a taste of what quantum computing is all about. What makes this possible is that we only need finite dimensional quantum mechanics, i.e., the vector spaces that represent the states of the system will only be of finite dimension. Such vector spaces consist of finite vectors with complex entries. These vectors will change by being them multiplied by operators or matrices. These matrices will be finite and have complex entries. We do not do any more quantum computing than what is needed for our final goal: Deutsch's algorithm. We stress that the reader does not need more than the ability to do matrix multiplication in order to understand this paper.

To motivate our use of vectors to describe states and matrices as ways of describing dynamics, we show that it is understandable if one looks at a basic toy models. Our models deal with childrens' marbles moving along the edges of a graph. Every computer scientist and logician who has taken a class in discrete structures knows how to represent a (weighted) graph as an adjacency matrix. We shall take this basic

Noson S. Yanofsky
Department of Computer and Information Science, Brooklyn College, CUNY, Brooklyn, NY 11210 and Computer Science Department, The Graduate Center, CUNY, New York, NY 10016, USA, e-mail: noson@sci.brooklyn.cuny.edu

J. van Benthem et al. (eds.), *Proof, Computation and Agency*, Synthese Library 352,
DOI 10.1007/978-94-007-0080-2_10,

idea and generalize it in several straightforward ways. While doing this, we shall present many concepts that are at the very core of quantum mechanics.

We begin with graphs that are without weights and progress to graphs that are weighted with real numbers, and finally to graphs that are weighted with complex numbers. With this in hand, we present a graph theoretic version of the double-slit experiment. This is *the* most important experiment in quantum mechanics. We conclude with a discussion of ways of combining systems to yield larger systems.

Throughout this chapter, we shall present an idea in a toy model, then generalize it to an abstract point and lastly discuss the connection with quantum mechanics before moving on to the next idea.

This paper is based on a forthcoming text *Quantum Computing for Computer Scientists* coauthored with Mirco Mannucci. The text was accepted for publication by Cambridge University Press and should see the light of day in the beginning of 2008. In the text we take the reader through the same material and go much further. The reader who appreciates this paper, will definitely gain from the text.

Acknowledgements I am grateful to Dr. Mirco Mannucci for many helpful discussions and cheery editing sessions.

10.1.1 Classical Deterministic Systems

We begin with a simple system described by a graph along with some children's marbles. Imagine the marbles as being on the vertices of the graph. The state of a system is described by how many marbles are on each vertex. For example, say that there are six vertices in the graph and a total of 27 marbles. We might place six marbles on vertex 0, two marbles on vertex 1 and the rest as described here:

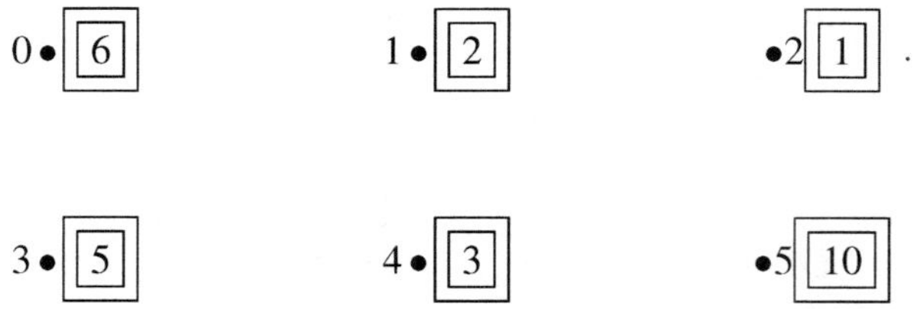

We shall denote this state as $X = [6, 2, 1, 5, 3, 10]^T$. The states of such a system will simply be a column vector of size 6.

We are not only interested in states of the system, but also in the way that the states change – or the "dynamics" of the system. This can be represented by a graph with directed edges. The dynamics might be described by the following directed graph:

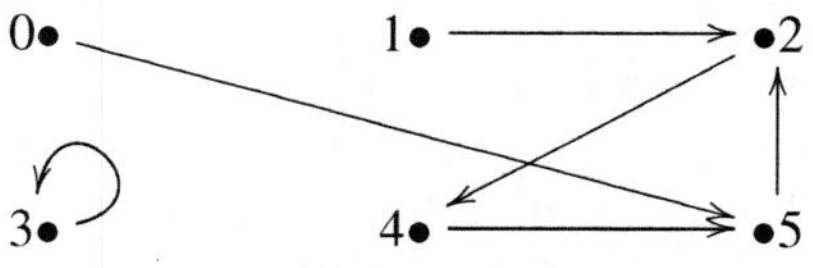

The idea is that if there exists an arrow from vertex i to vertex j, then in one time click, all the marbles on vertex i will move to vertex j. We place the following restriction on the types of graphs we shall be concerned with: graphs with exactly one outgoing edge from each vertex. This will correspond to the notion of a classical determinstic system. At each time click the marbles will have exactly one place to go. This graph is equivalent to the matrix, M (for "marbles"):

$$M = \begin{array}{c} \\ 0 \\ 1 \\ 2 \\ 3 \\ 4 \\ 5 \end{array} \begin{array}{c} \begin{array}{cccccc} 0 & 1 & 2 & 3 & 4 & 5 \end{array} \\ \begin{bmatrix} 0 & 0 & 0 & 0 & 0 & 0 \\ 0 & 0 & 0 & 0 & 0 & 0 \\ 0 & 1 & 0 & 0 & 0 & 1 \\ 0 & 0 & 0 & 1 & 0 & 0 \\ 0 & 0 & 1 & 0 & 0 & 0 \\ 1 & 0 & 0 & 0 & 1 & 0 \end{bmatrix} \end{array}$$

where $M[i, j] = 1$ if and only if there is an arrow from vertex j to vertex i[1]. Our restricted class of graphs are related to the restricted class of boolean matrices that have exactly one 1 in each column.

Let us say we multiply M by a state of the system $X = [6, 2, 1, 5, 3, 10]^T$. Then we have

$$MX = \begin{bmatrix} 000000 \\ 000000 \\ 010001 \\ 000100 \\ 001000 \\ 100010 \end{bmatrix} \begin{bmatrix} 6 \\ 2 \\ 1 \\ 5 \\ 3 \\ 10 \end{bmatrix} = \begin{bmatrix} 0 \\ 0 \\ 12 \\ 5 \\ 1 \\ 9 \end{bmatrix} = Y$$

What does this correspond to? If X describes the state of the system at time t, then Y is the state at time $t+1$, i.e., after one time click. We can see this clearly by looking at the formula for matrix multiplication:

$$Y[i] = (M \star X)[i] = \sum_{k=0}^{5} M[i, k] X[k].$$

In plain English, this states that the number of marbles that will reach vertex i after one time step is the sum of all the marbles that are on vertices with edges connecting

[1] Although most texts might have $M[i, j] = 1$ if and only if there is an arrow from vertex i to vertex j, we shall need it to be the other way for reasons which will become apparent later. The difference is trivial.

to vertex i. Notice that the top two entries of Y are zero. This corresponds to the fact that there are no arrows going to vertex 0 or 1.

In general if $X = [x_0, x_1, \ldots, x_{n-1}]^T$ is a column vector corresponding to having x_i marbles on vertex i, M is a n by n Boolean matrix, and if $MX = Y$ where $Y = [y_0, y_1, \ldots, y_{n-1}]^T$, then there are y_j marbles on vertex j in one time click. M is thus a way of describing how the state of the marbles can change from time t to time $t+1$.

As we shall soon see, (finite dimensional) quantum mechanics works in the same way. States of a system are represented by column vectors and the way in which the system changes in one time click is represented by matrices. Multiplying a matrix with a column vector yields a new state of the system. Quantum mechanics explores the way states of similar systems evolve over time.

Returning to our marbles, let's multiply M by itself. $MM = M^2$. However, since our entries are Boolean, we shall multiply the matrices as Boolean, i.e., $1+1 = 1 \vee 1 = 1$.

$$\begin{bmatrix} 0&0&0&0&0&0 \\ 0&0&0&0&0&0 \\ 0&1&0&0&0&1 \\ 0&0&0&1&0&0 \\ 0&0&1&0&0&0 \\ 1&0&0&0&1&0 \end{bmatrix} \begin{bmatrix} 0&0&0&0&0&0 \\ 0&0&0&0&0&0 \\ 0&1&0&0&0&1 \\ 0&0&0&1&0&0 \\ 0&0&1&0&0&0 \\ 1&0&0&0&1&0 \end{bmatrix} = \begin{bmatrix} 0&0&0&0&0&0 \\ 0&0&0&0&0&0 \\ 1&0&0&0&1&0 \\ 0&0&0&1&0&0 \\ 0&1&0&0&0&1 \\ 0&0&1&0&0&0 \end{bmatrix}$$

Looking at the formula for Boolean matrix multiplication

$$M^2[i,j] = \bigvee_{k=0}^{n-1} M[i,k] \wedge M[k,j]$$

we can see that this formula really shows us how to go from vertex j to vertex i in *two* time clicks.

And so we have that

$$M^2[i,j] = 1 \text{ if and only if there is a path of length 2 from vertex } j \text{ to vertex } i.$$

For an arbitrary k we have

$$M^k[i,j] = 1 \text{ if and only if there is a path of length } k \text{ from vertex } j \text{ to vertex } i.$$

In general, multiplying an n by n matrix by itself several times will correspond to the whether there is a path after several time clicks. Consider $X = [x_0, x_1, \ldots, x_{n-1}]^T$ to be the state where one has x_0 marbles on vertex 0, x_1 marbles on vertex 1, ..., x_{n-1} marbles on vertex $n-1$. Then, after k steps, the state of the marbles is Y where $Y = [y_0, y_1, \ldots, y_{n-1}]^T = M^k X$. In other words, y_j is the number of marbles on vertex j after k steps.

In quantum mechanics, if there are two or more matrices or operators that manipulate states, then the action of one followed by another is described by their matrix product. We shall take different states of systems and multiply the states by various matrices (of the appropriate type) to obtain other states. These other states will again be multiplied by other matrices until we attain the desired state.

10.1.2 Classical Probabilistic Systems

In quantum mechanics, neither the state of a system nor the dynamics of a system are deterministic. There is an indeterminacy in our knowledge of a state. Furthermore, the states change with probabilistic laws as opposed to deterministic laws. That means that states do not change in set ways. Rather, the laws are given by stating that states will change from one state to another state with a certain likelihood.

In order to capture these probabilistic scenarios, let us generalize what we did in the last subsection. Instead of dealing with a bunch of marbles moving around, we shall deal with a single marble. The state of the system will tell us the probabilities of the single marble being on each vertex. For a three-vertex graph, a typical state might look like this $X = [\frac{1}{5}, \frac{3}{10}, \frac{1}{2}]^T$. This will correspond to the fact that there is a one-fifth[2] chance that the marble is on vertex 0; a three-tenth chance that the marble is on vertex 1; and a half chance that the marble is on vertex 2. Since the marble must be somewhere on the graph, the sum of the probabilities is 1.

We must generalize the dynamics as well. Rather than exactly one arrow leaving each vertex, we will have several arrows leaving each vertex with non-negative real numbers between 0 and 1 as weights. These weights will describe the probability of the single marble going from one vertex to another in one time click. We shall restrict our attention to weighted graphs that satisfy the following two conditions: a) the sum of all the weights leaving a vertex is 1 and b) the sum of all the weights entering a vertex is 1. This will correspond to the fact that a marble must go someplace (there might be loops) and a marble must come from someplace. An example of such a graph is

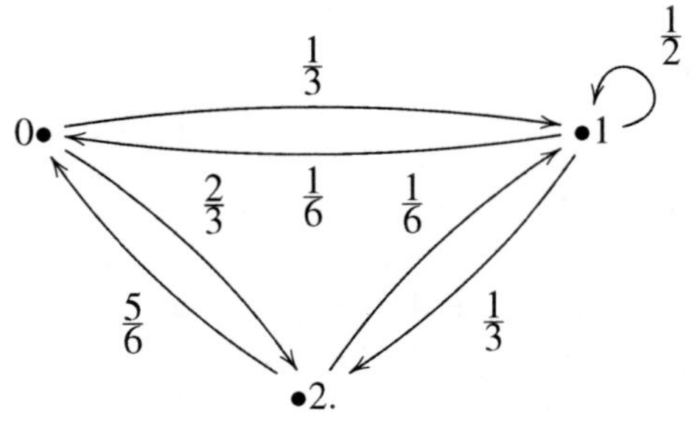

The matrix for this graph is

[2] Although the theory works with any $r \in [0,1]$, we shall deal only with fractions.

$$M = \begin{bmatrix} 0 & \frac{1}{6} & \frac{5}{6} \\ \frac{1}{3} & \frac{1}{2} & \frac{1}{6} \\ \frac{2}{3} & \frac{1}{3} & 0 \end{bmatrix}$$

The adjacency matrices for our graphs will have non-negative real entries where the sums of the rows and the sums of the columns are all 1. Such matrices are called "doubly stochastic matrices."

Let us see how the states interact with the dynamics. Suppose we have a state $X = \left[\frac{1}{6}, \frac{1}{6}, \frac{2}{3}\right]^T$. We will calculate how a state changes: $MX = Y$

$$\begin{bmatrix} 0 & \frac{1}{6} & \frac{5}{6} \\ \frac{1}{3} & \frac{1}{2} & \frac{1}{6} \\ \frac{2}{3} & \frac{1}{3} & 0 \end{bmatrix} \begin{bmatrix} \frac{1}{6} \\ \frac{1}{6} \\ \frac{2}{3} \end{bmatrix} = \begin{bmatrix} \frac{21}{36} \\ \frac{9}{36} \\ \frac{6}{36} \end{bmatrix}$$

Notice that the sum of the entries of Y is 1. If we have X expressing the probability of the position of a marble, and M expressing the probability of the way the marble moves around, then $MX = Y = \left[\frac{21}{36}, \frac{9}{36}, \frac{6}{36}\right]^T$ is to be interpreted as expressing the probability of the marble's location after moving. In other words, if X is the probability of the marble at time t, then MX is the probability of the marble at time $t+1$.

If M is an n by n doubly stochastic matrix and X is an n by 1 column vector whose entries sum to 1, then $M^k X = Y$ can be interpreted as expressing the probability of the position of a marble after k time clicks. That is, if $X = [x_0, x_1, \ldots, x_{n-1}]^T$ means that there is an x_i chance that a marble is on vertex i, then $M^k X = Y = [y_0, y_1, \ldots, y_{n-1}]^T$ means that after k time clicks, there is a y_j chance that the marble is on vertex j.

We are not constrained to multiply M by itself. We may also multiply M by another doubly stochastic matrix. Let M and N be two n by n doubly stochastic matrices. $M \star N$ will then describe a probability transition of going from time t to $t+1$ to $t+2$.

In quantum computers, quantum systems are generally in a probabilistic state. Manipulating the system will correspond to multiplying the state by matrices. Each time click will correspond to one matrix multiplication. At the end of the computation, the resulting vector will describe the state of the system.

Before moving on to the next section, we shall examine an interesting example. This shall be known as the "probabilistic double slit experiment." Consider the picture in figure 10.1 of a shooting gun.

Fig. 10.1 Double slit experiment with bullets.

There are two slits in the wall. The shooter is a good enough shot to always get the bullets through one of the two slits. There is a 50–50 chance of which slit the bullet will go through. Once a bullet is through a slit, there are three targets to the right of each slit that the bullet can hit with equal probability. The middle target can get hit in one of two ways: from the top slit going down, or from the bottom slit going up. It is assumed that it takes the bullet one time click to go from the gun to the wall and one time click to go from the wall to the targets. The picture correspond to the following weighted graph.

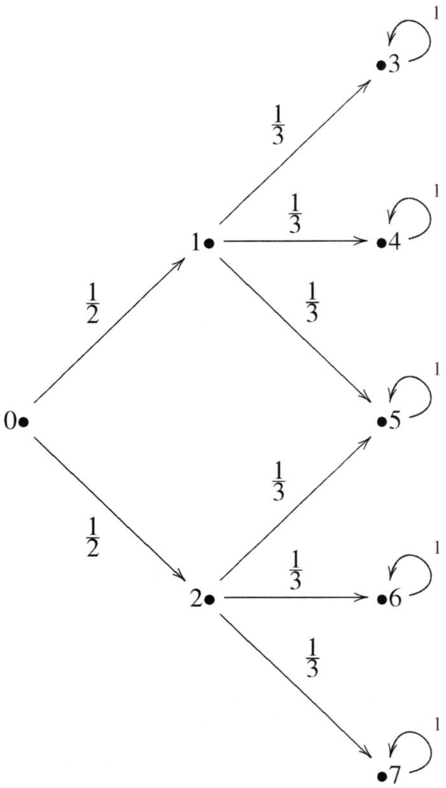

Notice that the vertex marked 5 can receive bullets from either of the two slits. Corresponding to this graph is the matrix B (for "bullets")

$$B = \begin{bmatrix} 0 & 0 & 0 & 0 & 0 & 0 & 0 & 0 \\ \frac{1}{2} & 0 & 0 & 0 & 0 & 0 & 0 & 0 \\ \frac{1}{2} & 0 & 0 & 0 & 0 & 0 & 0 & 0 \\ 0 & \frac{1}{3} & 0 & 1 & 0 & 0 & 0 & 0 \\ 0 & \frac{1}{3} & 0 & 0 & 1 & 0 & 0 & 0 \\ 0 & \frac{1}{3} & \frac{1}{3} & 0 & 0 & 1 & 0 & 0 \\ 0 & 0 & \frac{1}{3} & 0 & 0 & 0 & 1 & 0 \\ 0 & 0 & \frac{1}{3} & 0 & 0 & 0 & 0 & 1 \end{bmatrix}$$

In words, B describes the way a bullet will move after one time click.[3]

Let us calculate the probabilities for the bullet's position after two time clicks.

[3] The matrix B is not a doubly stochastic matrix. The sum of the weights entering vertex 0 is not 1. The sum of weights leaving vertices 3, 4, 5, 6, and 7 are more than 1. This fact should not bother you. We are interested in demonstrating the way probabilities behave with respect to these matrices.

$$B \star B = B^2 = \begin{bmatrix} 0 & 0 & 0 & 0 & 0 & 0 & 0 & 0 \\ 0 & 0 & 0 & 0 & 0 & 0 & 0 & 0 \\ 0 & 0 & 0 & 0 & 0 & 0 & 0 & 0 \\ \frac{1}{6} & \frac{1}{3} & 0 & 1 & 0 & 0 & 0 & 0 \\ \frac{1}{6} & \frac{1}{3} & 0 & 0 & 1 & 0 & 0 & 0 \\ \frac{1}{3} & \frac{1}{3} & \frac{1}{3} & 0 & 0 & 1 & 0 & 0 \\ \frac{1}{6} & 0 & \frac{1}{3} & 0 & 0 & 0 & 1 & 0 \\ \frac{1}{6} & 0 & \frac{1}{3} & 0 & 0 & 0 & 0 & 1 \end{bmatrix}$$

So B^2 indicates the probabilities of the bullet's position after two time clicks.

If we are sure that we start with the bullet in position 0, i.e., $X = [1,0,0,0,0,0,0,0]^T$, then, after two time clicks, the state of the bullets will be

$$B^2X = [0,0,0,\frac{1}{6},\frac{1}{6},\frac{1}{3},\frac{1}{6},\frac{1}{6}]^T$$

The key idea is to notice that $B^2[5,0] = \frac{1}{6} + \frac{1}{6} = \frac{1}{3}$ because the bullets start from position 0, then there are two possible ways of the bullet getting to position 5. The possibilities sum to $\frac{1}{3}$. This is what we would expect. We shall revisit this example in the next subsection where strange things start happening!

10.1.3 Quantum Systems

We are now ready to leave the world of classical probabilities and enter the world of the quantum. One of the central facts about quantum mechanics is that complex numbers play a major role in the calculations. Probabilities of states and transitions are not given as a real numbers p between 0 and 1. Rather, they are given is a complex numbers c such that $|c|^2$ is a real number[4] between 0 and 1.

What is the difference how the probabilities are given? What does it matter if a probability is given as a real number between 0 and 1, or as a complex number whose modulus squared is a real number between 0 and 1? The difference is – and this is the core of quantum theory – that real number probabilities can be added to obtain larger real numbers. In contrast, complex numbers can cancel each other and lower their probability. In detail, if p_1 and p_2 are two real numbers between 0 and 1, then $(p_1 + p_2) \geq p_1$ and $(p_1 + p_2) \geq p_2$. Now let's look at the complex case. Let c_1 and c_2 be two complex numbers with their squares of modulus $|c_1|^2$ and $|c_2|^2$. $|c_1 + c_2|^2$ need not be bigger then $|c_1|^2$ and it also does not need to be bigger than $|c_2|^2$.

---

[4] We remind the reader that if $c = a + bi$ is a complex number, then its modulus is $|c| = \sqrt{a^2 + b^2}$ and $|c|^2 = a^2 + b^2$.

For example[5] , if $c_1 = 5+3i$ and $c_2 = -3-2i$, then $|c_1|^2 = 34$ and $|c_2|^2 = 13$ but $|c_1+c_2|^2 = |2+i|^2 = 5$. 5 is less than 34 and 5 is less than 13.

This possibility of canceling out complex numbers corresponds to something called "interference" in quantum mechanics. One complex number might interfere with another. It is one of the most important ideas in quantum theory.

Let us generalize our states and graphs from the previous subsection. Rather than insisting that the sum of the entries in the column vector is 1, we insist that the sum of the modulus squared of the entries is 1. This makes sense since we are considering the probability as the modulus squared.

For dynamics, rather than talking about graphs with real number weights, we shall talk about graphs with complex number weights. Instead of insisting that the adjacency matrix of such a graph be a doubly stochastic matrix, we ask instead that the adjacency matrix be unitary.[6]

For example, consider the graph

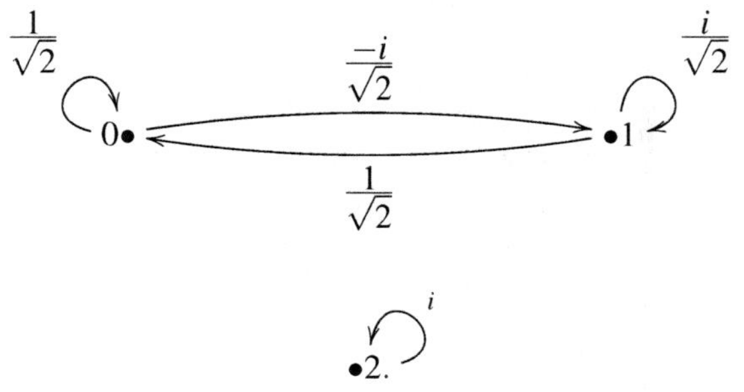

The corresponding unitary adjacency matrix is

$$U = \begin{bmatrix} \frac{1}{\sqrt{2}} & \frac{1}{\sqrt{2}} & 0 \\ \frac{-i}{\sqrt{2}} & \frac{i}{\sqrt{2}} & 0 \\ 0 & 0 & i \end{bmatrix}.$$

Unitary matrices are related to doubly stochastic matrices as follows. The modulus squared of the all the complex entries in U forms a doubly stochastic matrix. The i,jth element in U is denoted $U[i,j]$, and its modulus squared is denoted $|U[i,j]|^2$. By abuse of notation, we shall denote the entire matrix of modulus squares as $|U[i,j]|^2$:

$$|U[i,j]|^2 = \begin{bmatrix} \frac{1}{2} & \frac{1}{2} & 0 \\ \frac{1}{2} & \frac{1}{2} & 0 \\ 0 & 0 & 1 \end{bmatrix}.$$

[5] The important point here is that the modulus squared is positive. For simplicity of calculations, we have chosen easy complex numbers.

[6] Let us just remember: a matrix U is unitary if $U \star U^\dagger = I = U^\dagger \star U$. The adjoint of U, denoted as $U^\dagger$, is defined as $U^\dagger = (\overline{U})^T = \overline{(U^T)}$ or $U^\dagger[j,k] = \overline{U[k,j]}$.

It is easy to see that this is a doubly stochastic matrix.

From this graph-theoretic point of view, it is easy to see what unitary means: the conjugate transpose of the U matrix is

$$U^\dagger \quad = \quad \begin{bmatrix} \frac{1}{\sqrt{2}} & \frac{i}{\sqrt{2}} & 0 \\ \frac{1}{\sqrt{2}} & \frac{-i}{\sqrt{2}} & 0 \\ 0 & 0 & -i \end{bmatrix}.$$

This matrix corresponds to the graph

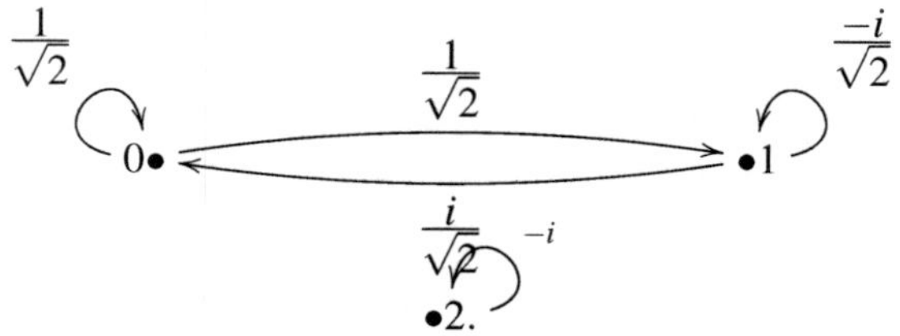

If U is the matrix that takes a state from time t to time $t+1$, then $U^\dagger$ is the matrix that takes a state from time t to time $t-1$. If we multiply U and $U^\dagger$, we get the identity matrix I_3 which corresponds to the graph

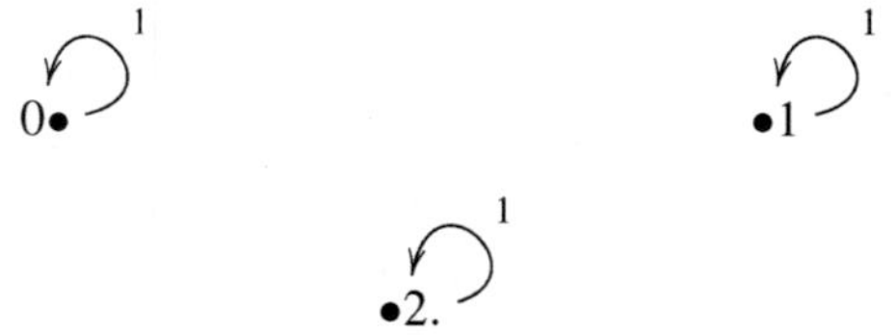

This means that if we perform some operation and then "undo" the operation, we will find ourselves in the same state as we began with probability 1. It is important to note that unitary does not only mean invertable. It means invertable in a very easy way, i.e., the inverse is the dagger of the matrix. This "invertibility" is again an important issue in quantum mechanics. Most of the dynamics will be invertible (except measurements).

In order to see the interference phenomenon and the related "superposition" phenomenon, we will revisit the double-slit experiment from the last subsection.

Rather than looking at bullets, which are relatively large objects and hence adhere to the laws of classical physics, we shall look at microscopic objects such as photons that follow the laws of quantum physics. Rather than having a gun, we shall have a laser shoot photons through two slits as in Fig. 10.2.

Again we shall make the assumption that a photon will pass through one of the two slits. Each slit has a 50% chance of the photon going through it. To the left of each slit there are three measuring devices. It is assumed that it takes one time click to go from the laser to the slits and one time click to go from the slits to the targets.

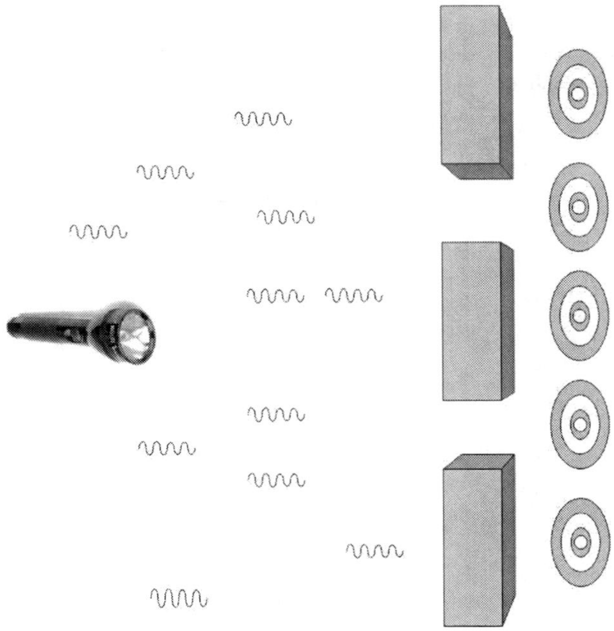

Fig. 10.2 Double slit experiment with photons.

We are not interested in how large the slits are or how far the measuring positions are from the slits. Physicists are very adapt at calculating many different aspects of this experiment. We are only interested in the set-up.

Consider the following weighted graph that describes the set-up of the experiment:

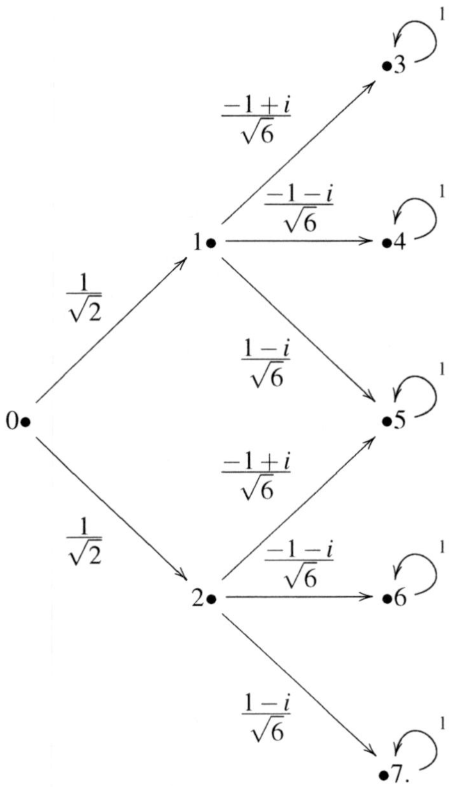

The modulus squared of $\frac{1}{\sqrt{2}}$ is $\frac{1}{2}$, which corresponds to the that there is a 50–50 chance of the photon going through either slit. $\left|\frac{\pm 1 \pm i}{\sqrt{6}}\right|^2 = \frac{1}{3}$ which corresponds to the fact that whichever slit the photon goes through, there is a $\frac{1}{3}$ of a chance of its hitting any of the three measuring positions to the right of that slit.[7]

The adjacency matrix for this graph is P (for "photons")[8]

[7] The actual complex number weights are not our interest here. If we wanted to calculate the actual numbers, we would have to measure the width of the slits, the distance between the slits, the distance from the slits to the measuring devices etc. However, our goal here is to clearly demonstrate the interference phenomenon. And so we chose the above complex numbers simply because the modulus squared are exactly the same as the bullets case.

[8] This matrix is not a unitary matrix. Looking carefully at row 0, one can immediately see that P is not unitary. In our graph, there is nothing entering vertex 0. The reason why this matrix fails to be unitary is because we have not put in all the arrows in our graph. There are many more possible ways the photon can travel in a real-life physical situation. In particular, the photon might go from the right to the left. The diagram and matrix would become too complicated if we put in all the transitions. We are simply trying to demonstrate the interference phenomenon and we can accomplish that even with a matrix that is not quite unitary.

$$P = \begin{bmatrix} 0 & 0 & 0 & 0\,0\,0\,0\,0 \\ \frac{1}{\sqrt{2}} & 0 & 0 & 0\,0\,0\,0\,0 \\ \frac{1}{\sqrt{2}} & 0 & 0 & 0\,0\,0\,0\,0 \\ 0 & \frac{-1+i}{\sqrt{6}} & 0 & 1\,0\,0\,0\,0 \\ 0 & \frac{-1-i}{\sqrt{6}} & 0 & 0\,1\,0\,0\,0 \\ 0 & \frac{1-i}{\sqrt{6}} & \frac{-1+i}{\sqrt{6}} & 0\,0\,1\,0\,0 \\ 0 & 0 & \frac{-1-i}{\sqrt{6}} & 0\,0\,0\,1\,0 \\ 0 & 0 & \frac{1-i}{\sqrt{6}} & 0\,0\,0\,0\,1 \end{bmatrix}.$$

The modulus squared of the P matrix is exactly the same as the bullets matrix i.e., $|P[i,j]|^2 = B$. Let us see what happens if we calculate the transitions matrix after *two* time clicks.

$$P^2 = \begin{bmatrix} 0 & 0 & 0 & 0\,0\,0\,0\,0 \\ 0 & 0 & 0 & 0\,0\,0\,0\,0 \\ 0 & 0 & 0 & 0\,0\,0\,0\,0 \\ \frac{-1+i}{\sqrt{12}} & \frac{-1+i}{\sqrt{6}} & 0 & 1\,0\,0\,0\,0 \\ \frac{-1-i}{\sqrt{12}} & \frac{-1-i}{\sqrt{6}} & 0 & 0\,1\,0\,0\,0 \\ 0 & \frac{-1+i}{\sqrt{6}} & \frac{1-i}{\sqrt{6}} & 0\,0\,1\,0\,0 \\ \frac{-1-i}{\sqrt{12}} & 0 & \frac{-1-i}{\sqrt{6}} & 0\,0\,0\,1\,0 \\ \frac{-1+i}{\sqrt{12}} & 0 & \frac{1-i}{\sqrt{6}} & 0\,0\,0\,0\,1 \end{bmatrix}.$$

How do we interpret this in terms of probability? Let us look at the modulus squared of each of the entries.

$$|P^2[i,j]|^2 = \begin{bmatrix} 0 & 0 & 0 & 0\,0\,0\,0\,0 \\ 0 & 0 & 0 & 0\,0\,0\,0\,0 \\ 0 & 0 & 0 & 0\,0\,0\,0\,0 \\ \frac{1}{6} & \frac{1}{3} & 0 & 1\,0\,0\,0\,0 \\ \frac{1}{6} & \frac{1}{3} & 0 & 0\,1\,0\,0\,0 \\ \mathbf{0} & \frac{1}{3} & \frac{1}{3} & 0\,0\,1\,0\,0 \\ \frac{1}{6} & 0 & \frac{1}{3} & 0\,0\,0\,1\,0 \\ \frac{1}{6} & 0 & \frac{1}{3} & 0\,0\,0\,0\,1 \end{bmatrix}.$$

This matrix is almost exactly the same as B^2 but with one glaring difference. $B^2[5,0] = \frac{1}{3}$ because of the two ways of starting at position 0 and ending at position 5. We added the nonnegative probabilities $\frac{1}{6} + \frac{1}{6} = \frac{1}{3}$. However with a photon that follows the laws of quantum mechanics, the complex numbers are added as opposed to their probabilities.

$$\frac{1}{\sqrt{2}}\left(\frac{-1+i}{\sqrt{6}}\right)+\frac{1}{\sqrt{2}}\left(\frac{1-i}{\sqrt{6}}\right) \quad = \quad \frac{-1+i}{\sqrt{12}}+\frac{1-i}{\sqrt{12}} \quad = \quad \frac{0}{\sqrt{12}} \quad = \quad 0.$$

Thus giving us $|P^2[5,0]|^2 = 0$. In other words, although there are two ways of a photon going from vertex 0 to vertex 5, there will be no photon at vertex 5.

How is one to understand this phenomenon? Physicists have a simple explanation for interference: waves. A familiar observation such as a pebble thrown into a pool of water will easily convince us that waves interfere, sometimes reinforcing each other, sometimes cancelling each other. In our experiment, photons are going through both slits at one time and they are canceling each other out at the middle measuring device. Thus, the double-slit experiment points to the wave-like nature of light.

The experiment can be done with only one photon shot out from vertex 0. This ensures that there will not be another wave for it to cancel out with. And yet, when only one photon goes through a slit, there is still an interference phenomenon. What is going on here?

The naive probabilistic interpretation of the position of the photon following the bullet metaphor of the previous section, is thus not entirely adequate. Let the state of the system be given by $X = [c_0, c_1, \ldots, c_{n-1}]^T \in \mathbb{C}^n$. It is incorrect to say that the probability of the photon being in position k is $|c_k|^2$. Rather, a system in state X means that the particle is in *all* positions simultaneously. It is only after we measure the photon that the chances of it being found in position k is $|c_k|^2$. The photon (or its associated wave) goes through the top slit *and* the bottom slit simultaneously. And when the photon exits both slits, it can cancel *itself* out. A photon is not in *a* singe position, rather it is in *many* positions or a "superposition".

This might cause some justifiable disbelief. After all, we do not see things in a superposition of states. Our everyday experience tells us that things are in one position or (exclusive or!) another position. How can this be? The reason why we see particles in exactly one position is because we have performed a measurement. When we measure something at the quantum level, the quantum object that we have measured is no longer in a superposition of states, rather it collapses to a single classical state. So we have to redefine what a state of a quantum system means: A system in state X means that *after measuring* the photon it will be in position k with probability $|c_k|^2$.

What are we to make of these strange ideas? Are we really to believe them? Richard Feynman in discussing the double-slit experiment [3] (Vol. III, p. 1–1) waxes lyrical:

> We choose to examine a phenomenon which is impossible, *absolutely* impossible, to explain in any classical way, and which has in it the heart of quantum mechanics. In reality, it contains the *only* mystery. We can not make the mystery go away by "explaining" how it works. We will just tell you how it works.

It is exactly this superposition of states that is the real power behind quantum computing. Classical computers are in one state at every moment. Quantum computers can be put in a superposition of states. Imagine putting a computer in many different classical states at one time and then processing with *all* the states. This is the ultimate in parallel processing! Such a computer can only be conceived in the quantum world.

10.1.4 Combining Systems

Quantum mechanics can also deal with systems that have more than one part. In this section we learn how to combine several systems into one. We shall talk about combining classical probabilistic systems. However, whatever is stated about classical probabilistic systems is also true for quantum systems.

Consider two different marbles. Imagine that a red marble follows the probabilities of the graph whose corresponding adjacency matrix is

$$M = \begin{bmatrix} 0 & \frac{1}{6} & \frac{5}{6} \\ \frac{1}{3} & \frac{1}{2} & \frac{1}{6} \\ \frac{2}{3} & \frac{1}{3} & 0 \end{bmatrix}.$$

Consider also a blue marble that follows the transitions given by the graph

$G_N =$ 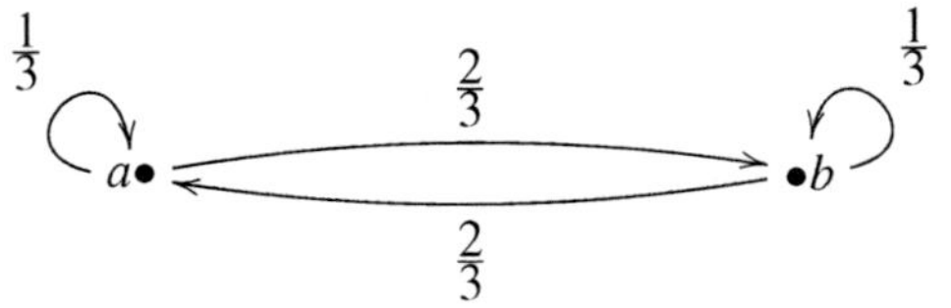

i.e., the matrix

$$N = \begin{bmatrix} \frac{1}{3} & \frac{2}{3} \\ \frac{2}{3} & \frac{1}{3} \end{bmatrix}.$$

What does a state for a *two* marble system look like? Since the red marble can be on one of three vertices and the blue marble can be on one of two vertices, there are $3 \times 2 = 6$ possible states of the combined system. This is the tensor product of a 3 by 1 vector with a 2 by 1 vector. A typical state might look like:

$$X = \begin{array}{c} 0a \\ 0b \\ 1a \\ 1b \\ 2a \\ 2b \end{array} \begin{bmatrix} \frac{1}{18} \\ 0 \\ \frac{2}{18} \\ \frac{1}{3} \\ 0 \\ \frac{1}{2} \end{bmatrix},$$

which would correspond to the fact that there is a

$\frac{1}{18}$ chance of the red marble being on vertex 1 and the blue marble being on vertex a,
0 chance of the red marble being on vertex 1 and the blue marble being on vertex b,
⋮
$\frac{1}{2}$ chance of the red marble being on vertex 3 and the blue marble being on vertex b.

Now we may ask, how does a system with these *two* marbles change? What is its dynamics? Imagine that the red marble is on vertex 1 and the blue marble is on vertex a. We may write this as "$1a$." What is the probability of the state going from state $1a$ to state $2b$? Obviously, the red marble must go from vertex 1 to vertex 2 and (multiply) the blue marble must go from vertex a to vertex b. The probability is $\frac{1}{3} \times \frac{2}{3} = \frac{2}{9}$. In general, for a system to go from state ij to a state $i'j'$ we must multiply the probability of going from state i to state i' with the probability of going from state j to state j'.

$$ij \xrightarrow{M[i',i] \times N[j',j]} i'j' \,.$$

For the changes of all the states, we have to do this for all the entries. We are really giving the tensor product of two matrices[9]:

[9] Formally, the tensor product of matrices is a function

$$\otimes : \mathbb{C}^{m\times m} \times \mathbb{C}^{n\times n} \longrightarrow \mathbb{C}^{m\times m} \otimes \mathbb{C}^{n\times n} = \mathbb{C}^{mn\times mn}$$

and it is defined as: $(A \otimes B)[j,k] = A[j/n, k/m] \times B[j \text{ MOD } n, k \text{ MOD } m]$.

$$M \otimes N \quad = \quad \begin{array}{c} \\ 0 \\ \\ 1 \\ \\ 2 \end{array} \begin{array}{c} \begin{array}{ccc} 0 & 1 & 2 \end{array} \\ \begin{bmatrix} 0\begin{bmatrix} \frac{1}{3} & \frac{2}{3} \\ \frac{2}{3} & \frac{1}{3} \end{bmatrix} & \frac{1}{6}\begin{bmatrix} \frac{1}{3} & \frac{2}{3} \\ \frac{2}{3} & \frac{1}{3} \end{bmatrix} & \frac{5}{6}\begin{bmatrix} \frac{1}{3} & \frac{2}{3} \\ \frac{2}{3} & \frac{1}{3} \end{bmatrix} \\ \frac{1}{3}\begin{bmatrix} \frac{1}{3} & \frac{2}{3} \\ \frac{2}{3} & \frac{1}{3} \end{bmatrix} & \frac{1}{2}\begin{bmatrix} \frac{1}{3} & \frac{2}{3} \\ \frac{2}{3} & \frac{1}{3} \end{bmatrix} & \frac{1}{6}\begin{bmatrix} \frac{1}{3} & \frac{2}{3} \\ \frac{2}{3} & \frac{1}{3} \end{bmatrix} \\ \frac{2}{3}\begin{bmatrix} \frac{1}{3} & \frac{2}{3} \\ \frac{2}{3} & \frac{1}{3} \end{bmatrix} & \frac{1}{3}\begin{bmatrix} \frac{1}{3} & \frac{2}{3} \\ \frac{2}{3} & \frac{1}{3} \end{bmatrix} & 0\begin{bmatrix} \frac{1}{3} & \frac{2}{3} \\ \frac{2}{3} & \frac{1}{3} \end{bmatrix} \end{bmatrix} \end{array}.$$

$$= \quad \begin{array}{c} \\ 0a \\ 0b \\ 1a \\ 1b \\ 2a \\ 2b \end{array} \begin{array}{c} \begin{array}{cccccc} 0a & 0b & 1a & 1b & 2a & 2b \end{array} \\ \begin{bmatrix} 0 & 0 & \frac{1}{18} & \frac{2}{18} & \frac{5}{18} & \frac{10}{18} \\ 0 & 0 & \frac{2}{18} & \frac{1}{18} & \frac{10}{18} & \frac{5}{18} \\ \frac{1}{9} & \frac{2}{9} & \frac{1}{6} & \frac{2}{6} & \frac{1}{18} & \frac{2}{18} \\ \frac{2}{9} & \frac{1}{9} & \frac{2}{6} & \frac{1}{6} & \frac{2}{18} & \frac{1}{18} \\ \frac{2}{9} & \frac{4}{9} & \frac{1}{9} & \frac{2}{9} & 0 & 0 \\ \frac{4}{9} & \frac{2}{9} & \frac{2}{9} & \frac{1}{9} & 0 & 0 \end{bmatrix} \end{array}$$

The graph that corresponds to this matrix, $G_M \times G_N$ is called the Cartesian product of two weighted graphs. In quantum theory, the states of two systems are combined using the tensor product of two vectors and the dynamics of two systems are combined by using the tensor product of two matrices. The tensor product of the matrices will then act on the tensor product of the vectors.

In general, the Cartesian product of an n-vertex graph with an n'-vertex graph is an $(n \times n')$-vertex graph. If we have an n-vertex graph G and we are interested in m different marbles moving in this system, we would need to look at the graph

$$G^m \quad = \quad \underbrace{G \times G \times \cdots \times G}_{m \text{ times}}$$

which will have n^m vertices. If M_G is the associated adjacency matrix, then we will be interested in

$$M_G^{\otimes m} \quad = \quad \underbrace{M_G \otimes M_G \otimes \cdots \otimes M_G}_{m \text{ times}}$$

which will be a n^m by n^m matrix.

One might think of a bit as a two vertex graph with a marble on the 0 vertex or a marble on the 1 vertex. If one were then interested in m bits, one would need a 2^m vertex graph or equivalently a 2^m by 2^m matrix. So there is exponential growth of the resources needed for the amount of bits discussed.

This exponential growth for a system is actually one of the main reasons why Richard Feynman started thinking about quantum computing. He realized that because of this exponential growth, it would be hard for a classical computer to simulate such a system. He asked whether a quantum computer with its ability to do massive parallel processing, might be able to simulate such a system.

10.2 Basic Quantum Theory

Armed with the intuition, we tackle the formal statement of quantum theory. A disclaimer is in order: We are only presenting a small part of finite dimensional quantum physics. There is no way that we can give more than a minute fraction of this magnificent subject in these few pages. It is a sincere hope that this will inspire the reader to go on and study more. For a mathematician, the best book to start reading about quantum mechanics is, of course, Dirac's classic text [2]. However, there are many other primers available, e.g., [1, 4, 5, 7]. The more advanced mathematician might want to look at [6].

10.2.1 States

An n dimensional quantum system is a system that can be observed in one of n possible states. Examples of such systems are

- a particle can be in one of n positions;
- a system might have one of n energy levels;
- a photon might have one of n polarization directions.

For clarity, lets talk of the first example. Lets say we have a particle that can be in one of n positions.

The states of such a system shall be represented by column vectors of n complex numbers. We shall denote these vectors with the "ket" $|\ \rangle$[10] notation:

$$|\varphi\rangle = [c_0, c_1, \ldots, c_j, \ldots, c_{n-1}]^T.$$

How is one to interpret these kets? Let us look at simple cases. The state

$$|\psi\rangle = [0, 1, \ldots, 0, \ldots, 0]^T$$

is to be thought of as saying that our particle will be found in position 1. The state

$$|\psi'\rangle = [0, 0, \ldots, 1, \ldots, 0]^T$$

is to be interpreted that the particle is in position j. These two states are examples of what are called "pure states".

How is one to interpret an arbitrary

[10] "Ket" is the second half of "bracket". However we shall not use the "bra" part in our exposition.

$$|\varphi\rangle = [c_0, c_1, \ldots, c_j, \ldots, c_{n-1}]^T?$$

Let S be the sum of the squares of modulus of the c_j, i.e.,

$$S = |c_0|^2 + |c_1|^2 + \ldots + |c_{n-1}|^2.$$

This is the length of the vector $|\varphi\rangle$ squared. Then $|\varphi\rangle$ is to be interpreted that if one was to measure the state described by $|\varphi\rangle$ we would find the particle in position 0 with probability $|c_0|^2/S$, in position 1 with probability $|c_1|^2/S$, in position 2 with probability $|c_2|^2/S$, ..., in position $n-1$ with probability $|c_{n-1}|^2/S$ Such states are called "superpositions". They say that the particle is in more than one "position" at a time. It is important to stress that $|\varphi\rangle$ means that the particle is in *all* positions simultaneously. It does not mean that the particle is in some single position and the c_j are giving us probabilities of which position.

These superpositions can be added: if

$$|\varphi\rangle = [c_0, c_1, \ldots, c_j, \ldots, c_{n-1}]^T \text{ and } |\varphi'\rangle = [c'_0, c'_1, \ldots, c'_j, \ldots, c'_{n-1}]^T,$$

then

$$|\varphi\rangle + |\varphi'\rangle = [c_0 + c'_0, c_1 + c'_1, \ldots, c_j + c'_j, \ldots, c_{n-1} + c'_{n-1}]^T.$$

Also, if there is a complex number $c \in \mathbb{C}$, we can multiply a ket by this c:

$$c|\varphi\rangle = [c \times c_0, c \times c_1, \ldots, c \times c_j, \ldots, c \times c_{n-1}]^T$$

These operation satisfy all the properties of being a complex vector space. So the states of an n dimensional quantum system are represented by the complex vector space $\mathbb{C}^n$.

Let us add a superposition to itself.

$$|\varphi\rangle + |\varphi\rangle = 2|\varphi\rangle = [c_0 + c_0, c_1 + c_1, \ldots, c_j + c_j, \ldots, c_{n-1} + c_{n-1}]^T$$

$$= [2c_0, 2c_1, \ldots, 2c_j, \ldots, 2c_{n-1}]^T.$$

The sum of the modulus squared is

$$S' = |2c_0|^2 + |2c_1|^2 + \ldots + |2c_{n-1}|^2$$
$$= 2^2|c_0|^2 + 2^2|c_1|^2 + \ldots + 2^2|c_{n-1}|^2 = 2^2(|c_0|^2 + |c_1|^2 + \ldots + |c_{n-1}|^2).$$

The chances of a particle being found in position j is

$$\frac{|2c_j|^2}{S'} = \frac{2^2|c_j|^2}{2^2(|c_0|^2 + |c_1|^2 + \ldots + |c_{n-1}|^2)} = \frac{|c_j|^2}{|c_0|^2 + |c_1|^2 + \ldots + |c_{n-1}|^2}.$$

In other words, the ket $2|\varphi\rangle$ describes the same physical system as $|\varphi\rangle$. This makes sense, after all when we add two of the same superpositions, we do not expect any interference. We expect that they should reinforce each other. A similar analysis shows that for an arbitrary $c \in \mathbb{C}$ we have that the ket $|\varphi\rangle$ and the ket $c|\varphi\rangle$ describe the same physical state. Geometrically, we can say that the vector $|\varphi\rangle$ and the extension $c|\varphi\rangle$ describe the same physical state. So the only thing that is important is the direction of $|\varphi\rangle$ not the length of $|\varphi\rangle$. We might as well work with a "normalized"

$|\varphi\rangle$, i.e.,

$$\frac{|\varphi\rangle}{||\varphi\rangle|}$$

which has length 1. In fact, in section 1, we only worked with vectors of length 1.

Given an n dimensional quantum system that is represented by $\mathbb{C}^n$ and an m dimensional quantum system represented by $\mathbb{C}^m$, we can combine these two systems to form one system. This one system is to represented by the tensor product of the two vector spaces:

$$\mathbb{C}^n \otimes \mathbb{C}^m \cong \mathbb{C}^{n \times m}.$$

If $|\varphi\rangle$ is in the first system and $|\varphi'\rangle$ is in the second system, then we represent the combined system as

$$|\varphi\rangle \otimes |\varphi'\rangle = |\varphi, \varphi'\rangle = |\varphi\varphi'\rangle.$$

It is important to realize that, in general, there are more elements in the tensor product of the two systems than in the union of each of the two systems. States in $\mathbb{C}^n \otimes \mathbb{C}^m$ that cannot be represented simply as an element in $\mathbb{C}^n$ and an element in $\mathbb{C}^m$ are said to be "entangled".

10.2.2 Dynamics

Quantum systems are not static. The states of the system are constantly changing. Changes, or "operators", on an n dimensional quantum system are represented by n by n unitary matrices. Given a state $|\varphi\rangle$ that represents a system at time t, then the system will be in state $U|\varphi\rangle$ at time $t+1$.

What does unitary really mean? If $U|\varphi\rangle = |\varphi'\rangle$ then we can easily form $U^\dagger$ and multiply both sides of the equation by $U^\dagger$ to get $U^\dagger U|\varphi\rangle = U^\dagger|\varphi'\rangle$ or $|\varphi\rangle = U^\dagger|\varphi'\rangle$. In other words, because U is unitary, there is a related matrix that can "undo" the action that U does. $U^\dagger$ takes the result of U's action and gets the original vector back. In the quantum world, most actions actions are "undoable" or "reversible" in such a manner.

If U operates on $\mathbb{C}^n$ and U' operates on $\mathbb{C}^m$, then $U \otimes U'$ will operate on $\mathbb{C}^n \otimes \mathbb{C}^m$ in the following way:

$$(U \otimes U')(|\varphi\rangle \otimes |\varphi'\rangle) = U|\varphi\rangle \otimes U'|\varphi'\rangle.$$

10.2.3 Observables

There are other types of operations that one can do to a n dimensional quantum system: one can observe, or "measure", the system. When we measure a system, it is no longer in a superposition. The superposition is said to "collapse" to a pure state.

$$|\varphi\rangle = [c_0, c_1, \ldots, c_j, \ldots, c_{n-1}]^T \rightsquigarrow |\varphi'\rangle = [0, 0, \ldots, 1, \ldots, 0].^T$$

The question is which of the n pure states will the state collapse to? The answer is that it is random. Let S be the sum of all the squares of the modulus, i.e.,

$$S = |c_0|^2 + |c_1|^2 + \cdots + |c_j|^2 + \cdots + |c_{n-1}|^2.$$

There is a $|c_0|^2/S$ of a chance of the superposition collapsing to the 0th pure state. There is a $|c_1|^2/S$ of a chance of the superposition collapsing to the 1th pure state. etc. There is no known mathematical way to decide which pure state the system will, in-fact, collapse to.

An observable, or "measurement" on an n dimensional system is represented by an n by n hermitian matrix. We remind the reader that a n by n matrix A is hermitian, or "self-adjoint" if $A^\dagger = A$. In other words, $A[j,k] = \overline{A[k,j]}$. Equivalently A is hermitian if and only if $A^T = \overline{A}$. For example, the matrix

$$\begin{bmatrix} 5 & 4+5i & 6-16i \\ 4-5i & 13 & 7 \\ 6+16i & 7 & -2.1 \end{bmatrix}$$

is hermitian.

For a matrix A in $\mathbb{C}^{n \times n}$, if there is a number c in $\mathbb{C}$ and a vector $|\psi\rangle$ in $\mathbb{C}^n$ such that

$$A|\psi\rangle = c|\psi\rangle$$

then c is called an "eigenvalue" of A and $|\psi\rangle$ is called an "eigenvector" of A associated to c. The eigenvalues of a hermitian matrix are all real numbers. Furthermore, distinct eigenvectors which have distinct eigenvalues of any hermitian matrix are orthogonal. The hermitian matrices that represent observables for an n dimensional system have the further property that there are n distinct eigenvalues and n distinct eigenvectors. That means that the set of eigenvectors form a basis for the entire complex vector space that represents the quantum system we are interested in. Hence if we have an observable A and $|\varphi\rangle$ an eigenvalue of A then $A|\varphi\rangle = c|\varphi\rangle$ for some $c \in \mathbb{C}$. $c|\varphi\rangle$ represents the same state as $|\varphi\rangle$ as we said before. So if the system is in an eigenvector of the basis, then the system will not change.

10.3 Architecture

In this section we are going to show how the ideas of quantum mechanics are going to influence our construction of quantum computers. In this paper we do not dwell on actual hardware implementations. Rather we shall look at the quantum generalizations of bits and logical gates. In Section 3.1 we go from bits to qubits. We also discuss the notation that is needed for this. In Section 3.2 we show how to look at classical computing as matrix manipulations. From the view afforded by this perspective, we easily generalize the notion of logical gate to quantum gate. There are many quantum gates, but we shall only look at a few that will be needed in the next section.

10.3.1 Bits and Qubits

What is a bit? A bit is an atom of information that represents one of two disjoint situations. An example of a bit is electricity going through a circuit or electricity not going through a circuit; a switch turned on or a switch turned off; a way of saying true or false. All these examples are saying the same thing: a bit is a way of describing a system whose set of states is of size two.

A bit can be represented by two 2 by 1 matrices. We shall represent 0 – or better the state $|0\rangle$ as

$$|0\rangle \quad = \quad \begin{matrix} 0 \\ 1 \end{matrix} \begin{bmatrix} 1 \\ 0 \end{bmatrix}$$

We shall represent a 1, or state $|1\rangle$ as:

$$|1\rangle \quad = \quad \begin{matrix} 0 \\ 1 \end{matrix} \begin{bmatrix} 0 \\ 1 \end{bmatrix}$$

Since these are two different representations, we have an honest-to-goodness bit.

A bit is either in state $|0\rangle$ or in state $|1\rangle$. That was sufficient for the classical world. But that is not sufficient for the quantum world. In that world we have situations where we are in one state *and* in the other state simultaneously. In the quantum world we have systems where a switch is in a superposition of states on *and* off. So we define a "quantum bit" or a "qubit" as a way of describing a quantum system of dimension two. We shall represent any such qubit as a two by one matrix with complex numbers:

$$\begin{matrix} 0 \\ 1 \end{matrix} \begin{bmatrix} c_0 \\ c_1 \end{bmatrix}$$

where $|c_0|^2 + |c_1|^2 = 1$. Notice that a classical bit is a special type of qubit. $|c_0|^2$ is to be interpreted as the probability that after measuring the qubit, it will be found in state $|0\rangle$. $|c_1|^2$ is to be interpreted as the probability that after measuring the qubit it will be found in state $|1\rangle$. Whenever we measure a qubit, it automatically becomes a bit. So we shall never "see" a general qubit. Nevertheless, they do exist and they are the core of our tale. The power of quantum computers is due to the fact that a system can be in many states at the same time.

Following the normalization procedure that we learned in the last section, any element of $\mathbb{C}^2$ can be converted into a qubit. For example, the vector

$$V = \begin{bmatrix} 5+3i \\ 6i \end{bmatrix}$$

has norm

$$|V| = \sqrt{\langle V, V\rangle} = \sqrt{[5-3i, -6i] \begin{bmatrix} 5+3i \\ 6i \end{bmatrix}} = \sqrt{34+36} = \sqrt{70}.$$

So V describes the same physical state as the qubit

$$\frac{V}{\sqrt{70}} = \begin{bmatrix} \frac{5+3i}{\sqrt{70}} \\ \frac{6i}{\sqrt{70}} \end{bmatrix}.$$

After measuring the qubit $\frac{V}{\sqrt{70}}$, the probability of being in state $|0\rangle$ is $\frac{34}{70}$, and the probability of being in state $|1\rangle$ is $\frac{36}{70}$.

It is easy to see that the bits $|0\rangle$ and $|1\rangle$ are the canonical basis of $\mathbb{C}^2$. So any qubit can be written as

$$\begin{bmatrix} c_0 \\ c_1 \end{bmatrix} = c_0 \cdot \begin{bmatrix} 1 \\ 0 \end{bmatrix} + c_1 \cdot \begin{bmatrix} 0 \\ 1 \end{bmatrix} = c_0|0\rangle + c_1|1\rangle.$$

Let us look at several ways of writing different qubits. $\frac{1}{\sqrt{2}}\begin{bmatrix} 1 \\ 1 \end{bmatrix}$ can be written as

$$\begin{bmatrix} \frac{1}{\sqrt{2}} \\ \frac{1}{\sqrt{2}} \end{bmatrix} = \frac{1}{\sqrt{2}}|0\rangle + \frac{1}{\sqrt{2}}|1\rangle = \frac{|0\rangle + |1\rangle}{\sqrt{2}}.$$

Similarly $\frac{1}{\sqrt{2}}\begin{bmatrix} 1 \\ -1 \end{bmatrix}$ can be written as

$$\begin{bmatrix} \frac{1}{\sqrt{2}} \\ \frac{-1}{\sqrt{2}} \end{bmatrix} = \frac{1}{\sqrt{2}}|0\rangle - \frac{1}{\sqrt{2}}|1\rangle = \frac{|0\rangle - |1\rangle}{\sqrt{2}}.$$

It is important to realize that

$$\frac{|0\rangle + |1\rangle}{\sqrt{2}} = \frac{|1\rangle + |0\rangle}{\sqrt{2}}.$$

These are both ways of writing $\begin{bmatrix} \frac{1}{\sqrt{2}} \\ \frac{1}{\sqrt{2}} \end{bmatrix}$. In contrast,

$$\frac{|0\rangle - |1\rangle}{\sqrt{2}} \neq \frac{|1\rangle - |0\rangle}{\sqrt{2}}$$

The first state is $\begin{bmatrix} \frac{1}{\sqrt{2}} \\ -\frac{1}{\sqrt{2}} \end{bmatrix}$ and the second one is $\begin{bmatrix} -\frac{1}{\sqrt{2}} \\ \frac{1}{\sqrt{2}} \end{bmatrix}$. However the two states are related:

$$\frac{|0\rangle - |1\rangle}{\sqrt{2}} = (-1)\frac{|1\rangle - |0\rangle}{\sqrt{2}}.$$

How are qubits to be implemented? While this is not our focus, some examples of qubit implementations are given:

- An electron might be in one of two different orbits around a nucleus of an atom. (Ground state and excited state.)
- A photon might be in one of two different polarized states.
- A subatomic particle might be in spinning in one of two different directions.

There will be enough quantum indeterminacy and quantum superposition effects within all these systems to represent a qubit.

Computers with only one bit of storage are not very interesting. Similarly, we will need quantum devices with more then one qubit. Consider a byte. A typical byte might look like

$$01101011.$$

We might also write it as

$$\begin{bmatrix} 1 \\ 0 \end{bmatrix}, \begin{bmatrix} 0 \\ 1 \end{bmatrix}, \begin{bmatrix} 0 \\ 1 \end{bmatrix}, \begin{bmatrix} 1 \\ 0 \end{bmatrix}, \begin{bmatrix} 0 \\ 1 \end{bmatrix}, \begin{bmatrix} 1 \\ 0 \end{bmatrix}, \begin{bmatrix} 0 \\ 1 \end{bmatrix}, \begin{bmatrix} 0 \\ 1 \end{bmatrix}.$$

Previously, we learned that in order to combine systems one should use the tensor product. We can describe the above byte as

$$|0\rangle \otimes |1\rangle \otimes |1\rangle \otimes |0\rangle \otimes |1\rangle \otimes |0\rangle \otimes |1\rangle \otimes |1\rangle.$$

As a qubit, this is an element of

$$\mathbb{C}^2 \otimes \mathbb{C}^2 \otimes \mathbb{C}^2 \otimes \mathbb{C}^2 \otimes \mathbb{C}^2 \otimes \mathbb{C}^2 \otimes \mathbb{C}^2 \otimes \mathbb{C}^2.$$

This vector space can be written as $(\mathbb{C}^2)^{\otimes 8}$. This is a complex vector space of dimension $2^8 = 256$. Since there is only one complex vector space (up to isomorphism) of this dimension, this space is isomorphic to $\mathbb{C}^{256}$.

Our byte can be described as in yet another way: As a $2^8 = 256$ row vector

$$\begin{array}{c} 00000000 \\ 00000001 \\ \vdots \\ 01101010 \\ 01101011 \\ 01101100 \\ \vdots \\ 11111110 \\ 11111111 \end{array} \begin{bmatrix} 0 \\ 0 \\ \vdots \\ 0 \\ 1 \\ 0 \\ \vdots \\ 0 \\ 0 \end{bmatrix}.$$

This is fine for the classical world. However, for the quantum world, a general qubit can be written as

$$\begin{array}{c} 00000000 \\ 00000001 \\ \vdots \\ 01101010 \\ 01101011 \\ 01101100 \\ \vdots \\ 11111110 \\ 11111111 \end{array} \begin{bmatrix} c_0 \\ c_1 \\ \vdots \\ c_{106} \\ c_{107} \\ c_{108} \\ \vdots \\ c_{254} \\ c_{255} \end{bmatrix}$$

where $\sum_{i=0}^{255} |c_i|^2 = 1$.

In the classical world, you need to write the state of each of the eight bits. This amounts to writing eight bits. In the quantum world, a state of eight qubits is given by writing 256 complex numbers. This exponential growth was one of the reasons why researchers started thinking about quantum computing. If you wanted to emulate a quantum computer with a 64 qubit register, you would need to store $2^{64} = 18,446,744,073,709,551,616$ complex numbers. This is beyond our current ability.

Let us practice writing two qubits in ket notation. The qubits

$$\begin{array}{c} 00 \\ 01 \\ 10 \\ 11 \end{array} \begin{bmatrix} 0 \\ 1 \\ 0 \\ 0 \end{bmatrix}$$

can be written as $|0\rangle \otimes |1\rangle$.

Since the tensor product is understood, we might also write these qubits as $|0,1\rangle$ or $|01\rangle$. The qubit corresponding to

$$\frac{1}{\sqrt{3}}\begin{bmatrix}1\\0\\-1\\1\end{bmatrix}$$

can be written as

$$\frac{1}{\sqrt{3}}|00\rangle - \frac{1}{\sqrt{3}}|10\rangle + \frac{1}{\sqrt{3}}|11\rangle = \frac{|00\rangle - |10\rangle + |11\rangle}{\sqrt{3}}.$$

The tensor product of two states is not commutative.

$$|0\rangle \otimes |1\rangle = |0,1\rangle = |01\rangle \neq |10\rangle = |1,0\rangle = |1\rangle \otimes |0\rangle.$$

The first ket says that the first qubit is in state 0 and the second qubit is in state 1. The second ket says that first qubit is in state 1 and the second state is in state 0.

10.3.2 Classical Gates

Classical logical gates are ways of manipulating bits. Bits go into logical gates and bits come out. We represent n input bits as a 2^n by 1 matrix and m output bits as a 2^m by 1 matrix. How should we represent our logical gates? A 2^m by 2^n matrix takes a 2^n by 1 matrix and outputs a 2^m by 1 matrix. In symbols:

$$(2^m \times 2^n)(2^n \times 1) = (2^m \times 1).$$

So bits will be represented by column vectors and logic gates will be represented by matrices.

Let us do a simple example. Consider the NOT gate. NOT takes as input one bit, or a 2 by 1 matrix, and outputs one bit, or a 2 by 1 matrix. NOT of $|0\rangle$ equals $|1\rangle$ and NOT of $|1\rangle$ equals $|0\rangle$. The matrix

$$\text{NOT} = \begin{bmatrix}0 & 1\\1 & 0\end{bmatrix}.$$

This matrix satisfies

$$\begin{bmatrix}0 & 1\\1 & 0\end{bmatrix}\begin{bmatrix}1\\0\end{bmatrix} = \begin{bmatrix}0\\1\end{bmatrix} \qquad \begin{bmatrix}0 & 1\\1 & 0\end{bmatrix}\begin{bmatrix}0\\1\end{bmatrix} = \begin{bmatrix}1\\0\end{bmatrix},$$

which is exactly what we want.

What about the other gates? The other gates will be given by the following matrices:

AND	*NAND*	*OR*	*NOR*
$\begin{bmatrix}1 & 1 & 1 & 0\\0 & 0 & 0 & 1\end{bmatrix}$	$\begin{bmatrix}0 & 0 & 0 & 1\\1 & 1 & 1 & 0\end{bmatrix}$	$\begin{bmatrix}1 & 0 & 0 & 0\\0 & 1 & 1 & 1\end{bmatrix}$	$\begin{bmatrix}0 & 1 & 1 & 1\\1 & 0 & 0 & 0\end{bmatrix}$

When we perform a computation, often we have to carry out one operation followed by another.

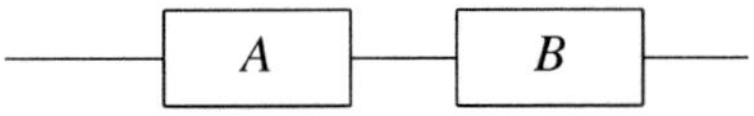

We shall call this performing "sequential" operations. If matrix A corresponds to performing an operation and matrix B corresponds to performing another operation, then the multiplication matrix $B \star A$ corresponds to performing the operation sequentially.

Besides sequential operations, there are "parallel" operations:

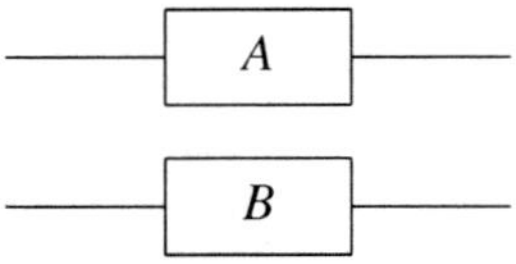

Here we are doing A to some bits and B to other bits. This will be represented by $A \otimes B$, the tensor product of two matrices.

Combination of sequential and parallel operations gates/matrices will be circuits. A starting point is the realization that the circuit

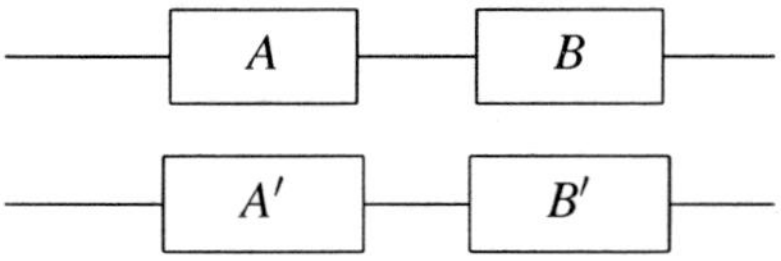

can be realized as

$$(B \star A) \otimes (B' \star A') = (B \otimes B') \star (A \otimes A').$$

We will of course have some really complicated matrices, but they will all be decomposable into the sequential and parallel composition of simple gates.

10.3.3 Quantum Gates

A quantum gate is simply any unitary matrix that manipulates qubits. There are some simple gates that are quantum gates.

The **Hadamard matrix**:

$$H = \frac{1}{\sqrt{2}} \begin{bmatrix} 1 & 1 \\ 1 & -1 \end{bmatrix} = \begin{bmatrix} \frac{1}{\sqrt{2}} & \frac{1}{\sqrt{2}} \\ \frac{1}{\sqrt{2}} & -\frac{1}{\sqrt{2}} \end{bmatrix}.$$

The Hadamard matrix is also its own inverse. As it turns out, the Hadamard matrix is one of the most important matrices in quantum computing.

Consider the following **controlled-not** gate.

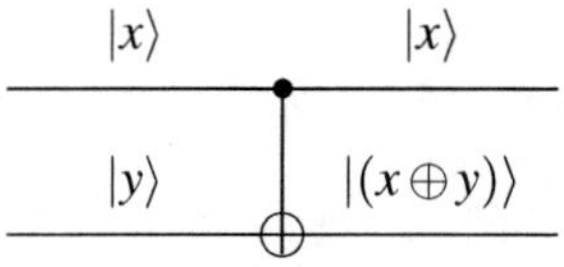

This gate has two inputs and has two outputs. The top input is the control bit. It controls what the output will be. If $|x\rangle = |0\rangle$, then the output of $|y\rangle$ will be the same as the input. If $|x\rangle = |1\rangle$ then the output of $|y\rangle$ will be the opposite. If we write the top qubit first and then the bottom qubit, then the controlled-not gate takes $|x,y\rangle$ to $|x, x\oplus y\rangle$ where $\oplus$ is the binary exclusive or operation.

The matrix that corresponds to this reversible gate is

$$\begin{array}{c} \\ 00 \\ 01 \\ 10 \\ 11 \end{array} \begin{array}{c} \begin{array}{cccc} 00 & 01 & 10 & 11 \end{array} \\ \begin{bmatrix} 1 & 0 & 0 & 0 \\ 0 & 1 & 0 & 0 \\ 0 & 0 & 0 & 1 \\ 0 & 0 & 1 & 0 \end{bmatrix} \end{array}.$$

One last piece of notation: we shall denote an observation (measurement) by the following "meter":

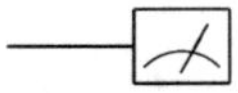

There are many other quantum gates, but we shall not need them for our work in the next section.

10.4 Deutsch's Algorithm

The simplest quantum algorithm is Deutsch's algorithm which is a nice short algorithm that solves a slightly contrived problem. This algorithm is concerned with functions from the set $\{0,1\}$ to the set $\{0,1\}$. There are four such functions which we might visualize as

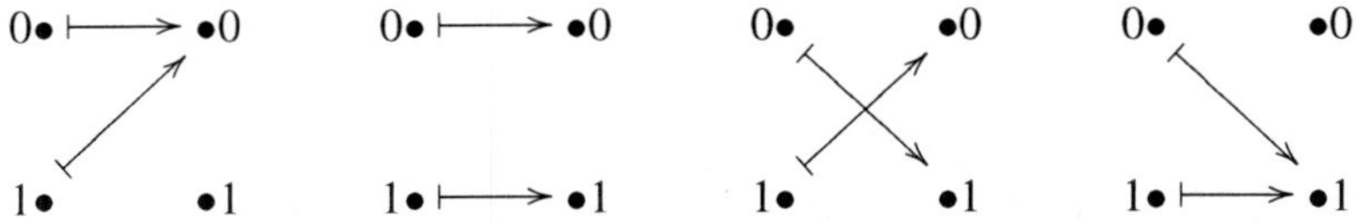

Call a function $f : \{0,1\} \longrightarrow \{0,1\}$, "balanced" if $f(0) \neq f(1)$). In contrast, call a function "constant" if $f(0) = f(1)$. Of the four functions, two are balanced and two are constant.

Deutsch's algorithm solves the following problem: Given a function $f : \{0,1\} \longrightarrow \{0,1\}$ as a black-box, where one can evaluate an input, but cannot "look inside" and "see" how the function is defined, tell if the function is balanced or constant.

With a classical computer, one would have to first evaluate f on an input, then evaluate f on the second input and then compare the outputs. The point is that with a classical computer, f must be evaluated twice. Can we do better?

A quantum computer can be in two states at one time. We shall use this superposition of states to evaluate both inputs at one time.

In classical computing, evaluating a given function f would correspond to performing the following operation

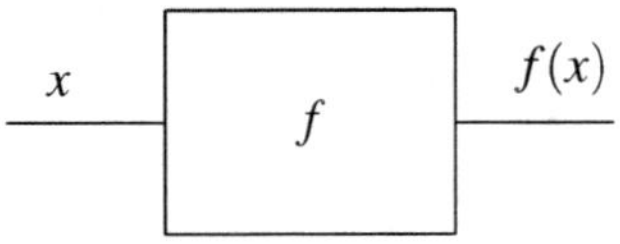

Such a function could be thought of as a matrix. The function

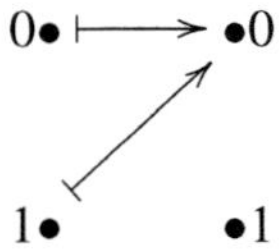

is equivalent to the matrix

$$\begin{array}{c c} & \begin{array}{cc} 0 & 1 \end{array} \\ \begin{array}{c} 0 \\ 1 \end{array} & \begin{bmatrix} 1 & 1 \\ 0 & 0 \end{bmatrix} \end{array}.$$

Multiplying either state $|0\rangle$ or state $|1\rangle$ on the right of this matrix would result in state $|0\rangle$. The column name is to be thought of as the input and the row name is to be thought of as the output.

However, this will not be good for a quantum system. For a quantum system we need a little something extra. A quantum system must be unitary (reversible). Given the output, we must be able to find the input. If f is the name of the function, then the following black-box U_f system will be the quantum gate that we shall employ to evaluate input:

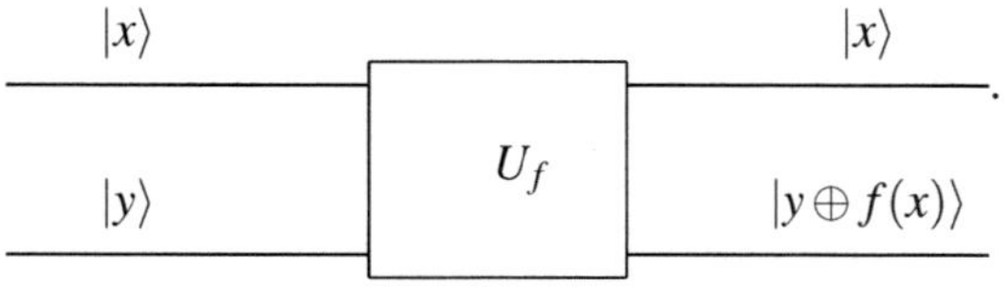

The top input, $|x\rangle$, will be the qubit value that one wishes to evaluate and the bottom input, $|y\rangle$, controls the output. The top output will be the same as the input qubit $|x\rangle$ and the bottom output will be the qubit $|y \oplus f(x)\rangle$ where $\oplus$ is XOR, the exclusive or operation. We are going to write from left to right the top qubit first and then the bottom. So we say that this function takes the state $|x,y\rangle$ to the state $|x, y \oplus f(x)\rangle$. If $y = 0$ this simplifies: $|x,0\rangle$ to $|x, 0 \oplus f(x)\rangle = |x, f(x)\rangle$. This gate is reversible by simply looking at the following system.

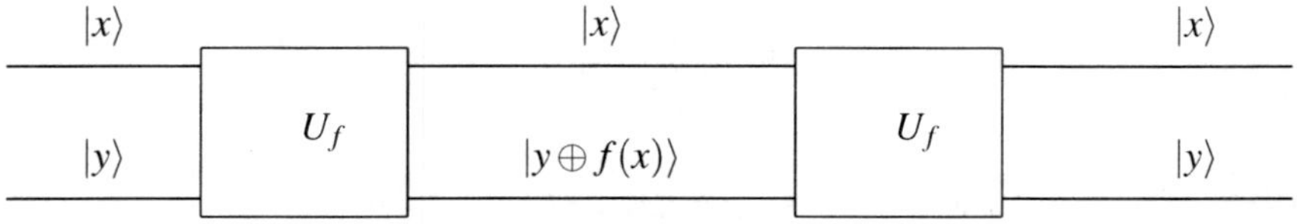

State $|x,y\rangle$ goes to $|x,y\oplus f(x)\rangle$ which further goes to

$$|x,(y\oplus f(x))\oplus f(x)\rangle = |x,y\oplus(f(x)\oplus f(x))\rangle = |x,y\oplus 0\rangle = |x,y\rangle.$$

In quantum systems, evaluating f is equivalent to multiplying a state of the input by a unitary matrix U_f. For the function

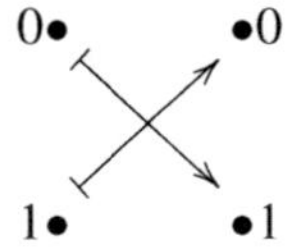

the corresponding unitary matrix is

$$\begin{array}{c} \\ 00 \\ 01 \\ 10 \\ 11 \end{array}\begin{array}{c} \begin{array}{cccc} 00 & 01 & 10 & 11 \end{array} \\ \begin{bmatrix} 0 & 1 & 0 & 0 \\ 1 & 0 & 0 & 0 \\ 0 & 0 & 1 & 0 \\ 0 & 0 & 0 & 1 \end{bmatrix} \end{array}.$$

Remember that the top column name correspond to the input $|x,y\rangle$ and the left-hand row name corresponds to the outputs $|x,y\oplus f(x)\rangle$. A 1 in the xy column and the $x'y'$ row means for input $|x,y\rangle$ the output will be $|x',y'\rangle$.

So we are given such a matrix that expresses a function but we cannot "look inside" the matrix to "see" how it is defined. We are asked to determine if the function is balanced or constant.

10.4.1 First Attempt

Let us take a first stab at a quantum algorithm to solve this problem. Rather than evaluating f twice, let us try our trick of superposition of states. Instead of having the top input to be either in state $|0\rangle$ or in state $|1\rangle$, we shall put the top input in state

$$\frac{|0\rangle + |1\rangle}{\sqrt{2}}$$

which is "half-way" $|0\rangle$ and "half-way" $|1\rangle$. The Hadamard matrix can place a qubit in such a state.

$$H|0\rangle = \begin{bmatrix} \frac{1}{\sqrt{2}} & \frac{1}{\sqrt{2}} \\ \frac{1}{\sqrt{2}} & -\frac{1}{\sqrt{2}} \end{bmatrix} \begin{bmatrix} 1 \\ 0 \end{bmatrix} = \begin{bmatrix} \frac{1}{\sqrt{2}} \\ \frac{1}{\sqrt{2}} \end{bmatrix} = \frac{|0\rangle + |1\rangle}{\sqrt{2}}.$$

The obvious (and not necessarily correct) state to put the bottom input is as state $|0\rangle$. And so we have:

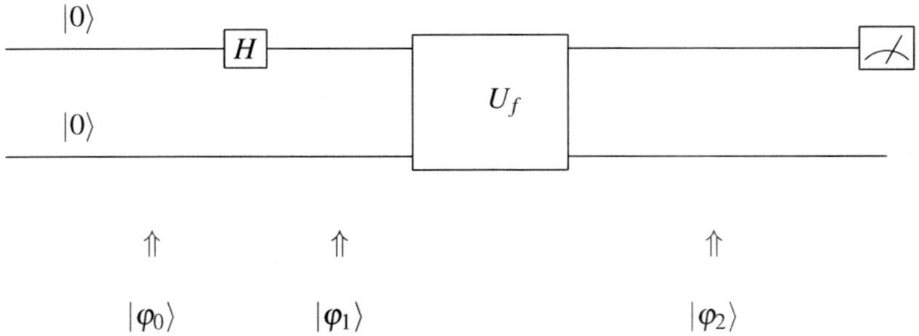

In terms of matrices this corresponds to

$$U_f(H \otimes I)(|0\rangle \otimes |0\rangle).$$

We shall carefully examine the states of the system at every point. The system starts in

$$|\varphi_0\rangle \quad = \quad |0\rangle \otimes |0\rangle \quad = \quad |0,0\rangle.$$

We then apply the Hadamard matrix only to the top input – leaving the bottom input alone – to get

$$|\varphi_1\rangle \quad = \quad \left[\frac{|0\rangle + |1\rangle}{\sqrt{2}}\right] |0\rangle \quad = \quad \frac{|0,0\rangle + |1,0\rangle}{\sqrt{2}}.$$

After multiplying with U_f we have

$$|\varphi_2\rangle \quad = \quad \frac{|0,f(0)\rangle + |1,f(1)\rangle}{\sqrt{2}}$$

For the function $0 \mapsto 1$ and $1 \mapsto 0$ the state $|\varphi_2\rangle$ would be

$$|\varphi_2\rangle = \begin{array}{c} \\ 00 \\ 01 \\ 10 \\ 11 \end{array} \begin{array}{c} \begin{array}{cccc} 00 & 01 & 10 & 11 \end{array} \\ \begin{bmatrix} 0 & 1 & 0 & 0 \\ 1 & 0 & 0 & 0 \\ 0 & 0 & 1 & 0 \\ 0 & 0 & 0 & 1 \end{bmatrix} \end{array} \begin{array}{c} 00 \\ 01 \\ 10 \\ 11 \end{array} \begin{bmatrix} \frac{1}{\sqrt{2}} \\ 0 \\ \frac{1}{\sqrt{2}} \\ 0 \end{bmatrix} = \begin{array}{c} 00 \\ 01 \\ 10 \\ 11 \end{array} \begin{bmatrix} 0 \\ \frac{1}{\sqrt{2}} \\ \frac{1}{\sqrt{2}} \\ 0 \end{bmatrix} = \frac{|0,1\rangle + |1,0\rangle}{\sqrt{2}}$$

If we measure the top qubit, there will be a 50–50 chance of finding it either in state $|0\rangle$ or in state $|1\rangle$. Similarly there is no real information to be gotten by measuring the bottom qubit. So the obvious algorithm does not work. We need another trick.

10.4.2 Second Attempt

Let us take another stab at solving our problem. Rather then leaving the bottom qubit in state $|0\rangle$, let us put it in the superposition state:

$$\frac{|0\rangle - |1\rangle}{\sqrt{2}} = \begin{bmatrix} \frac{1}{\sqrt{2}} \\ -\frac{1}{\sqrt{2}} \end{bmatrix}.$$

Notice the minus sign. Even though there is a negation, this state is also "half-way" in state $|0\rangle$ and "half-way" in state $|1\rangle$. The change of phase will help us get our desired results. We can get to this superposition of states by multiplying state $|1\rangle$ with the Hadamard matrix. We shall leave the top qubit as an ambiguous $|x\rangle$.

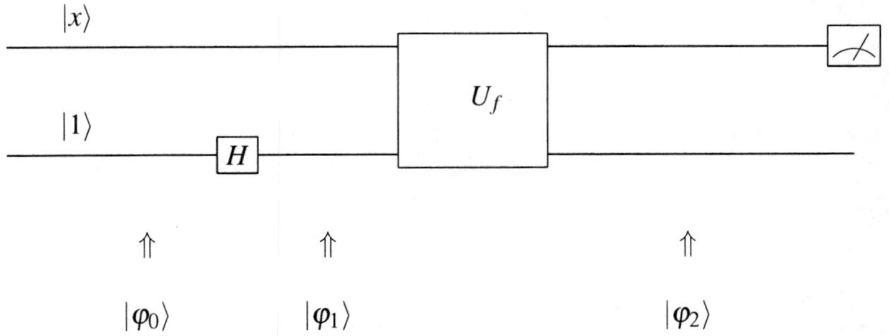

In terms of matrices, this becomes:

$$U_f(I \otimes H)|x,1\rangle.$$

Let us look carefully at how the states of the qubits change.

$$|\varphi_0\rangle \quad = \quad |x,1\rangle.$$

After the Hadamard matrix, we have

$$|\varphi_1\rangle \quad = \quad |x\rangle \left[\frac{|0\rangle - |1\rangle}{\sqrt{2}} \right] \quad = \quad \frac{|x,0\rangle - |x,1\rangle}{\sqrt{2}}.$$

Applying U_f we get

$$|\varphi_2\rangle \quad = \quad |x\rangle \left[\frac{|0 \oplus f(x)\rangle - |1 \oplus f(x)\rangle}{\sqrt{2}} \right] \quad = \quad |x\rangle \left[\frac{|f(x)\rangle - |\overline{f(x)}\rangle}{\sqrt{2}} \right]$$

where $\overline{f(x)}$ means the opposite of $f(x)$. And so we have

$$|\varphi_2\rangle = \begin{cases} |x\rangle \left[\frac{|0\rangle - |1\rangle}{\sqrt{2}}\right] & \text{if } f(x) = 0 \\ \\ |x\rangle \left[\frac{|1\rangle - |0\rangle}{\sqrt{2}}\right] & \text{if } f(x) = 1 \end{cases}.$$

Remembering that $a - b = (-1)(b - a)$ we might write this as

$$|\varphi_2\rangle = (-1)^{f(x)}|x\rangle \left[\frac{|0\rangle - |1\rangle}{\sqrt{2}}\right].$$

What would happen if we evaluate either the top or the bottom state? Again, this does not really help us. We do not gain any information. The top qubit will be in state $|x\rangle$ and the bottom qubit will – with equal probability – be either in state $|0\rangle$ or in state $|1\rangle$. We need something more.

10.4.3 Deutsch's Algorithm

Now let us combine both of these attempts to actually give Deutsch's algorithm. Deutsch's algorithm works by putting *both* the top and bottom qubits into a superposition. We will also put the results of the top qubit through a Hadamard matrix.

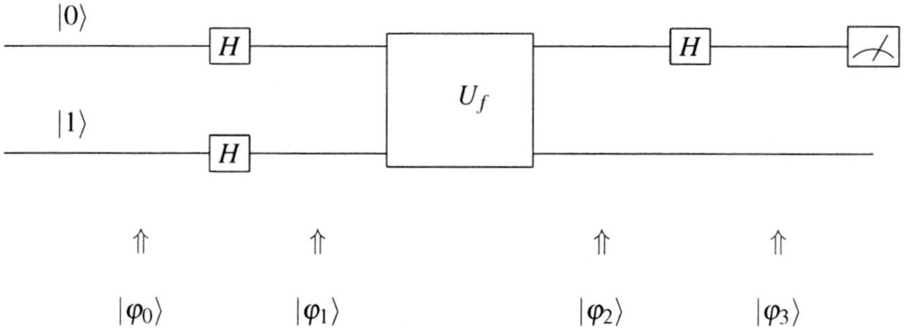

In terms of matrices this becomes:

$$(I \otimes H)U_f(H \otimes H)|0,1\rangle$$

At each point of the algorithm the states are as follows.

$$|\varphi_0\rangle = |0,1\rangle.$$

$$|\varphi_1\rangle = \left[\frac{|0\rangle + |1\rangle}{\sqrt{2}}\right]\left[\frac{|0\rangle - |1\rangle}{\sqrt{2}}\right] = \frac{+|0,0\rangle - |0,1\rangle + |1,0\rangle - |1,1\rangle}{2} =$$

$$\begin{matrix} 00 \\ 01 \\ 10 \\ 11 \end{matrix} \begin{bmatrix} +\frac{1}{2} \\ -\frac{1}{2} \\ +\frac{1}{2} \\ -\frac{1}{2} \end{bmatrix}.$$

We saw from our last attempt at solving this problem, that when we put the bottom qubit into a superposition and then multiply by U_f we will be in a superposition

$$(-1)^{f(x)}|x\rangle \left[\frac{|0\rangle - |1\rangle}{\sqrt{2}}\right].$$

Now with $|x\rangle$ in a superposition, we have

$$|\varphi_2\rangle = \left[\frac{(-1)^{f(0)}|0\rangle + (-1)^{f(1)}|1\rangle}{\sqrt{2}}\right]\left[\frac{|0\rangle - |1\rangle}{\sqrt{2}}\right].$$

Let us carefully look at

$$(-1)^{f(0)}|0\rangle + (-1)^{f(1)|1\rangle}.$$

If f is constant this becomes either

$$+1(|0\rangle + |1\rangle) \text{ or } -1(|0\rangle + |1\rangle)$$

(depending on being constantly 0 or constantly 1.) If f is balanced it becomes either

$$+1(|0\rangle - |1\rangle) \text{ or } -1(|0\rangle - |1\rangle)$$

(depending on which way it is balanced.) Summing up, we have that

$$|\varphi_2\rangle = \begin{cases} (\pm 1)\left[\frac{|0\rangle + |1\rangle}{\sqrt{2}}\right]\left[\frac{|0\rangle - |1\rangle}{\sqrt{2}}\right] & \text{if } f \text{ is constant} \\ (\pm 1)\left[\frac{|0\rangle - |1\rangle}{\sqrt{2}}\right]\left[\frac{|0\rangle - |1\rangle}{\sqrt{2}}\right] & \text{if } f \text{ is balanced} \end{cases}$$

Remembering that the Hadamard matrix is its own inverse and takes $\frac{|0\rangle + |1\rangle}{\sqrt{2}}$ to $|0\rangle$ and takes $\frac{|0\rangle - |1\rangle}{\sqrt{2}}$ to $|1\rangle$, we apply the Hadamard matrix to the top qubit to get

$$|\varphi_3\rangle = \begin{cases} (\pm 1)|0\rangle \left[\frac{|0\rangle - |1\rangle}{\sqrt{2}}\right] & \text{if } f \text{ is constant} \\ (\pm 1)|1\rangle \left[\frac{|0\rangle - |1\rangle}{\sqrt{2}}\right] & \text{if } f \text{ is balanced} \end{cases}$$

Now we simply measure the top qubit. If it is in state $|0\rangle$, then we know that f is a constant function, otherwise it is a balanced function. And we did this all with only one evaluation as opposed to the two evaluations that a classical algorithm demands.

Notice that although the ± 1 tells us even more information, namely which of the two balanced functions or which of the two constant functions, we are not privy to this information. Upon measuring, if the function is balanced, we will measure $|1\rangle$ regardless if the state was $(-1)|1\rangle$ or $(+1)|1\rangle$.

In conclusion, we have solved a problem that a classical computer would require two function evaluations. Deutsch's quantum algorithm solved the same problem with one function evaluation. Other quantum algorithms are built-up with similar ideas to the ones presented here.

References

1. Chester M. *Primer of Quantum Mechanics (Physics)*. Dover Publications, Mineola, NY, 2003.
2. Dirac P. A. M. *The Principles of Quantum Mechanics (The International Series of Monographs on Physics)*. Oxford University Press, USA, 1982.
3. Feynman R. P. *Feynman Lectures On Physics (3 Volume Set)*. Addison Wesley Longman, Boston, MA, 1970.
4. Martin J. L. *Basic Quantum Mechanics (Oxford Physics Series)*. Oxford University Press, USA, 1982.
5. Polkinghorne J. *Quantum Theory, A Very Short Introduction*. Oxford University Press, USA, 2002.
6. Sudbery A. *Quantum Mechanics and the Particles of Nature: An Outline for Mathematicians*. Cambridge University Press, Cambridge, 1986.
7. White R. L. *Basic Quantum Mechanics (McGraw-Hill physical and Quantum Electronics Series)*. McGraw-Hill, New York, NY, 1966.

Part IV
Logic, Agency and Games

Chapter 11
Logic Games: From Tools to Models of Interaction

Johan van Benthem

11.1 From Products to Activities: Logic in Games

Logical dynamics Logic is often taken to be about propositions, truth, and proofs: abstract objects in Heaven, and their Platonic properties. But the discipline arose in Antiquity by studying *activities* on Earth: dialogue and argumentation. And the very terminology of logic still has a double meaning. "Statement" is both a dynamic activity and the static product of that activity, "proof" is a procedure of establishing a claim and a formal record of that procedure, etc. These activities are usually kept in the background, as mainly didactical motivation. Placing the dynamics at centre stage in logical theory is the program of "logical dynamics" [37].

Logic games The best source in modern logic for structured activities are logic games. These have been around since the Middle Ages, with the "Obligatio" debates of authors like Walter Burleigh. Mathematical logic games were defined in the 1950s and 1960s by Lorenzen [25], Ehrenfeucht-Fraïssé [12], Hintikka [17], and Parikh (cf. [29]) – not accidentally in conjunction with the new wave in game theory. Today, two-person games are in wide use for logical tasks of evaluation of propositions in a model, comparing models, building models for assertions, or constructing proofs for claims in argumentation. And new varieties are still appearing. For a survey with many references, see [39].

This paper is based on the lecture notes series [38]. It presents logic games in one setting, to show their role as a model of "intelligent interaction" which fits well with game theory. Our presentation is elementary, and mostly in the nature of a survey. We use existing results, plus an occasional new observation connecting up relevant strands, to paint a total picture. A word of warning to the reader is in order here. This paper is not a didactical introduction to logic games, but we hope to motivate the

Johan van Benthem
Institute for Logic, Language and Computation (ILLC), University of Amsterdam, Amsterdam, The Netherlands and Spring Quarters: Department of Philosophy, Stanford University, Stanford, CA 94305, USA, e-mail: johan@science.uva.nl

J. van Benthem et al. (eds.), *Proof, Computation and Agency*, Synthese Library 352,
DOI 10.1007/978-94-007-0080-2_11,

reader to seek further information in the many references given here. Conversely, for those who already know the subject, we hope to add some fresh perspectives and links to other areas of research.

Before doing all this, let us first look at some basic examples.

11.2 The Basic Logic Games

11.2.1 Gamification

In principle, any logical task can be "gamified", by pulling it apart into roles for two players whose dynamic interaction tests the notion involved. Interaction involves dependence, and gamification works once we have an interplay between a universal quantifier **A** ("Abelard", "Adam", "Alter") and an existential **E** ("Eloise", "Eve", "Ego"). Leibniz already explained the meaning of quantifier forms $\forall\varepsilon\exists\delta$ expressing dependence in mathematical settings in terms of a game:

> A challenger $\forall$ chooses some number ε at his discretion,
> and the defender $\exists$ has to produce a suitable response δ.

In what follows, we give some sketches of major logic games, referring to the literature for more detailed exposition and references: cf. [39, 45]. Our aim is to make the reader aware of the ubiquity of these games, and give some examples for the general discussions later on in this paper.

11.2.2 Argumentation

Perhaps the oldest example of a logic game is argumentation: one makes a claim against an opponent, upholding it in the face of objections. We all experience its game-like character of having to say the right thing at the right time, and also, the bitter taste of defeat when we have talked ourselves into a corner, contradicting our earlier statements. The latter are the typical losing stages in argumentation – being at the same time wins for the other player. Precise dialogue games for argumentation, in the style of Lorenzen, are found in [32].

Here is the key game-theoretic feature of argumentation – which probably led to the Greeks discovering logical patterns of reasoning in the first place. Roughly speaking,

> *Logically valid propositions* φ will be precisely those
> for which their proponent Eve has a *winning strategy*:

that is, a way of choosing her conversational moves against opponent Adam which guarantees that she never loses – no matter how Adam attacks her claim.

Thus, Eve's winning strategy is a dynamic counterpart of a logical *proof* for the proposition φ. Now there may be more than one proof for a claim, and this reflects the diversity of rational behavior. Players may have more than one strategy to win a debate. This adversarial strategy-based style of analysis holds across a wide range of logic games. (For representative studies of modern argumentation theory, cf. [13, 54].)

But players are on a par in games, and there is no need to glorify one over the other. It is their interaction which really matters. In particular, since not all φ are valid, there are argumentation games where Adam has a winning strategy for involving Eve in self-contradictions. Section 11.2.4 below takes up what these look like.

11.2.3 Obligation

Most conversation is not about argumentation. We tend to believe things people tell us – even implausible ones like "I love you" – as long as they are *consistent*. Only in special settings we will be challenged to prove our assertions, say, in a juridical procedure, or in teaching mathematics. But maintaining consistency is itself a major logical task! Medieval training disputations often had a form like this:

> Eve has to maintain consistency when Adam confronts her with successive assertions φ, each of which she has to accept or reject. In the former case, φ is added to her cumulative store of commitments, otherwise $\neg\varphi$ is added. Eve loses if at any stage, her commitments become inconsistent.

One may ask whether this sort of logic exam is fair. And indeed, in principle,

> Eve always has a *winning strategy*, being a string of YES/NO answers to any sequence of propositions, keeping her commitments consistent.

To see this, Eve's strategies for passing the exam are correlated with logical models: situations that make all her commitments true, and indeed any model will help her pass the test. Strategic insights into consistency are, alas, also logical skills that help us misinform and lie. For details of medieval disputations, involving Eve's initial knowledge, and an option of giving a third response of "doubt" – cf. [11].

11.2.4 Model Checking

Arguably the most basic logic game today occurs in a different setting. Let a first-order assertion φ be made about a model $\mathbf{M}$ and variable assignment s. A game *game* $(\varphi, \mathbf{M}, s)$ of semantic evaluation or "model checking" lets a "Verifier" Eve claim the truth of φ, while a "Falsifier" Adam defends its falsity:

With atomic φ one checks who is right about $\mathbf{M}, s$, and that player wins. Disjunctions are a *choice* for Eve, and play goes on with the disjunct chosen. Conjunctions are an initial choice for Adam. Negations trigger a *role switch*, with all turns interchanged between players. Existential quantifiers $\exists x\varphi$ are a move for Eve who picks a witness object d in $\mathbf{M}$, and play continues with the formula $\varphi(d)$. Universal quantifiers are Adam's choice of a challenge object.

The game for the formula $\forall x \exists y.\, x \neq y$ in a model with two distinct objects s, t is this (writing $\mathbf{A}$ for "Adam" and $\mathbf{E}$ for "Eve"):

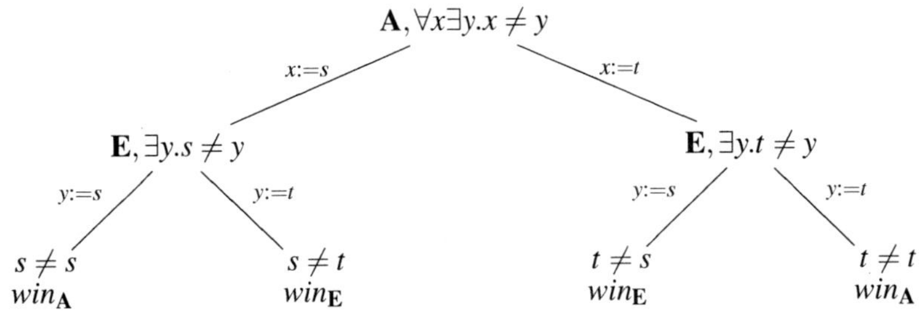

Note that $\mathbf{E}$ has a winning strategy here. Essentially, this perspective pulls one standard logical formula, say $\forall x \exists y \forall z \exists u \varphi(x,y,z,u)$ apart into an interactive alternation $\forall_1 x \exists_2 y \forall_1 z \exists_2 u \varphi(x,y,z,u)$. A good account of evaluation games is [19], while [34] has variants for fixed-point languages in computer science. Again, the central notion involves winning strategies [17]:

Fact 11.1 *For all* $\mathbf{M}, s, \varphi$, *the following assertions are equivalent:*

1. $\mathbf{M}, s \models \varphi$,
2. Eve has a winning strategy in game $(\varphi, \mathbf{M}, s)$.

Adam has a winning strategy if φ is false in $\mathbf{M}$. Different winning strategies encode different *reasons* for the truth or falsity of an assertion φ. "Reasons" are unusual logical objects – but with quantifier combinations such as a true $\forall x \exists y \forall z \exists u \varphi(x,y,z,u)$, one can think of them as bunches of *Skolem functions* $f(x), g(x,z)$ of the right arities providing Eve with her winning response in φ to Adam's successive choices.

11.2.5 Model Construction

The evaluation task of checking if $\mathbf{M}, s \models \varphi$ starts from a given model and formula. But in the earlier task of checking consistency, only assertions are given, with the satisfiability question if one can find a model for these. This suggests a *model construction game* between a "Builder" Eve who tries to create a model making some initial assertions true and others false, and a "Critic" Adam who raises objections, making sure that every building task gets scheduled. In particular, Critic can force

Builder to choose when a disjunct is to be made true, and he can keep calling new instances of initial universal quantifiers as Builder puts new objects into the model under construction. Builder loses at any stage if her current schedule tells her to make the same formula both true and false.

A precise format for the game arises by gamifying *semantic tableaus* [47], with decomposition rules for logical operators as game moves. The result is

Fact 11.2 *The following assertions are equivalent in tableau construction games:*

1. *A given set of first-order formulas Σ has a model,*
2. *Builder has a winning strategy in the construction game for Σ.*

Unlike first-order evaluation games, whose depth is bounded by the operator depth of the initial formula φ, construction games can have infinite runs. The reason is that some first-order formulas have only infinite models, making Builder go on forever. Again, there is a match between Builder's winning strategies and different *models* (if any) for Σ. More sophisticated construction games are used in [21].

If Builder's winning strategies are like models, what about Critic's? The latter are guaranteed ways of blocking any construction attempt. It can be shown that these are essentially *proofs* of the negation of the initial assertion. Thus, the construction game is like the earlier argumentation game, when we reformulate the roles. Critic tries to prove some initial assertion, while Builder is looking for a *counter-example*. This answers our question in Section 11.2.2 about Adam. His winning strategies correlate with counter-examples to the claim by Proponent. Thus, argumentation and model construction games are really two takes on the same logical process.

11.2.6 Model Comparison

In addition to model checking and satisfiability testing, other basic tasks for a modern logical system have to do with its expressive power. We measure the latter by seeing which models can be distinguished by our language. The most widely used logic game performs just this task. *Ehrenfeucht-Fraïssé* games cast Eve as a "Duplicator" who claims that two models $\mathbf{M}, \mathbf{N}$ are similar, while the "Spoiler" Adam claims they are different. Each round of the game starts with pointed models $(\mathbf{M}, \mathbf{a}) - (\mathbf{N}, \mathbf{b})$, where $\mathbf{a}, \mathbf{b}$ are tuples of objects chosen according to the following procedure:

> In each round, Spoiler chooses a model $\mathbf{M}$ or $\mathbf{N}$, and an object x in it, Duplicator then chooses a corresponding object y in the other model; and the link $x - y$ is then added to the current match $\mathbf{a} - \mathbf{b}$

Duplicator loses whenever the function from $\mathbf{M}$ into $\mathbf{N}$ defined by the current match of objects is no longer a partial isomorphism between the two models. She wins those runs of the game where this failure never occurs. We can play over a fixed finite number of rounds, or forever. An excellent textbook on Ehrenfeucht-Fraïssé games is [10]. Here is the key result about the method:

Fact 11.3 *For all models* **M**, **N**, *the following are equivalent:*

1. *Duplicator has a winning strategy in a k-round comparison game,*
2. **M**, **N** *satisfy the same first-order sentences up to quantifier depth k.*

In games with no finite bound, Duplicator's winning strategies match up with *potential isomorphisms* between **M**, **N**: a notion of structural similarity for models. But as always in logic games, the viewpoint of the other player is of independent interest. In the k-round model comparison game, a winning strategy for Spoiler is essentially a first-order formula φ of depth $\leq k$ which is true in **M** and false in **N**. We will make this more precise in Section 11.3. This reflects the original goal of testing the expressive power of our language. In this dual view of the game, then, Spoiler is the "positive" player, claiming the language is rich, while Duplicator claims it is poor. A state-of-the-art exposition of model comparison games is [36].

11.2.7 Other Logic Games

New logic games are still emerging today. For example, [20] study when a given abstract relational algebra **A** is representable as an algebra **S** of binary set relations over some set of individual objects U. Thus, we want to build a model $(U, \mathbf{S})$ standing in some isomorphism-like relationship E to the model **A**. Mixing ideas from model construction and model comparison games, the authors let players create "networks", as stages of a representation in progress, with Builder responding to challenges made by Critic. An abstract relational algebra **A** is representable iff Builder has a winning strategy in this representation game. The result is a perspicuous new axiomatization of the representable relational algebras.

After this survey of logic games, we now turn to their general features.

11.3 The Unifying Role of Strategies

11.3.1 Strategies as a Unifying Notion

Our games all had an "Adequacy Theorem" stating that some standard notion obtains (truth, satisfiability, potential isomorphism) iff some designated player has a winning strategy. Thus, the typically game-theoretic notion of a strategy becomes a unifying idea across logic. This leads to surprising connections. For example, in one and the same type of game, viz. that for model construction, we may encounter *proofs* as winning strategies for Critic, and *models* as winning strategies for Builder. Thus, logically very different notions turn out cousins after all. Sometimes also, strategies are new citizens asking for recognition in logic, such as "semantic reasons" for truth or falsity in first-order evaluation games. These analogies suggest that underlying logic, there is a calculus of strategies, for combining them and proving their basic properties. We will return to this issue in Sections 11.6 and 11.11.

11.3.2 ∃-sickness, and Its Cure

Despite their crucial role, strategies are often hidden in Adequacy Theorems. For example, truth amounts to the existence of a winning strategy for Verifier in an evaluation game, but we are not told how. Indeed, there is a wide-spread disease of *∃-sickness*: the wilful hiding of available specific information under existential quantifiers. Sure symptoms are indefinite articles "a", or modal affixes "-ility": cf. "having a strategy" or "winnab-ility". ∃-sickness also afflicts completeness theorems relating "provability" to validity, instead of a more informative match from proofs to semantic structures (but see the strong completeness theorems in [3]). An early case is in [5], who pointed out the self-inflicted problems of tense logic taking the past tense in "Lida fell down the stairs" as "at *some time* in the past", thereby losing the particular episode we have in mind. Our point here is not that we want some constructive reading of the existential quantifier. Indeed, curing ∃-sickness may well involve classical logic. The point is rather the hiding of relevant information when it is in fact explicitly available.

Fortunately, the disease is often cured with a little exercise. Our first illustration shows how to make an existential quantifier explicit by analyzing a standard proof - in this case, adequacy for model comparison games – making the strategies explicit:

Theorem 11.1 *There is an explicit correspondence between*

1. *Winning strategies for* **S** *in the k-round comparison game for* **M**, **N**
2. *First-order sentences* φ *of quantifier depth k with* $\mathbf{M} \models \varphi$, *not* $\mathbf{N} \models \varphi$

Proof. From 2 to 1. Every such "difference formula" φ of quantifier depth k defines a winning strategy for Spoiler in a k-round game between arbitrary models. Each round $k-m$ starts with a match between linked objects chosen so far which differ on some subformula ψ of φ with quantifier depth $k-m$. By straightforward Boolean analysis, **S** then finds some existential subformula $\exists x.\ \alpha$ of ψ with a matrix formula α of quantifier depth $k-m-1$ on which the two models disagree. **S**'s next choice is a witness in that model of the two where $\exists x.\ \alpha$ holds.

From 1 to 2. Each winning strategy σ for Spoiler induces a distinguishing formula of proper depth. Let **S** make his first choice d in model **M** according to σ – and write down an existential quantifier for d. Our formula under construction will be true in **M**, and false in **N**. We know that each choice of Duplicator for a corresponding object e in **N** gives a winning position for **S** in all remaining $(k-1)$-round games starting from an initial match $d-e$. By the inductive hypothesis, these induce distinguishing formulas of depth $k-1$. Now, the Finiteness Lemma for first-order logic over a fixed finite relational signature says that, for any fixed set of free variables and fixed quantifier depth, only finitely many non-equivalent formulas exist. In particular, only finitely many of the above distinguishing formulas can occur modulo logical equivalence. Some of these will start with "their" first quantifier in **M** (say $A_1,\ldots,A_r$) others in **N** (say $B_1,\ldots,B_s$). The total distinguishing formula for strategy σ is then the **M**-true assertion $\exists x. (A_1 \wedge \ldots A_r \wedge \neg B_1 \ldots \neg B_s)$.

Thus, Spoiler's winning strategies in a comparison game correspond to formulas, logical objects of prime interest. For Duplicator, the objects corresponding to her winning strategies might be called "analogies", of a finite quality measured by the game length k. They are cut-off versions of the *potential isomorphisms* widely used in logical model theory.

Of course, even Theorem 11.1 is still $\exists$-sick! But the outer existential quantifier in its formulation may be harmless, in that its instantiation is *the proof.*

Our second illustration shows another way of high-lighting strategies, by analyzing the available *number of them.* Consider verification games for propositional logic. Here is an elementary, but perhaps, unusual observation on counting strategies.

Fact 11.4 *One can count the number of verifying (falsifying) strategies, say* $\#(\mathbf{V},\varphi)$, $\#(\mathbf{F},\varphi)$ *for any propositional formula* φ *as follows:*

$$\begin{array}{ll}
\#(\mathbf{V},p) = 1 & \#(\mathbf{F},p) = 1 \\
\#(\mathbf{V},\neg\varphi) = \#(\mathbf{F},\varphi) & \#(\mathbf{F},\neg\varphi) = \#(\mathbf{V},\varphi) \\
\#(\mathbf{V},\varphi\vee\psi) = 2\#(\mathbf{V},\varphi)\cdot\#(\mathbf{V},\psi) & \#(\mathbf{F},\varphi\vee\psi) = \#(\mathbf{F},\varphi)\cdot\#(\mathbf{F},\psi) \\
\#(\mathbf{V},\varphi\wedge\psi) = \#(\mathbf{V},\varphi)\cdot\#(\mathbf{V},\psi) & \#(\mathbf{F},\varphi\wedge\psi) = 2\#(\mathbf{F},\varphi)\cdot\#(\mathbf{F},\psi)
\end{array}$$

Proof. The rationale for these clauses is immediate from the standard definition of strategies in game trees as functions assigning unique moves to players' turns.

Such counting is much more complex with first-order evaluation games.

11.3.3 Strategies: Actions or Powers?

Strategies are stepwise instructions for players to act. This detailed level of game structure was suppressed by existential

quantifiers of "having a strategy". But upon reflection, one person's $\exists$-sickness may be another's sanity! In games, we are sometimes not interested in details of moves and actions, but just in the *control* that players have over possible *outcomes.* For example, that Eve has a winning strategy really says that it is within her power – whatever Adam does – to make sure that the game ends in some specific set of runs or outcomes, designated as "winning". And players may have further powers over outcomes in games: say, via losing strategies, or strategies that guarantee long runs. Such powers provide a natural level for describing influence in social settings [30].

This coarser level of control raises a general question. At what level do we want to describe games – in terms of their global outcomes, or more detailed local actions? This choice is one of many general game-theoretic issues lying behind logic games.

11.3.4 From Logic Games to Game Theory

Logic games, though a very specialized class of activities, high-light issues which concern all games: playing at cards, competing in markets, or engaging in warfare (not *that* far removed from Academia). Some of these issues have arisen independently in game theory, others seem new. (A compact lucid source on game theory is [27].) In our next sections, we look at some pervasive ones – with heuristics inspired by the evaluation games of Section 11.2.4. An obvious bridge from logic to game theory are Adequacy Theorems like the one relating the logical notion of truth with the game-theoretic notion of a strategy. Unimaginative people interpret such results as a "kiss of death" for the game-theoretic stance, as it "just restates" what we know from standard logic already. But the opposite is true for the unprejudiced reader!

11.4 Game Equivalences and Game Languages

11.4.1 When Are Two Games the Same?

To raise our first question in a simple manner, consider the propositional distribution law for conjunction over disjunction:

$$p \wedge (q \vee r) \leftrightarrow (p \wedge q) \vee (p \wedge r)$$

The two finite trees in the following figure correspond to evaluation games for the two propositional formulas involved, letting **A** stand for Falsifier and **E** for Verifier. This picture raises the following intuitive, and yet fundamental, question

When are two games the same?

In particular, does the logical validity of propositional Distribution mean that the pictured games are the same in some natural sense?

$p \wedge (q \vee r)$ $(p \wedge q) \vee (p \wedge r)$

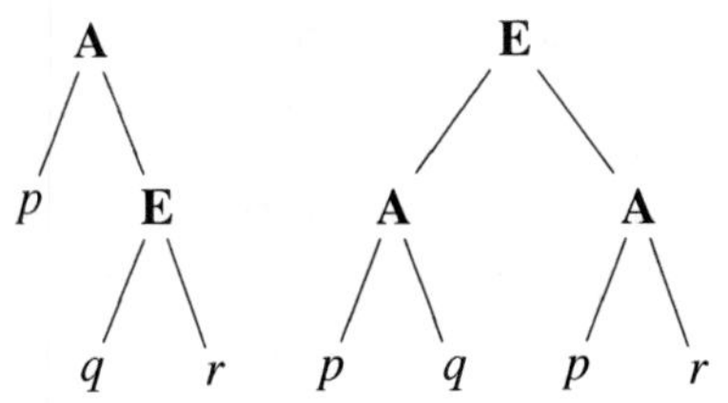

Clearly, the answer depends on our viewpoint:

I *If we focus on turns and moves, then the two games are not equivalent:* they differ in "etiquette" (who gets to play first) and in choice structure.

This is the level of what game theorists call "extensive games", with the familiar tree pictures involving details of choices and actions. But there are also "strategic forms" of games, where we are just interested in listing outcomes that players can control. For example, the fact that the order of players' turns is reversed by applying Distribution is immaterial then. At the latter level, the answer to our question becomes "Yes":

II *Both players have the same powers of achieving outcomes in both games*:
A can force the outcome to fall in the sets $\{p\}, \{q, r\}$
E can force the outcome to fall in the sets $\{p, q\}, \{p, r\}$.

Here, a player's *powers* are those sets U of outcomes for which she has a *strategy* making sure that the game will end inside U, no matter what the other player does. On the left, **A** has strategies "left" and "right", yielding powers $\{p\}$ and $\{q, r\}$. Player **E** also has two strategies, yielding powers $\{p, q\}, \{p, r\}$. On the right, **E** has two strategies "left" and "right", which give the same powers for **E** as on the left. By contrast, player **A** now has four strategies, which may be written ad-hoc as

"left: L, right: L", "left: L, right: R", "left: R, right: L", "left: R, right: R"

The first and fourth give the same powers for **A** as on the left, while the second and third strategy produce merely weaker powers subsumed by the former.

11.4.2 Game Equivalences and Game Languages

The general issue here are natural equivalences between games, setting coarser or finer levels of detail. Reference [41] uses an analogy with theories of computation, and *process equivalences* such as modal bisimulation. In particular, outcome-control views are like black-box input-output views of processes, while extensive games lie closer to views of computation endowing processes with richer internal states, including choices. Thus, structure theory of games is much like general multi-agent process theory.

As usual in mathematics, however, invariance relations between structures are one side of a coin. The other side are the matching properties of games one wants to define. For example, for strategic games, it does not matter in which schedule **E**, **A** took their turns: for extensive games, this is a relevant property. Such properties are expressed in *game languages* appropriate to the chosen "equivalence level". For example, [41] shows that a good language for describing extensive game forms of perfect information is modal logic, plus fixed-point extensions like the modal μ-calculus. But the appropriate language for describing players' outcome powers only are modal logics over neighbourhood semantics, in the spirit of [28].

11.4.3 Logic and Games: The Plot Thickens

We have reached a delicate point in our story. We started with specific games for logical notions. But now, logic enters in a different guise, as different description levels for games correspond to different languages and associated logics. Thus, in one of those happy Hegelian inversions, in addition to *logic games*, there are also *game logics* describing games in general.

11.5 Players' Powers and Modal Forcing Languages

In this section, we focus on the input–output level of outcomes and players' powers – broadening the connections between logic and game theory. We start with the global level for games, since it seems most congenial to logic games (see the more elaborate discussion in Section 11.9.2).

11.5.1 Determinacy

In evaluation games, Verifier has a winning strategy if the relevant formula φ is true, and Falsifier has a winning strategy if φ is false. This means that these games have the following important game-theoretical property:

Fact 11.5 *Evaluation games are determined: one of the two players in game* $(\varphi, \mathbf{M}, s)$ *must always have a winning strategy.*

A general proof uses *Zermelo's Theorem* which says all zero-sum two-player games of *finite depth* are determined. Indeed all infinite games with topologically open winning conditions for one player are determined by the Gale-Stewart Theorem. This explains why our games of model construction or model comparison are determined, even though runs may be infinite. Critic (Spoiler) have open winning conditions, as Builder's (Duplicator's) failures always arise at some finite stage. Finally, Martin's Theorem says that all infinite games are determined with winning conditions in the Borel Hierarchy of sets. Non-open Borel winning conditions occur with some games in computer science. Examples include *fairness* of runs for interactive game systems.

With non-Borel conventions, infinite games can become non-determined. We display one example, to make a general point about players' powers later on (Section 11.8.2). Take any free ultrafilter $\mathbf{U}^*$ on the natural numbers. Two players pick successive neighbouring closed intervals, of any finite sizes, producing a succession like this:

$$\mathbf{A} : [0, n_1], \text{ with } n_1 > 0, \quad \mathbf{E} : [n_1 + 1, n_2], \text{ with } n_2 > n_1 + 1, \quad \text{etc.}$$

E wins if the union of all her intervals is in $\mathbf{U}^*$, otherwise, **A** wins. It is easy to see that winning sets in this game are not open for either player.

Fact 11.6 *The interval selection game is not determined.*

Proof. If player **A** had a winning strategy, **E** could use that with a one-step delay to copy **A**'s responses to her own moves disguised as her moves. Both resulting sets of intervals (disjoint up to some finite initial segment!) would then have their unions in $\mathbf{U}^*$: which cannot be, as $\mathbf{U}^*$ was free. Likewise, **E** has no winning strategy.

There is a flourishing literature on determined games in descriptive set theory (cf. [26]), but we will concentrate on more general game issues here.
Remark This "strategy-stealing" argument is of interest per se, and should be formalizable in a suitable logic for explicit strategies.

11.5.2 Powers and Representation

There is more to players' powers, even in logic games, than just abilities to win. Consider the game in Section 11.2.4 for $\forall x \exists y.\ x \neq y$ in a first-order model with two distinct objects s, t – where we now number outcomes:

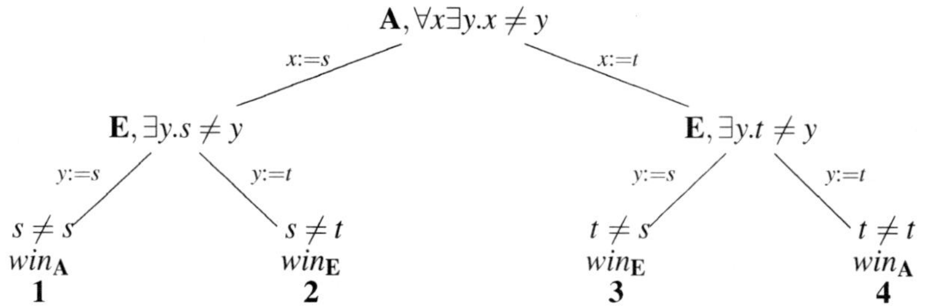

That $\forall x \exists y.Rxy$ is true is reflected in player **E**'s winning strategy "choose the object different from that chosen by **A**". But players have more strategies in this game, and calculating as in Section 11.4.1 we get their true powers:

A can force the sets $\{1,2\}, \{3,4\}$
E can force the sets $\{1,3\}, \{1,4\}, \{2,3\}, \{2,4\}$

Power families like these satisfy the following properties, which are reminiscent of the definition of an ultrafilter, but "spread out" over two players:

Monotonicity If Y is a power of **i** and $Y \subseteq Z$, then Z is also a power of **i**.
Consistency If Y is a power of **A** and Z a power of **E**, then Y, Z overlap.
Determinacy If Y is not a power of **A** (**E**), then its set complement $\overline{Y}$ in the total space of outcomes is a power of **E** (**A**).

These are also all the relevant properties:

Fact 11.7 *An assignment of non-empty subsets of some set to two players represents their powers in some finite game iff these powers satisfy Monotonicity, Consistency, and Determinacy.*

In non-determined games, the third condition drops out – after which we can do a similar representation result for monotone consistent families, in which their powers are realized in finite games of imperfect information ([40]; see also Section 11.10). A typical non-determined case is this specification:

$$\text{powers of } \mathbf{A}\text{:}\quad \{1,2\}, \{3,4\} \qquad \text{powers of } \mathbf{E}\text{:}\quad \{1,3\}, \{2,4\}$$

Alternatively, one can represent such families of non-determined powers using infinite games of perfect information – witness Section 11.8.

Digression The proof of Fact 11.7 allows for identifying different outcome states of a game. This can happen in first-order evaluation games for states with the same variable assignment, or in chess, for the same board configurations with different histories. If we insist on unique outcome states, additional conditions hold, reflecting a closer tie between strategies and powers. In particular, (a) the intersection of any two inclusion minimal power sets of two players is a singleton, and (b) each singleton outcome set can be obtained as such an intersection. A representation result for this case seems open (cf. [45]). This issue will return for logic games in Section 11.9.

11.5.3 Modal Forcing Languages

Games at the level of players' powers have a natural *modal game logic*. We just illustrate this – but cf. [28] for motivation, and [7, 8] for modal logic in general. The language has proposition letters, Boolean operators, and *forcing modalities* $\{G,\mathbf{i}\}\varphi$ saying that player **i** has a strategy for playing G which guarantees a set of outcomes all of which satisfy φ. Here, φ may express winning, or any property of states. Next, games are associated with modal models **M**, where states may interpret game-external proposition letters. We set

$$\mathbf{M},s \models \{G,\mathbf{i}\}\varphi \text{ iff there exists a power } X \text{ for player } \mathbf{i} \text{ in } G \text{ played from state } s \text{ such that for all } x \in X : \mathbf{M},x \models \varphi$$

To bring this in line with modal semantics in its "neighbourhood" version, one might use binary state-to-set *forcing relations* $\rho_{\mathbf{i}}^{G}\, s,Y$, and set

$$\text{there exists a set } X \text{ with } \rho_{\mathbf{i}}^{G}\, s,Y \text{ and for all } x \text{ in } X,\ \mathbf{M},x \models \varphi$$

The main effect of this at the level of validities is the following:

Fact 11.8 *Modal logic with the forcing interpretation satisfies all principles of the minimal modal logic K except for distribution of* $\{\}$ *over disjunctions.*

In particular, $\{G,\mathbf{i}\}\varphi$ is *upward monotone*:

$$\text{if} \models \varphi \to \psi, \text{ then} \models \{G,\mathbf{i}\}\varphi \to \{G,\mathbf{i}\}\psi$$

But distribution over disjunctions is *not valid*:

$$\{G,\mathbf{i}\}\varphi \vee \psi \to \{G,\mathbf{i}\}\varphi \vee \{G,\mathbf{i}\}\psi$$

This is precisely the point of forcing. Other players may have powers that keep us from determining results precisely. I may have a winning strategy, but it may still be up to *you* exactly *where* my victory is going to take place. For instance, in the game of Section 11.5.2, **E** can force $\{2,3\}$, but neither $\{2\}$ nor $\{3\}$. Two further axioms relate powers of different players: matching the earlier Consistency and Determinacy.

Finally, this modal language has a matching notion of *bisimulation* between game models **M**, which leaves truth of all modal forcing formulas invariant.

Definition A *power bisimulation* between two game models $\mathbf{M},\mathbf{N}$ is a binary relation E between states satisfying the following two conditions, for all **i**:

1. if xEy, then x, y satisfy the same proposition letters.
2. if xEy and $\rho_{\mathbf{i}}^{\mathbf{M}} x, U$ then there exists a set V with $\rho_{\mathbf{i}}^{\mathbf{N}} y, V$ and $\forall v \in V \exists u \in U : uEv$; and vice versa.

We will use this notion later. The theory of game logic is much like that of standard modal logic. Cf. [30, 43] for further details and results.

11.6 Extensive Games and Modal Action Logics

11.6.1 Extensive Games as Modal Process Models

Moving beyond players' global powers, extensive games have a finer level of states, turns, and individual moves. Such games are like ordinary models for a standard modal language with moves as atomic actions, and some special predicates, like **end** for being an end point, or $\mathbf{turn_i}$ for turns of player **i**. Consider again a simple game as in Section 11.5.2:

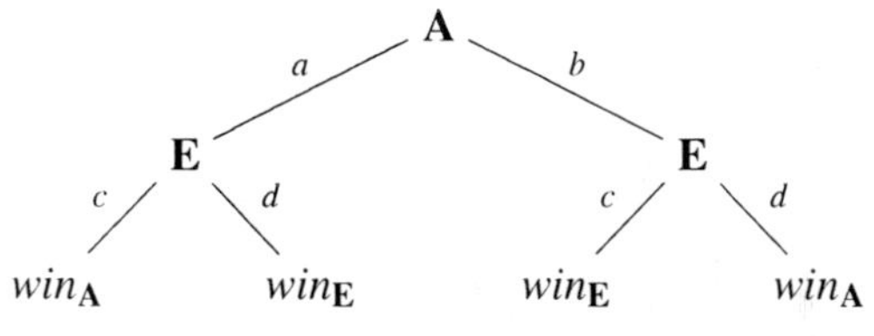

The assertion that **E** has a winning strategy can be expressed in detail here by this modal formula true in the root, which records player's intermediate moves:

$$[a]\langle b\rangle win_{\mathbf{E}} \wedge [b]\langle c\rangle win_{\mathbf{E}}$$

Alternatively, these models are the labelled transition systems of computer science, with a state space for computations by several interactive processors. A systematic study of extensive games as process models for modal languages is made in [41]. Here we just remark that the usual process of "solving" a game by means of a Zermelo-style colouring algorithm (cf. the discussion of determinacy in Section 11.5.1) really amounts to stage-wise computing a smallest (or greatest) fixed-point definition for **E**'s winning positions. For example, to define the above forcing modality $\{G,\mathbf{E}\}\varphi$ in more detail, one can use the following modal recursion:

$$\{\mathbf{E}\}\varphi \leftrightarrow ((\mathbf{end} \wedge \varphi) \vee (\mathbf{turn_E} \wedge \bigvee_a \langle a\rangle\{\mathbf{E}\}\varphi) \vee (\mathbf{turn_A} \wedge \bigwedge_a [a]\{\mathbf{E}\}\varphi))$$

In terms of the modal μ-calculus, this says (using greatest fixed-points) that

$$\{\mathbf{E}\}\varphi \leftrightarrow \nu p.((\mathbf{end} \wedge \varphi) \vee (\mathbf{turn_E} \wedge \bigvee_a \langle a\rangle p) \vee (\mathbf{turn_A} \wedge \bigwedge_a [a]p))$$

This means that all games can be solved by model checking some assertion in an appropriate game logic over them. Moreover, basic game-theoretic arguments, like the proof of Zermelo's Theorem or the Gale-Stewart Theorem can be formalized in modal fixed-point logics. Thus, one particular logic game from Section 11.2, viz. model checking, may help us understand games in general!

11.6.2 Dynamic Logic as Strategy Calculus

Modal languages also have another use germane to games: they can define players' *strategies* explicitly. For, in extensive game trees, strategies for player **i** are nothing but *binary relations* which are functions on the turns for **i**, while including all possible moves at other players' turns. Thus, they are of the same type as the above actions a,b. More generally, think of a standard dynamic logic with action expressions that can be formed from atomic ones using

> choice $\cup$, sequential composition ;, relational converse $^{\cup}$, finite iteration *, and the usual test program (φ)? for propositions φ.

For instance, then, the winning strategy for **E** in the above game may be defined as follows (with $\top$ for an assertion which is always true):

$$((\langle a^{\cup}\rangle\top)?;d) \cup ((\langle b^{\cup}\rangle\top)?;c)$$

This means that propositional dynamic logic [16] can be used as a general calculus of strategies, of the sort mentioned in Section 11.3.1. Thus, one particular modal game logic can help us understand logic games in general.

11.7 Game Constructions

The preceding two sections were about game-internal properties, and ways of expressing these in modal languages. But we can also take an external view of games, more like in Process Algebra or Category Theory. For a start, games do not occur by themselves, they live in families. A natural general theme are then game-forming constructions. Logic games provide many instances of this.

11.7.1 Logical Game Operations

For a start, the evaluation games of Section 11.2.4 provide the following game-theoretical take on the basic logical operations:

1. conjunction and disjunction are *choices* for players: $G \cap H$ is **A**'s choice, $G \cup H$ is **E**'s choice of G or H.
2. negation is a *role switch*, leading to the dual game G^d with all the turns and win markings reversed in G.

But clearly, choice and dual are completely general operations forming new games out of old. In particular, switch stands for the basic rational ability to put oneself into another person's position, which has been cited by psychologists and philosophers as central to human agency. Here is another such operation which operates inside evaluation games. Consider the rule for an existentially quantified formula $\exists x.\ \psi(x)$:

E must pick an object d in **M**, and play then continues with $\psi(d)$

Properly understood, the existential quantifier $\exists x$ does not serve as a game operation here: it clearly denotes an independent atomic game of "object picking" by Verifier. (The same point occurs in [2].) The general game operation in this clause hides behind the phrase "continues", which signals

3. *sequential composition* of games $G; H$.

These are just a few natural operations that form new games out of old. To arrive at others, consider the intuitive idea of "conjunction" of games. So far we have two candidates. *Boolean conjunction* $\cap$ forces a choice right at the start, with the game not chosen never accessed. Sequential composition ; may lead to play of both games, if the first is ever completed. But now consider the plight of academics playing games "family" and "career". Most of us compromise via a new operation:

4. *parallel composition* of games $G \parallel H$.

This means playing a stretch in one game, then switching to the other, and so on – running around and trying to do the best in both. Logic games also provide examples of such interleaving (see Section 11.9.6). But even this is just one plausible parallel game construction: we might also proceed simultaneously in both games, and so on. Indeed, no natural complete operational repertoire is known either in logic or in game theory.

11.7.2 Game Algebra of Sequential Operations

Game operations suggest game algebra: a calculus of equivalent game expressions. For example, intuitively, the above choices for the two players are related by a De Morgan duality under role switch:

$$G \cap H = (G^d \cup H^d)^d$$

Like composing of binary relations, game composition is associative. Another intuitive validity is left-distribution for composition over choice

$$(G \cup H); K = (G; K) \cup (H; K)$$

By contrast, the right-distribution law is not valid:

$$G; (H \cup K) = (G; H) \cup (G; K)$$

E's choice on the left-, but not on the right-hand side may depend on the outcome of first playing game G. Another intuitively valid principle concerns role switch:

$$(G; H)^d = G^d; H^d$$

These intuitions may be made precise using the notions of Section 11.5 [43]. The game equivalence that fits best with first-order validity looks at players' powers for determining outcomes. One can compute these inductively for complex games using choice, dual, and composition. We define forcing relations

$\rho_{\mathbf{i}}^{G} x, Y$ player **i** has a strategy in game G which makes sure that G ends in a state in Y when started from state x

Fact 11.9 *The following equivalences hold for forcing relations:*

$$\begin{aligned}
&\rho_{\mathbf{E}}^{G \cup G'} x, Y \textit{ iff } \rho_{\mathbf{E}}^{G} x, Y \textit{ or } \rho_{\mathbf{E}}^{G'} x, Y \\
&\rho_{\mathbf{A}}^{G \cup G'} x, Y \textit{ iff } \exists Z, Z' : \rho_{\mathbf{A}}^{G} x, Z \textit{ and } \rho_{\mathbf{A}}^{G'} x, Z' \textit{ and } Y = Z \cup Z' \\
&\rho_{\mathbf{E}}^{G^d} x, Y \quad \textit{iff } \rho_{\mathbf{A}}^{G} x, Y \\
&\rho_{\mathbf{A}}^{G^d} x, Y \quad \textit{iff } \rho_{\mathbf{E}}^{G} x, Y \\
&\rho_{\mathbf{E}}^{G;G'} x, Y \textit{ iff } \exists Z : \rho_{\mathbf{E}}^{G} x, Z \textit{ and } \forall z \in Z : \rho_{\mathbf{E}}^{G'} z, Y \\
&\rho_{\mathbf{A}}^{G;G'} x, Y \textit{ iff } \exists Z : \rho_{\mathbf{A}}^{G} x, Z \textit{ and } \forall z \in Z : \rho_{\mathbf{A}}^{G'} z, Y
\end{aligned}$$

Using superset closure of powers, the second clause simplifies to

$$\rho_{\mathbf{A}}^{G \cup G'} x, Y \text{ iff } \rho_{\mathbf{A}}^{G} x, Y \text{ and } \rho_{\mathbf{A}}^{G'} x, Y$$

Remark If we also assume *determinacy*, in the earlier sense that for each set Y, either one of the players can force Y, or the other can force $W - Y$, then we just need to define forcing relations for player **E**, because all powers for player **A** follow automatically by observing that $\rho_{\mathbf{A}}^{G^d} x, Y$ iff not $\rho_{\mathbf{E}}^{G} x, W - Y$.

Now, take an algebraic language of game expressions starting with variables, and operations $\cup$, ^{d}, ;. In addition, ι is the *idle game* staying at the same state.

Definition Two game expressions G, H are equivalent (written $G = H$) if they have the same power relations for players in all game models. We also write $G \leq H$ in case of a similar valid inclusion between the respective powers.

11.7.3 Excursion: A Complete System

Algebraic game validity [38] validates the preceding observations. There is a simple complete equational system for this algebra of the game-theoretic analogues of the usual first-order operations [15]. We display it here, to show a surprising fact behind our approach. Just underneath standard first-order logic, there lies a systematic game logic!

Theorem 11.2 *Basic Game Algebra consists of the following principles:*

1. *the laws of De Morgan algebra for choice and dual*
2. $G;(H;K) = (G;H);K$ *associativity*
   $(G \cup H);K = (G;K) \cup (H;K)$ *left-distribution*
   $(G;H)^d = G^d;H^d$ *dualization*
3. $G \leq H \to K;G \leq K;H$ *right-monotonicity*
4. $G;\iota = G = \iota;G$

De Morgan algebra is essentially Boolean Algebra minus special laws for 0, 1. For our discussion in Section 11.9, we note that basic game algebra is *decidable*.

11.7.4 Dynamic Game Logic

Basic game algebra can be embedded into the decidable *dynamic game logic* which extends the earlier modal forcing language with modalities $\{G, \mathbf{i}\}\varphi$ with complex game terms G. Typical for these systems is the interplay of two ingredients in one language: a dynamic component with expressions G for games, and a static component with propositions φ about states of play. This is like dynamic logics in computer science which manipulate program expressions and propositions about computational states together. For more on dynamic game logic, cf. [28, 30, 31, 39].

11.7.5 Logics of Parallel Game Operations

Parallel game constructions have been studied extensively in game semantics for *linear logic* whose tensor product involves interleaved games where Adam can switch between games. Under this interpretation, linear logic is a complete axiomatization of several central sequential and parallel game constructions. Cf. [1, 9, 14] for details. A complete dynamic logic of players' powers in products that model simultaneous games has been proposed in [49].

11.8 Finite Versus Infinite Games

11.8.1 The Importance of Infinite Runs

The emphasis so far has been on finite games and their outcome states. But several logic games in Section 11.2 support *infinite runs*, witness model construction, model comparison, and even model checking for first-order languages with fixed-point operators. The same move was behind the shift from Zermelo's Theorem to the Gale-Stewart Theorem, where players produce infinite runs, marked as winning or losing. And game theory also has more than just finite matrices or tree pictures. Infinite games model situations with ongoing behaviour, such as iterated Prisoner's Dilemma in studying the possible emergence of social cooperation. Such ongoing behaviour is just as important as finite termination. A good analogy comes from computer science, using our process analogy for games. *Terminating programs* are meant to find some value, or finish some task. But there are also crucial non-terminating programs like *operating systems* which ensure the proper functioning of some device: the longer the better. (Such infinite computations are at centre stage in modern *co-algebra*, cf. [53].) Likewise, in linguistics, language games for conversational tasks should terminate. But there is also the Great Game of Language – with discourse as the "operating system of cognition". This should keep functioning forever. Both kinds of game, finite and infinite, make sense.

11.8.2 Extending the Game Logic Perspective

Infinite games can still be studied by the logical techniques of Sections 11.3, 11.4, 11.5, 11.6 and 11.7. Outcomes are now the runs themselves. Then, both power and action levels still make sense. Here are two illustrations.

First, consider computation of powers. Recall the interval selection game of Section 11.5.1, showing that non-determined games exist. Further interesting information can be extracted by looking into players' powers. We can then prove in general that

Fact 11.10 *Identifying infinite runs that are equal up to finite initial segments, both players have the same powers in the interval selection game.*

In particular, this perfect information game is about the simplest infinite realization of a non-determined power specification (cf. Section 11.5.2), viz.

$$\text{powers of } \mathbf{A}: \{1,2\} \qquad \text{powers of } \mathbf{E}: \{1,2\}$$

For a finite game realization with imperfect information, see Section 11.10.

Our second example is *temporal logic of games*. Infinite games need more expressive logics, to formalize game-theoretic arguments. For example, the proof of the Gale-Stewart Theorem involves the following result true for all games:

Weak Determinacy Either player **E** has a winning strategy, or player **A** has a strategy which forces infinite branches on which player **E** never has a winning strategy.

The usual proof for the Gale-Stewart Theorem then runs as follows. Given that **E**'s winning set is open, this Weak Determinacy implies that the branch forced by player **A** is a loss for **E** – so **A** has a winning strategy.

Fact 11.11 *Weak Determinacy can be defined in a branching temporal language, evaluating formulas at pairs* $\langle h,t\rangle$ *of a current branch h and point t on it:*

$$\mathbf{M},h,t \models \{G,\mathbf{E}\}\varphi \vee \{G,\mathbf{A}\}\mathrm{A}\neg\{G,\mathbf{E}\}\varphi$$

Here, $\{G,\mathbf{i}\}\varphi$ is a *modal-temporal forcing modality* extending the a-temporal one of Section 11.5.3. It says at point t that player **i** has a strategy ensuring that only runs result with the current history h up to t as an initial segment, which satisfy the temporal logic formula φ. Temporal logic comes in explicitly once more through the use of the standard operator A ("always") in the right-hand disjunct. This says that a statement is *always true on the current branch*.

This temporal logic seems very powerful. Most reasonable winning conventions can be given in this format – such as the earlier *safety* and *liveness* properties, as well as winning conventions of infinite logic games. Also, like dynamic game logic (Section 11.7.4), temporal game logic involves a merge of internal and external game languages.

Infinite games are complex structures. In particular, they have huge unwieldy spaces of strategies – as high-lighted in the "folk theorems" of game theory on a plethora of equilibria. Another general logical issue then is *finitization* (cf. Section 11.9). To which extent can we know infinite strategies in a full infinite game through their finite approximations? This may be a lost cause in general, but things look brighter if games have finite branching width, and we can use principles like König's Lemma.

11.9 From Game Logics to Logic Games

Having discussed game logics for a while, let us now return to logic games proper, and see what additional light is cast on these by Sections 11.6, 11.7, and 11.8. We will see how all earlier themes make sense, and suggest new results and questions.

11.9.1 Questioning Game Equivalence

For a start, take the issue of "equivalence" between logic games. The literature is full of statements to the effect that one logic game is "equivalent" to another. For example, [22] says all games are equivalent to full infinite game trees, as we can disregard all undesired runs by calling them losses for **E**. But this presupposes just one notion of game equivalence, and one biased in favour of just one player. Other authors "reformulate" logic games without stating in what sense the new versions are equivalent. For example, [6] define infinite model comparison games inverting the schedule of Section 11.2.6:

> The game starts with one finite partial isomorphism between two models. Each round lets Duplicator **D** choose some family F of partial isomorphisms, followed by a selection by Spoiler **S** of one f in F. In the next round, **D** must select a set F^+ again, **S** then chooses a partial isomorphism in F^+ again, and so on. The back-and-forth property to be maintained by **D** is: *For every object a in one model, there exists an object b in the other model such that* $f \cup \{(a,b)\} \in F^+$ *– and likewise in the other direction.*

Fact 11.12 *The inverted model comparison game is equivalent to standard Ehrenfeucht-Fraïssé games at the level of players' powers.*

The trick is like the distributive law of Section 11.4.1, inverting scopes of logical

operators. In each round, **D** offers **S** a panorama of all choices he could make, plus her own responses to them. **S** then selects his own move plus **D**'s pre-packaged response – thereby setting the new stage. In human terms, **D** behaves like a colleague of mine, who tries to speed up department meetings by saying: "Now you're going to say A, and I will say B – or, you're going to say C, and then I will say D – etc."

In terms of Section 11.5 versus Section 11.6, logic games seem biased toward outcomes and powers, rather than the fine-structure of actions and turns. Most intuitive equivalences between such games can be justified in this manner.

11.9.2 The Outcome Perspective in Logic Games

Our insights about general game logic have interesting implications in specific cases. Consider our running example of evaluation games for first-order logic. We give three concrete illustrations.

Richer denotations Any game *game* $(\mathbf{M}, s, \varphi)$ assigns a much more structured denotation to a formula φ than just a truth value, viz. the complete power structure of the two players. We can think of these as their forcing relations $\rho_{\mathbf{i}}^{G}\, s, Y$ computed over the model $\mathbf{M}$ in the sense of Section 11.7.2. This suggests that there are several natural levels to assigning meaning, even for standard logical languages.
Power bisimulation At the level of powers, the right notion of equivalence between two models $\mathbf{M}$, $\mathbf{N}$ is the power bisimulation of Section 11.5.3. Translating this back into standard first-order terms yields a variant of standard potential isomorphism (Section 11.2.6). This time, states to be related are not tuples of objects from the models, but variable assignments over them. The bisimulation is then a family of links satisfying obvious back-and-forth conditions for shifting values for variables. Well-understood, this also seems a natural notion for first-order logic in general.
Representation of general games Perhaps most strikingly, evaluation games seem adequate for games in general!

Fact 11.13 *Any extensive game is outcome-equivalent to one where players evaluate an associated game-logical assertion.*

Proof. We just give an illustration. The game in Section 11.6.1 is clearly outcome-equivalent to an evaluation game for its associated modal formula $[a]\langle d\rangle win_{\mathbf{E}} \wedge [b]\langle c\rangle win_{\mathbf{E}}$. Modulo outcome equivalence, we can re-arrange any finite extensive game to one with a uniform alternating schedule, while making all runs of equal length. This allows us to write up a matching modal formula in iterated $[\,]\langle\,\rangle$ form, whose evaluation game proceeds like the original game.

Here is a more technical representation result showing how logic games are complete for game logics in one more sense [42]:

Theorem 11.3 *The basic algebra of sequential operations on arbitrary games is precisely the game algebra of first-order or modal evaluation games.*

Thus, any non-valid principle of basic game algebra (cf. Section 11.7.3) already has a counter-example in first-order evaluation games. More delicate issues arise when we demand uniqueness of outcomes (cf. Section 11.5.2): this causes no loss of generality in first-order evaluation games, but it does in modal ones [43].

11.9.3 Fine-Structure: The Action Level in Logic Games

Despite the noted power bias, it also makes sense to look at the less well-studied fine-structure of logic games at their action level. Evidently, in this setting, far fewer games will be identified.
Finer levels of denotation Consider again first-order evaluation games. We now get much finer notions of denotation, leading to a subset of game equivalences, viz. those which leave the move and turn structure intact as far as ordinary modal logic cares about them. Some axioms of basic game algebra survive this: De Morgan laws,

or left-distribution of composition over choice. Other axioms fail, as they merely preserved powers – e.g., the distribution law of Section 11.4.1 reversing players' turns.

Open problem Determine the complete basic game algebra for modal equivalence between extensive games.

This analysis suggests that there are many natural levels of equivalence for first-order formulas. And this again ties up neatly with a persistent philosophical tradition of looking for various levels of identifying "propositions" (cf. [24]).

Strategy calculus in dynamic logic Individual actions also emerge in players' strategies in logic games (cf. Section 11.3). These strategies unified such diverse notions as formulas, "reasons", proofs, models, or semantic analogies. Underlying all of these is the dynamic logic of Section 11.7.4. This gives a fresh look at known notions. For example, in first-order evaluation games, the item closest to strategies are Skolem functions. Dynamic logic suggests a calculus of *definable Skolem functions*, taken as relations (a natural generalization). Its major operations are choice, composition, and iteration of binary relations, allowing, amongst others, the standard sequential program constructions *IF P THEN* π *ELSE* π' and *WHILE P DO* π (cf. [41]).

Such an explicit format for strategies also makes sense in dynamic game logic (Section 11.7.4). That calculus was $\exists$-sick in the sense of Section 11.4.2, since it just says that a strategy exists without naming it. But we can remedy this with a modality

$$\{G, \mathbf{i}, \sigma\}\varphi$$

stating that, in game G, strategy σ for player **i** forces a set of outcomes satisfying proposition φ. Dynamic game logic in this guise is still to be developed.

11.9.4 Digression: Strategy Calculus in Type-Theoretic Format

Formats other than dynamic logic may be attractive, too. Type theories (cf. [4]) manipulate statements of the form $\sigma : G$ interpreted as "σ is a proof of assertion G", or "σ is an object having property G". Strategy calculi might then manipulate statements

$\sigma : G$ σ is a winning strategy for player **E** in game G

Such interpretations are given for linear logic games (cf. Section 11.7.5) in [3]. Here is a simple example. Consider the following sequent derivation for a propositional validity, whose steps are well-known valid inference rules:

$$\dfrac{\dfrac{\dfrac{\dfrac{A \Rightarrow A}{A,B \Rightarrow A} \quad \dfrac{B \Rightarrow B}{A,B \Rightarrow B}}{A,B \Rightarrow A \wedge B}}{A,B \Rightarrow (A \wedge B) \vee C} \quad \dfrac{\dfrac{\dfrac{C \Rightarrow C}{A,C \Rightarrow C}}{A,C \Rightarrow (A \wedge B) \vee C}}{}}{\dfrac{A, B \vee C \Rightarrow (A \wedge B) \vee C}{A \wedge (B \vee C) \Rightarrow (A \wedge B) \vee C}}$$

We want to make the strategy calculus behind this derivation explicit. This requires the identification of strategy combinations supporting the proof steps. Here is a concrete format of analysis, written with some ad-hoc notation:

$$\begin{array}{ll}
x:A \Rightarrow x:A \quad y:B \Rightarrow y:B & \\
x:A,y:B \Rightarrow x:A \quad x:A,y:B \Rightarrow y:B & z:C \Rightarrow z:C \\
x:A,y:B \Rightarrow (x,y):A \cap B & x:A,z:C \Rightarrow z:C \\
x:A,y:B \Rightarrow \langle L,(x,y)\rangle:(A\cap B)\cup C & x:A,z:C \Rightarrow \langle R,z\rangle:(A\cap B)\cup C
\end{array}$$

$$x:A,u:(B\cup C) \Rightarrow \textbf{IF}\, head(u)=L \;\textbf{THEN}\, \langle x, tail(u)\rangle \,\textbf{ELSE}\, tail(u) : (A\cap B)\cup C$$

$$v:A\cap(B\cup C) \Rightarrow \textbf{IF}\, head((v)_2)=L \;\textbf{THEN}\, \langle (v)_1, tail((v)_2)\rangle \,\textbf{ELSE}\, tail((v)_2) : (A\cap B)\cup C$$

This derivation composes strategies in complex games from those in sub-games. The operations for this task are completely general, not depending on proof theory:

storing strategies for a player who is not to move	$\langle , \rangle$
using a strategy from a list	$()_i$
computing the first recommendation of a strategy	$head()$
as well as the remaining strategy	$tail()$
making a choice dependent on some information	*IF THEN ELSE*

As strategies encode very different logical objects: proofs, models, analogies, etc., the above derivation can stand for quite different things. It is a recipe for constructing proofs, and the operations then encode what goes on as logical operations get added. But it also describes how any winning strategy for Verifier in an evaluation game for $A \wedge (B \vee C)$ can be turned into a winning strategy in $(A \wedge B) \vee C$. Not all logic games support operations of choice, though – and hence the above recipe may not make much sense (yet) for games of model construction, or model comparison.

11.9.5 Operations on Logic Games

Logic games are not isolated activities, they can be combined (cf. Section 11.2.7). And then, they support not just sequential game operations, but also parallel ones beyond those found in the computational literature. Nice examples are the Wadge Game in descriptive set theory [26] and the interleaved fixed-point games found in the proof of the Stage Comparison Theorem of Moschovakis 1974. The operational structure behind logic games should be a good testing ground for general game algebra. Here, we only offer one illustration showing the interest of such matters [38].

Relating evaluation and comparison games In this excursion, we relate two major logic games, viz. Ehrenfeucht games of model comparison with Hintikka games of evaluation. These have the same "back and forth" idea in their object-picking moves. We make this precise in terms of game operations, proving the following informal equation

$$\mathbf{E} = \mathbf{H}^2$$

The "squaring" operation here is interleaving games, and we can even correlate strategies in the two games directly. Recall the Adequacy Theorem for finite-depth Ehrenfeucht-Fraïssé games of Section 11.2.6, ∃-cured in Section 11.4.2. This suggests an explicit link between strategies across comparison and evaluation games for models $\mathbf{M}, \mathbf{N}$. First-order formulas φ of quantifier depth k between $\mathbf{M}, \mathbf{N}$ drove winning strategies for Spoiler in the k-round comparison game between $\mathbf{M}, \mathbf{N}$. But we can do away with this intermediary! Let $\mathbf{M} \models \varphi, \mathbf{N} \models \neg\varphi$. This induces a winning strategy for Verifier in an evaluation game $game(\varphi, \mathbf{M})$ *plus* one for Falsifier in $game(\varphi, \mathbf{N})$:

Theorem 11.4 *There exists an effective correspondence between*

1. *Winning strategies for Spoiler in the k-round comparison game,*
2. *Pairs of winning strategies for Verifier and Falsifier in some k-round evaluation game, played in opposite models.*

Proof. Without loss of generality, formulas can be assumed to be constructed from atoms with negations, disjunctions, and existential quantifiers only. *From 2 to 1.* Let an H-pair of depth k consist of a formula φ of quantifier depth k plus a winning strategy σ for $\mathbf{V}$ in the φ-game in one of the models, and a winning strategy τ for $\mathbf{F}$ in the φ-game in the other model. We sketch how to *merge* σ, τ. Spoiler looks at the two evaluation games. Suppose $\mathbf{V}$ wins φ in $\mathbf{M}$, and $\mathbf{F}$ wins φ in $\mathbf{N}$. If φ is a negation $\neg\psi$, Spoiler switches to the obvious strategies for $\mathbf{F}$ and $\mathbf{V}$ w.r.t. ψ. (Note that this is internal computation: the opponent in the comparison game does not see any action yet.) If φ is a disjunction $\psi \vee \xi$, Spoiler uses his $\mathbf{V}$-strategy in the one model to choose a disjunct. His $\mathbf{F}$-strategy in the other model will also win against that disjunct. Proceeding in this way, the formula is broken down until an existential subformula $\exists x\psi$ is reached. Spoiler then uses his $\mathbf{V}$-strategy σ in the model where it lives, say $\mathbf{M}$, to pick a witness d. This model $\mathbf{M}$ and object d are his opening move in the first round of the Ehrenfeucht game. Next, what remains for Spoiler is still a winning strategy σ^- for ψ in $\mathbf{M}$ after this first move. Now, let Duplicator respond with any object e in the other model $\mathbf{N}$. This choice can also be seen as a move by Verifier in the evaluation game for $\exists x\psi$ in $\mathbf{N}$. Now we know that Falsifier still has a winning strategy τ^- for ψ in $\mathbf{N}$ after this first move. So, by induction, we still have an H-pair of depth k-1, which can be merged into a follow-up winning strategy for Spoiler in the $(k-1)$-round comparison game between $\mathbf{M}$ and $\mathbf{N}$. The total effect is a k-round $\mathbf{S}$-strategy. This argument yields an algorithm for Spoiler's computation.

From 1 to 2. This direction seems harder, as we have to "decompose" one object: Spoiler's winning strategy, into two separate ones that must form a suitable H-pair.

One proof of this follows our earlier construction of a difference formula of depth k from an **S**-strategy (Section 11.4.2). This formula induces two evaluation strategies effectively. Let us describe "splitting" of a comparison strategy directly. Consider any winning strategy for **S** in the k-round comparison game between two models $\mathbf{M}, \mathbf{N}$. In the first move, **S** chooses, say, model **M** and object d. Our desired formula will then start with an existential quantifier, and **V** has the winning strategy in **M**. Let Duplicator now make any response e in **N**. We know that Spoiler still has a $(k-1)$-round winning strategy in the two expanded models $(\mathbf{M}, d), (\mathbf{N}, e)$. Inductively, we can find H-pairs of depth $k-1$ for each choice e that Duplicator makes. Moreover, by the earlier Finiteness Lemma, only finitely many logically non-equivalent formulas can be involved in these pairs. Then, one over-all existential quantification over a suitable conjunction of formulas of depth k-1 defines our desired H-pair of depth k. In particular, if it is **V** who has the winning strategy of a relevant H-pair φ in the model **M**, put itself in the conjunction; otherwise, put its negation.

Operations like interleaving are just the beginning of what may be called a logic of game *architecture*.

11.9.6 Finite Versus Infinite

Logic games can have finite or infinite depth, leading to issues of finitization (Section 11.8). Say, model construction games can go on forever. Critic's winning strategies made Builder lose on each run, by a finite stage.

Fact 11.14 *Critic's winning strategies in the Construction game are finite objects.*

Proof. How can this be? The reason is that the game tree for the model construction game is finitely branching. Hence by König's Lemma, there is some finite level at which Critic has already blocked every construction attempt. This closed game tree is a finite object. These strategies are associated with finite objects, viz. proofs.

Such a reduction may fail in infinite comparison games, where Spoiler may be able to win against Duplicator, blocking each run at some finite stage, without there being a finite object encoding this. There must be a formula of *infinitary* first-order logic witnessing the relevant difference between the two models being compared, but there need not be a standard first-order one.

Finitization can be very useful. Hirsch and Hodkinson [20] show the following for their game of Section 11.2.7, using finite branching for Builder's (though not Critic's!) moves: *Builder can win the infinite game if she can win all finite cut-offs*, and her winning strategy is easy to piece together from these. Then representability becomes equivalent to a set of first-order assertions expressing Builder's being able to win all finite cut-offs, which leads to a perspicuous axiomatization. Incidentally, in this game, by the same reasoning, Critic's winning strategy, if available, must be a *finite object* again. It is a sort of proof that the given algebra is not representable.

11.10 From Game Theory to Logic Games

Logic games are special, in that players' behaviour is much more constrained than in ordinary game theory. In real games, players have *preferences* between outcomes beyond winning or losing, they operate under *uncertainties* about what really happens during a move (think of card games), and there can be more than two players, leading to *coalitions*! Importing these concerns into logic games makes them more realistic, even though there is hardly any theory of the resulting activities.

We will just skim a few issues here – but there are natural motivations for enriching logic games, even from the standpoint of reasoning and other logical core business.

11.10.1 Preferences

Preferences in logic games allows for finer behaviour. Consider the game of Section 11.6.1, where **A** now has a slight preference for one site of defeat over the other (we write values for **A**, **E** in that order under the outcome nodes):

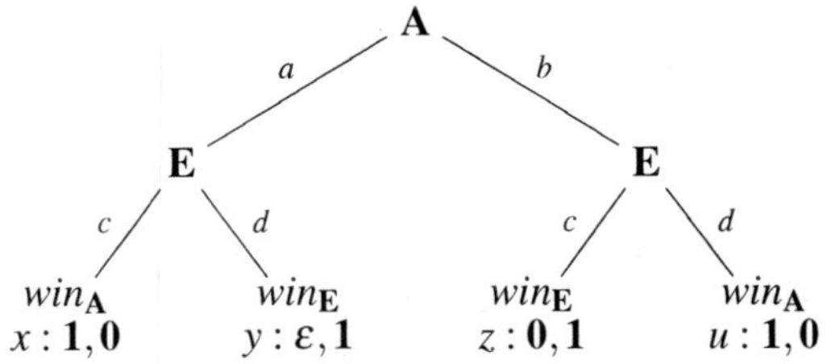

What will happen now? **A** can assume that, whatever he does, **E** will go for her most preferred outcome y or z. So, as he himself prefers y to z, he will choose "left", forcing **E** to end up in y. Thus the game ends in the run "left-right" with outcome y. This pair of strategies σ, τ is in Nash Equilibrium: neither player can gain by deviating from his strategy, assuming the other strategy in the pair does not change.

This story assumes a notion of "rationality", as in the solution method of *Backward Induction* computing game values via a Maximin rule. Here is how this works on the evaluation games of Section 11.4.1, with a preference structure as indicated below:

$p \wedge (q \vee r)$

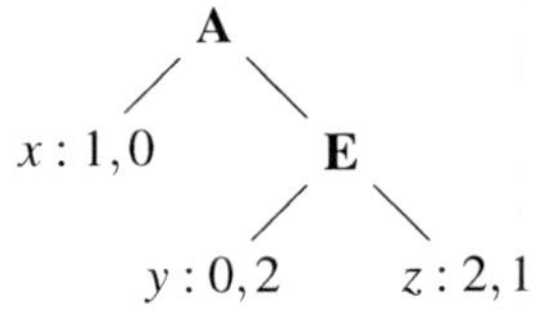

$(p \wedge q) \vee (p \wedge r)$

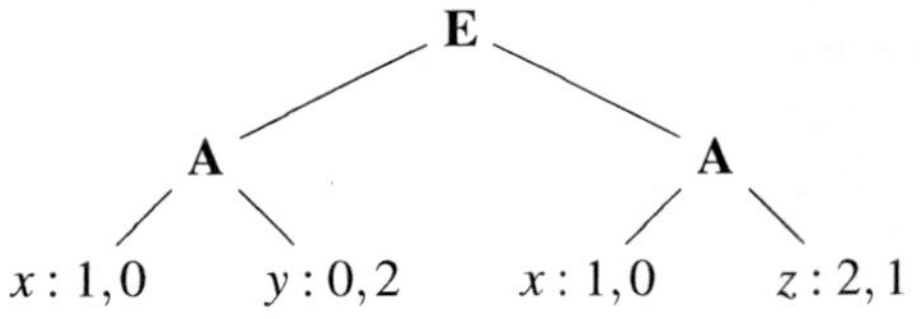

We display all pairs (**A**-value, **E**-value) computed bottom-up:

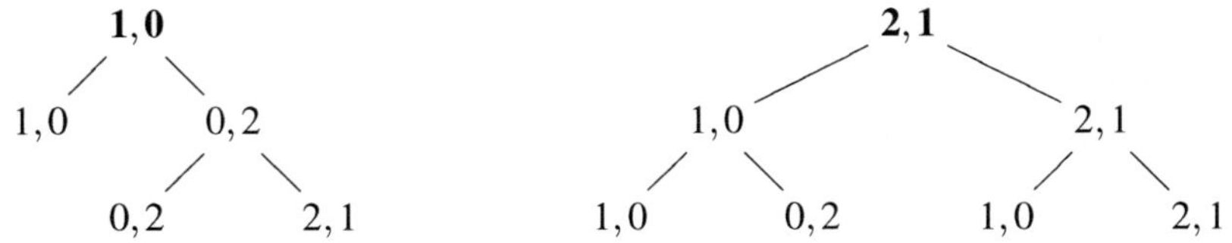

These trees correspond to different outcomes for the joint behaviour of the players. We predict outcome x on the left, but z on the right.

There are many new issues of general game logic for games with preferences. In general, these require combinations of modal logics for moves with preference logics (cf. [50–52]). Here we just look at one issue which also affects logic games: viz. how players evaluate the outcome of a game.

Game algebra with preferences Call two game expressions equivalent if their Nash equilibrium solutions are the same for every concrete realization including players' utilities. The preceding example showed that the basic game algebra of Section 11.7.3 no longer qualifies: propositional distribution fails. Vice versa, invalid equivalences may hold for special preference values. Assuming rationality, games A and $A \cup B$ are preference-equivalent whenever **E** prefers A to B. Of special interest are antagonistic zero-sum games, where **A** evaluates all outcomes opposite to **E**, as in logic games of winning and losing. Then some standard logic remains. Assume that different outcomes correspond to different preferences – otherwise, we might just as well identify them for game-theoretic purposes. Then it is easy to check that

Fact 11.15 *With zero-sum preferences, Boolean Absorption* $A = A \cap (A \cup B)$ *is valid, whereas with general preferences, it is not.*

Open problem Find the complete basic game algebra under the above defined Nash equilibrium equivalence in case of

1. arbitrary preferences,
2. zero-sum preferences.

Logic games with preferences Values and preferences also make sense in logic itself, witness many-valued logics, or preference models for logics of belief and conditionals. Another concrete source refers to game dynamics. Let us introduce resource structure into logic games, and say that, other things being equal, *players prefer outcome nodes lying on a shorter path from the root*. This assumes that players want to get to a winning node as soon as possible, or if no such node exists, to a losing node with the least effort. Then we can make more definite predictions about the course of games, and adapt definitions of validity and truth accordingly. A final example are the more realistic logical dialogue games of [44] where arguments may lose "force" from the moment they were first put on the table.

11.10.2 Solving Games

Making these connections does raise conceptual problems. In logic games, central notions are encoded in winning strategies, and in that case, the precise behaviour of the other player is unimportant: it will always be in Nash equilibrium with the winning strategy. But in game theory, a strategy models a type of behaviour vis-á-vis other players, and hence it is the *strategy profile for all players* that we are after. Nevertheless, multi-agent interactive behaviour seems crucial to logic, too. A proof is a long-term way of responding to objections, an isomorphism is a never-ending way of simulating one model in another, and sticking to the truth means emerging victorious no matter where your opponents in life try to push you. From a different perspective of resource-sensitivity, this also seems a major point of [14]. For similar ideas in computer science, see [2]. But then, we do seem to need a major change of perspective in logic. Based on preferences, game theory shows that Nash equilibria always exist for finite games if we are willing to admit *mixed strategies*, where pure actions are played with certain probabilities. What would be the logical point admitting of such *probabilistic solutions*?

11.10.3 Imperfect Information

In standard logic game, at each stage, players know exactly where they are. But in real games, players may have imperfect information as to where they are in the game tree. For example, in card games we do not know the complete distribution of the cards. Still players must move despite this partial ignorance.

Peculiarities of imperfect information games Games like this diverge from logic games in important respects [19]. Consider the evaluation game for the first-order formula $\forall x \exists y.\ x \neq y$ in Section 11.4.2. Now assume that Verifier is ignorant of the object chosen by Falsifier in his opening move. In game-theoretic notation, the new tree looks as follows, with a dotted line indicating **E**'s uncertainty:

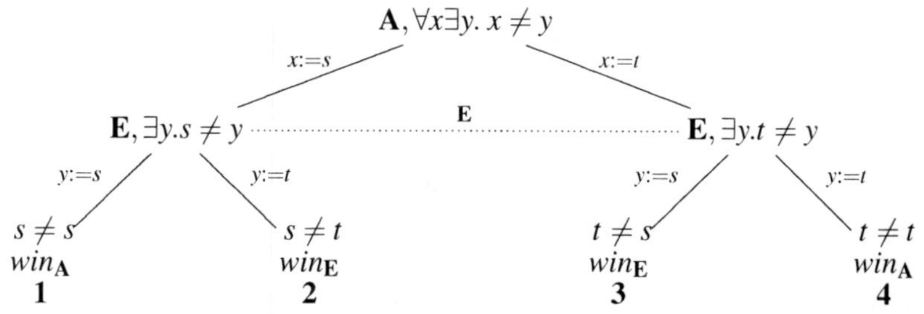

This game is quite different from its version with perfect information. In particular, if we allow only *uniform strategies* that can be played without resolving the uncertainty – as seems reasonable – **E** has only 2 of her original 4 strategies left in this game: “left” and “right”. Then *determinacy is lost*: neither player has a winning strategy!

Games with imperfect information like this still support game logics as in Sections 11.6 and 11.7 above, at both power and action levels (cf. [40]). For instance,

Fact 11.16 *Player* **E***'s situation in the central nodes of the preceding game can be defined by the following formulas of an epistemic-dynamic action logic:*

1. $K_{\mathbf{E}}(\langle y := t\rangle win_{\mathbf{E}} \vee \langle y := s\rangle win_{\mathbf{E}})$
   E *knows that some move will make her win, picking either s or t*
2. $\neg K_{\mathbf{E}}(\langle y := t\rangle win_{\mathbf{E}}) \wedge \neg K_{\mathbf{E}}(\langle y := s\rangle win_{\mathbf{E}})$
   there is no particular move of which **E** *knows that it will make her win.*

This is the well-known *de re – de dicto* distinction from philosophical logic. For instance, I may know that the ideal partner for me is walking out right there in the street, without ever finding out which one of these people was that person.

On the other hand, we can also describe games like this at the global level of powers. For example, with uniform strategies, players' powers in the above game are as follows:

$$\text{powers of } \mathbf{A}: \ \{1,2\}, \{3,4\}, \qquad \text{powers of } \mathbf{E}: \ \{1,3\}, \{2,4\}$$

The analysis of Section 11.5.2 can be extended to this situation. Families of powers satisfy Monotonicity and Consistency, though not Determinacy. And conversely, the former two conditions suffice for representability of given powers for two players as those realized in some game of imperfect information [40].

Logic games with imperfect information There is one exception to the above. Imperfect information has been added to logic games in the work of Hintikka and Sandu on *IF* logic (cf. [18] for a most recent version plus intended applications). In the notation of *IF* logic, the preceding game is written as

$$\forall x \exists y/\mathbf{x}.\ x \neq y$$

expressing that the choice of the existential witness for y must not depend on the universal challenge for x. As in Section 11.4, a good test question for a notion of games like this is: when do two given *IF* formulas define the same games? For instance,

Is the game $\forall x \exists y/\mathbf{x}.\ Rxy$ “equivalent to” the game $\exists y \forall x.\ Rxy$?

A popular answer is *YES*, as **E** has the same winning powers in both. But in games of imperfect information, the powers of one player do not automatically tell us all about the other. Hence **A** may not agree that these two games are the same. And indeed, the Thompson transformations of game theory [27] say that the correct equivalent is rather this game, with switched scheduling:

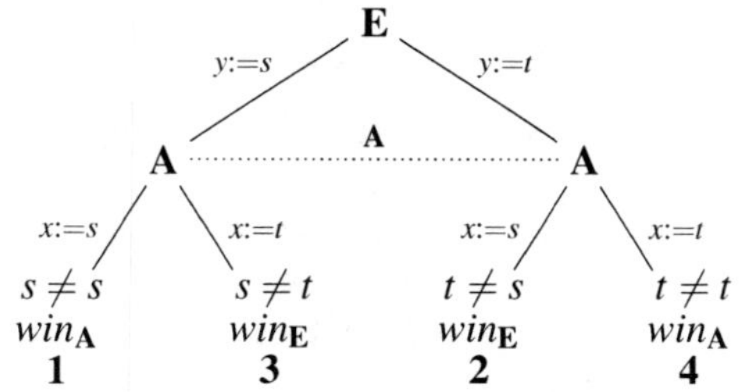

Stating this equivalence in terms of *IF* logic, the real situation involves a much nicer quantifier exchange:

Fact 11.17 $\forall x \exists y/\mathbf{x}.\ x \neq y$ *is uniformly outcome-equivalent to the slash formula representing the above game, which has the form* $\exists y \forall x/\mathbf{y}.\ x \neq y$.

IF games have generated a lot of controversy (cf. [46] for an epistemic analysis). Even so, they show that introducing imperfect information into logic games is exciting and perhaps even useful. In particular, this may provide a normal form for validities in all imperfect information games, the way first-order evaluation games did for the general algebra of perfect information games (Section 11.9.2).

One might try similar moves with the other logic games mentioned in Section 11.2, provided motivations are found. One might speculate about proofs where participants have forgotten some things that has been said. Or, one can play model comparison games with a fixed number of *pebbles*, representing some finite memory [23], where players will not be able to distinguish the same assignments of objects to these pebbles, even when they occur at different stages of the game.

Uncertainty about the future Finally, even in games of perfect information, where we know our position in the game tree at any time, we have "forward ignorance" of the future course of play, since we need not know the strategy of the other player. Nash equilibrium makes some predictions, but in general, we are in deliberation about future actions and choices, based on beliefs about ourselves and others. There is a flourishing literature on this, invoking belief revision and counterfactual reasoning (cf. [33, 48]). The resulting game logics put ideas from philosophical logic on top of the mathematical logic of game structure. Such issues, too, make sense for studying logic games, but we forego them here.

11.10.4 More Agents and Coalitions

Finally, many games involve more than two players, and hence the possibility of genuine *coalitions*: cf. [30] for a study with tools from modal logic. Many agents are the reality in conversation and debate, and they also make sense in logic games when thinking about teams of players – viewing Adam and Eve rather as some sort of Bourbaki. This suggestion is in the air today among logicians. The "team logic" of [35] is a serious contender.

11.11 Conclusions

This paper has shown how logic games are an interesting subclass of the totality of all games, raising new issues of game logic beyond standard game theory precisely since they are somewhat better delineated. On the other hand, the study of logic games might also benefit from importing ideas from general game theory. Either way, we think all this supports the idea of viewing logic as a study of the dynamics, rather than just the statics of agents involved in statement, reasoning, and communication.

Acknowledgements I would like to thank my students in the "Logic in Games" courses at ILLC Amsterdam and other venues since 1999. I also thank Rohit Parikh for his reader's comments. Finally, I thank Fernando Velázquez-Quesada and Abhisekh Sankaran for their generous help in preparing this manuscript for publication.

References

1. Abramsky S. Semantics of interaction: An introduction to game semantics. In P. Dybjer and A. Pitts, editors, *Proceedings 1996 CLiCS Summer School*, pages 1–31. Isaac Newton Institute, Cambridge University Press, Cambridge, 1996.
2. Abramsky S. Socially responsive, environmentally friendly logic. In T. Aho and A.-V. Pietarinen, editors, *Truth and Games: Essays in Honour of Gabriel Sandu*. Acta Philosophica Fennica, Helsinki, 2006.
3. Abramsky S., and Jagadeesan R. Games and full completeness for multiplicative linear logic, *J. Symbolic Log.*, 59(2): 543–574, 1994.
4. Barendregt H. Lambda calculi with types. In S. Abramsky, D. Gabbay, and T. Maibaum, editors, *Handbook of Logic in Computer Science*, pages 117–309. Clarendon Press, Oxford, 1992.
5. Barwise J., and Perry J. *Situations and Attitudes*. The MIT Press, Cambridge, MA, 1983.
6. Barwise J., and van Benthem J. Interpolation, preservation, and pebble games, *J. Symbolic Log.*, 64(2):881–903, 1999.
7. Blackburn P., de Rijke M., and Venema Y. *Modal Logic*. Cambridge University Press, Cambridge, 2001.
8. Blackburn P., van Benthem J., and Wolter F. editors. *Handbook of Modal Logic*. Elsevier Science Publishers, Amsterdam, 2006.
9. Blass A.A game semantics for linear logic, *Ann. Pure Appl. Log.*, 56:183–220, 1992.
10. Doets K. *Basic Model Theory*. CSLI Publications, Stanford, CA, 1996.
11. Dutilh-Novaes C. *Medieval Obligationes as Argumentation Games*. Philosophical Institute, University of Leiden, Leiden, 2002.
12. Ehrenfeucht A. Application of games to some problems of mathematical logic, *Bull. Acad. Pol. Sci., Cl. III.*, 5:35–37, 1957.
13. Gabbay D.M., Johnson R.H., Ohlbach H.-J., and Woods J. editors. *Handbook of the Logic of Argument and Inference*. Elsevier Publishers, Amsterdam, 2002.
14. Girard J.-Y. *The Meaning of Logical Rules, Parts I, II*. manuscripts, University of Sophia-Antipolis, Nice, 1997.
15. Goranko V. *The Basic Algebra of Game Equivalences*. Preprint 2000–2012, ILLC, Amsterdam, 2000.
16. Harel D., Kozen D., and Tiuryn K. *Dynamic Logic*. MIT Press, Cambridge, MA, 2000.

17. Hintikka J. Logic, *Language-Games and Information: Kantian Themes in the Philosophy of Logic*. Clarendon Press, Oxford, 1973.
18. Hintikka J. Hyperclassical logic (a.k.a. IF logic) and its implications for logical theory, *Bull. Symbolic Log.*, 8(3):404–423, 2002.
19. Hintikka J., and Sandu G. Game-theoretical semantics. In J. van Benthem and A. ter Meulen, editors, *Handbook of Logic and Language*, pages 361–410. Elsevier, Amsterdam, 1997.
20. Hirsch R., and Hodkinson I. *Relational Algebras by Games*. Elsevier Science Publishers, Amsterdam, 2002.
21. Hodges W. *Building Models by Games*. Cambridge University Press, Cambridge, 1985.
22. Hodges W. *An Invitation to Logical Games, Lecture Notes*. Department of mathematics, Queen Mary's College, London, 1999.
23. Immerman N., and Kozen D. Definability with bounded number of bound variables. In *Proceedings 2nd IEEE Symposium on Logic in Computer Science*, pages 236–244, 1987.
24. Lewis D. General semantics. In D. Davidson and G. Harman, editors, *Semantics of Natural Language*, pages 169–218. Reidel, Dordrecht, 1972.
25. Lorenzen P. *Einfuhrung in die Operative Logik und Mathematik*. Springer, Berlin, 1955.
26. Löwe B. Consequences of the axiom of Blackwell determinacy. *Bull. Irish Math. Soc.*, 49:43–69, 2002.
27. Osborne M., and Rubinstein A. *A Course in Game Theory*. MIT Press, Cambridge, MA, 1994.
28. Parikh R. The logic of games and its applications. *Ann. Discrete Math.*, 24:111–140, 1985.
29. Parikh R. D-structures and their semantics. In J. Gerbrandy, M. Marx, M. de Rijke, and Y. Venema, editors, *JFAK: Essays Dedicated to Johan van Benthem on the Occasion of his 50th Birthday*, Vossiuspers, Amsterdam University Press, Amsterdam, 1999. http://www.illc.uva.nl/j50/
30. Pauly M. *Logic for Social Software*. Dissertation DS-2001-10, Institute for Logic, Language and Computation, University of Amsterdam, 2001.
31. Pauly M., and van der Hoek W. Modal logic of information and games. In P. Blackburn, J. van Benthem, and F. Wolter, editors, *Handbook of Modal Logic*. Elsevier, Amsterdam, 2006.
32. Rahman S., and Rueckert H. New perspectives in dialogical logic. *Synthese*, 127: 2001. Guest issue.
33. Stalnaker R. Extensive and strategic form: Games and models for games. *Res. Econ.*, 53(2):93–291, 1999.
34. Stirling C. Bisimulation, modal logic, and model checking games. *Log. J. IGPL. Spec. Issue Temporal Log.*, 7(1):103–124, 1999.
35. Väänänen J. *Team Logic*. Mathematical Institute, University of Helsinki, Helsinki, 2006.
36. Väänänen J. *Models and Games*. Cambridge University Press, Cambridge, 2011.
37. van Benthem J. *Exploring Logical Dynamics*. CSLI Publications, Stanford, CA, 1996.
38. van Benthem J. Games and strategies inside elementary logic. In *Invited lecture at the 7th Asian Logic Conference, Hsi-Tou*. Manuscript, Institute for Logic, Language and Computation, University of Amsterdam, Amsterdam, 1999.
39. van Benthem J. *Logic in Games, Lecture Notes*. ILLC Amsterdam & Department of Philosophy, Stanford University, Stanford, CA, 1999–2002. http://staff.science.uva.nl/~johan/Teaching/
40. van Benthem J. Games in dynamic-epistemic logic. *Bull. Econ. Res.*, 53(4):219–248, October 2001.
41. van Benthem J. Extensive games as process models. *J. Log. Lang. Inf.*, 11:289–313, 2002.
42. van Benthem J. Logic games are complete for game logics. *Stud. Log.*, 75:183–203, 2003.
43. van Benthem J. *Powers and Representation in Games*. Manuscript, Institute for Logic, Language and Computation, University of Amsterdam, Amsterdam, 2003.
44. van Benthem J. De kunst van het vergaderen. In W. van der Hoek, editor, *Liber Amicorum "John-Jules Charles Meijer 50"*, pages 5–7. Onderzoeksschool SIKS, Utrecht, 2004.
45. van Benthem J. Open problems in logic and games. In S. Artemov, H. Barringer, A. d'Avila Garcez, L. Lamb, and J. Woods, editors, *Essays in Honour of Dov Gabbay*, pages 229–264. King's College Publications, London, 2005.

46. van Benthem J. The epistemic logic of if games. In R. Auxier and L. Hahn, editors, *The Philosophy of Jaakko Hintikka (Schilpp Series)*, pages 481–513. Open Court Publishers, Chicago, IL, 2006.
47. van Benthem J. Logical construction games. In T. Aho and A.-V. Pietarinen, editors, *Truth and Games: Essays in Honour of Gabriel Sandu*, pages 123–138. Acta Philosophica Fennica, Helsinki, 2006.
48. van Benthem J. Dynamic logic of belief revision, *J. Appl. Non-Class. Logic*, 17(2): 129–155, 2007.
49. van Benthem J., Ghosh S., and Liu F. Modeling simultaneous games with concurrent dynamic logic. In *Proceedings of the First LORI Workshop*, Beijing, 2007.
50. van Benthem J., Girard P., and Roy O. *All Other Things Being Equal, a Logic of Ceteris Paribus Preference*. Institute for Logic, Language and Computation, University of Amsterdam, Amsterdam, 2007.
51. van Benthem J., and Liu F. Dynamic logics of preference upgrade, *J. Appl. Non-Class. Logic*, 17(2): 125–155, 2007.
52. van Benthem J., van Otterloo S., and Roy O. Preference logic, conditionals, and solution concepts in games. In H. Lagerlund, S. Lindström, and R. Sliwinski, editors, *Modality Matters*, pages 61–76. University of Uppsala, Uppsala, 2006.
53. Venema Y. Modal logic and algebra. In P. Blackburn, J. van Benthem, and F. Wolter, editors, *Handbook of Modal Logic*. Elsevier, Amsterdam, 2006.
54. Walton D. N., and Krabbe E. C. W. *Commitment in Dialogue: Basic Concepts of Interpersonal Reasoning*. State University of New York Press, Albany, NY, 1995.

Chapter 12
In memory of Jasu Magan Bhana Panchia (1963–1991): Iterated Belief Revision in Dynamic Doxastic Logic

Krister Segerberg

12.1 Introduction to DDL

In propositional dynamic doxastic logic (DDL) there are two kinds of nonclassical operators: doxastic and dynamic. The doxastic operators are **B** and **K**, with the duals **b** and **k**, respectively. The dynamic ones are $[*\varphi]$, for every purely Boolean formula φ, with the dual $\langle *\varphi \rangle$. Borrowing from modal terminology, we refer to $\mathbf{B}, \mathbf{K}$ and $[*\varphi]$ as box operators and $\mathbf{b}, \mathbf{k}$ and $\langle *\varphi \rangle$ as diamond operators. The former are primitive, the latter are defined by abbreviation (if $\triangle$ is a box operator, then its dual is defined as $\neg\triangle\neg$).This object language has been introduced in order to analyse certain theories about belief change. The following unofficial readings should be helpful for understanding the motivation:

$\mathbf{B}\theta$	"the agent believes that θ",
$\mathbf{b}\theta$	"it is consistent with the agent's beliefs that θ",
$\mathbf{K}\theta$	"the agent is doxastically committed to θ",
$\mathbf{k}\theta$	"it is consistent with the agent's doxastic commitments that θ",
$[*\varphi]\theta$	"after the agent has revised his beliefs with φ it is definitely the case that θ",
$\langle *\varphi \rangle\theta$	"after the agent has revised his beliefs with φ it is perhaps the case that θ".

In practice it is convenient to read "the agent knows that θ" for $\mathbf{K}\theta$ and "it is consistent with the agent's knowledge that θ" for $\mathbf{k}\theta$. However, insofar it suggests that the agent's doxastic commitments cannot be mistaken, this jargon is misleading. In other words, in this paper we are not speaking of knowledge in a Platonic or Cartesian sense but rather of what would be called knowledge in everyday life or perhaps "knowledge" (with scare quotes).To complete the above informal description of our object language, assume a denumerable set of propositional letters and

Krister Segerberg
Uppsala University, Uppsala, Sweden, e-mail: krister.segerberg@filosofi.uu.se

J. van Benthem et al. (eds.), *Proof, Computation and Agency*, Synthese Library 352,
DOI 10.1007/978-94-007-0080-2_12,

a truth-functionally complete set of Boolean operators. There are two restrictions on the language: (i) the dynamic operators can be applied only to purely Boolean formulæ, that is, formulæ containing at most Boolean operators, and (ii) the star operator $*$, too, applies only to purely Boolean formulæ. By contrast, a dynamic operator may be applied to any formula whatever. Thus in the language of basic DDL (the only language considered in this paper)$[*\varphi][*\psi]\theta$ is always well-formed if φ, ψ and θ are purely Boolean, whereas $\mathbf{BB}\varphi$ and $\mathbf{B}[*\varphi]\theta$ and $[*\mathbf{B}\varphi]\theta$ are always ill-formed. These restrictions reflect three assumptions basic to our analysis: (i) that purely Boolean formulæ represent possible (partial) states of the environment, (ii) that the agent holds beliefs about the environment, and (iii) that the agent can process new information about the environment. It is of course possible to make other assumptions; the ones chosen here are the simplest.

12.2 Basic DDL

Our basic axiom system consists of four blocks: classical, modal, AGM proper, and extra. Each block contains certain postulates, that is, axioms (described by means of axiom schemata) and rules of inference.The axioms of the classical block are the tautologies of classical propositional logic. The single rule is modus ponens:

(MP) from φ and $\varphi \to \psi$ infer ψ.

The postulates of the modal block are, for each modal operator $\triangle$, enough for the smallest normal modal logic K, for example,

(NN) $\triangle(\varphi \wedge \psi) \leftrightarrow (\triangle\varphi \wedge \triangle\psi)$,
(N) $\triangle\top$,
(RC) from $\varphi \leftrightarrow \psi$ infer $\triangle\varphi \leftrightarrow \triangle\psi$.

The block AGM proper consists of (in some cases free) translations into DDL language of the original AGM-axioms (the original numbering is kept):

($*2$) $[*\varphi]\mathbf{B}\varphi$,
($*3$) $[*\top]\mathbf{B}\varphi \to \mathbf{B}\varphi$,
($*4$) $\mathbf{b}\top \to (\mathbf{B}\varphi \to [*\top]\mathbf{B}\varphi)$,
($*5$) $\mathbf{k}\varphi \to \langle *\varphi\rangle\mathbf{b}\top$,
($*6$) $\mathbf{K}(\varphi \leftrightarrow \psi) \to ([*\varphi]\mathbf{B}\theta \leftrightarrow [*\psi]\mathbf{B}\theta)$,
($*7$) $[*(\varphi \wedge \psi)]\mathbf{B}\theta \to [*\varphi]\mathbf{B}(\psi \to \theta)$,
($*8$) $\langle *\varphi\rangle\mathbf{b}\psi \to ([*\varphi]\mathbf{B}(\psi \to \theta) \to [*(\varphi \wedge \psi)]\mathbf{B}\theta)$.

The final extra block consists of postulates that seem to be (more or less) implicit in the original AGM theory:

($*0$) $\theta \leftrightarrow [*\varphi]\theta$, if θ is purely Boolean,
($*$D) $[*\varphi]\theta \to \langle *\varphi\rangle\theta$,
($*$F) $\langle *\varphi\rangle\theta \to [*\varphi]\theta$,
(K$*$) $\mathbf{K}\theta \to [*\varphi]\mathbf{K}\theta$,

(∗K) $[*\varphi]\mathbf{K}\theta \to \mathbf{K}\theta$,
(KB) $\mathbf{K}\varphi \to \mathbf{B}\varphi$.

The basic modal logic underlying the present system was studied in [4]. For an analysis in DDL of one-shot AGM, see [5].Among the several kinds of semantics available we choose what we call the onion semantics, originally due (under a different name) to David Lewis [3] and introduced into the theory of belief change by Adam Grove [2]. By a *basic frame* we understand a Stone space (U,T); that is, a compact, totally separated topological space U, with T its set of open sets. The clopen sets (that is, those subsets of U that are both closed and open) are also called the *propositions* of the frame; and closed sets are also called *theories*. We write clop(T) for the set of clopen sets.An *onion* is a nonempty set of theories ordered by set inclusion (NESTEDNESS) and closed under arbitrary nonempty intersection (AINT). It follows that every onion O will contain the set $\bigcap O$ as an element, often referred to as the *belief set* of O. Elements of an onion are called *fallbacks*, and elements of an onion other than its belief set are called *proper* fallbacks. An onion O is said to be *trivial* if its only element is the empty set; that is, if $O = \{\emptyset\}$. Thus the trivial onion has an empty belief set and lacks proper fallbacks.An important condition discussed by David Lewis is his famous Limit Condition

(LIMIT): If an onion overlaps a proposition, then there is a smallest element of the onion that intersects that proposition.

More carefully, if O is an onion and P is a clopen set such that $P \cap \bigcup O \neq \emptyset$, then there is an element $Z \in O$ such that $P \cap Z \neq \emptyset$ and, for all $X \in O$, if $P \cap X \neq \emptyset$ then $Z \subseteq X$. It is worth noting that this condition, which is slightly weaker than the condition that the relation of set inclusion be well-ordered, holds in our version of the Lewis/Grove semantics:

Observation. *In a Stone space, onions satisfy* LIMIT.

The set $\mathsf{C}\bigcup O$, where C denotes topological closure, is called the *commitment set* of an onion O. Unlike the belief set, the commitment set need not be an element of O. But it can be, and it is said to be *replete* if it is. An *onion frame* is a structure (U,T,H,R) such that (U,T) is a basic frame, H is a set of onions, R is a function from clop(T) to $H \times H$, and certain conditions are satisfied:

(o1) If $(O,O') \in R^P$, then $\bigcap O' = P \cap \bigcup O$. (REVISION)
(o2) $\mathsf{C}\bigcup O = \mathsf{C}\bigcup O'$, for all $O,O' \in H$. (CONSTANT COMMITMENT)
(o3) For every $O \in H$ there is some $O' \in H$ such that $(O,O') \in R^P$. (SERIALITY)
(o4) If $(O,O') \in R^P$ and $(O,O'') \in R^P$, then $O' = O''$. (FUNCTIONALITY)

We might refer to H as the *heap* of onions and R as the *revision function* of the frame. Notice that in the presence of condition (o2), an onion frame containing the trivial onion has no other elements.A *valuation* in a Stone space (U,T) is a function from the set of propositional letters to the set clp(T) of clopen subsets of the space. It is clear that a valuation V can always be liftedto a function $\overline{V}$ defined on the set of all purely Boolean formulæ: $\overline{V}(\mathbb{P}) = V(\mathbb{P})$, for all propositional letters

$\mathbb{P}, \overline{V}(\varphi \wedge \psi) = \overline{V}(\varphi) \cap \overline{V}\psi), \overline{V}(\varphi \vee \psi) = \overline{V}(\varphi) \cup \overline{V}\psi), \overline{V}(\neg\varphi) = U - \overline{V}(\varphi)$, etc.When it is clear what valuation V is understood, we will write $[\![\varphi]\!]$ for $\overline{V}(\varphi)$. Note that the notation $[\![\varphi]\!]$ is meaningful only if φ is a purely Boolean formula.An *onion model* is an onion frame together with a valuation; that is to say, (U,T,H,R,V) is an onion model if (U,T,H,R) is an onion frame and V is a valuation in (U,T). The notion of *truth* of a formula in an onion model, symbolized by the symbol $\models$, is defined relative to a pair (O,u), where $O \in H$ and $u \in U$. (Intuitively, O is, or represents, the belief state of the agent, while u is the state of the environment).Truth-conditions:

$(O,u) \models \varphi$ iff $u \in [\![\varphi]\!]$, for every purely Boolean formula φ,
$(O,u) \models \neg\varphi$ iff not $(O,u) \models \varphi$,
$(O,u) \models \varphi \wedge \psi$ iff $(O,u) \models \varphi$ and $(O,u) \models \psi$,(analogous conditions for other Boolean operators,)

$(O,u) \models \mathbf{B}\varphi$ iff $\bigcap O \subseteq [\![\varphi]\!]$,
$(O,u) \models \mathbf{K}\varphi$ iff $\bigcup O \subseteq [\![\varphi]\!]$,
$(O,u) \models [*\varphi]\theta$ iff, for all O', if $(O,O') \in R^{[\varphi]}$ then $(O',u) \models \theta$.

Theorem 12.1 *The given axiom system is strongly complete with respect to the class of onion frames.*

12.3 Iteration

According to the pattern of analysis employed in this paper (discrete) belief change is modelled as a change from one belief state (the old belief state) to another (the new belief state). In revising the old belief state by new information in the form of a proposition, it seems reasonable to build the new belief state with material supplied by the old onion and the proposition. Richer modellings may take further aspects into account, but the proposed restriction will make it possible to discern certain principles regulating what possibilities of iteration there are. As we are representing belief states by onions, this suggestion means that the elements of the new onion should be constructed

- from certain sets: elements of the old onion and the proposition,
- with the help of appropriate set theoretical operations.

Specifically, let O_* be the old onion waiting to be revised by a proposition P; we are wondering what a new onion O^* may look like. (We assume all onions satisfy the conditions (o1)–(o4) laid down above but also, in addition, that they are replete. Two consequences of this assumption are worth keeping in mind: revision (in this theory) always leads to a unique new onion O^*, and all onions contain their own commitment set as an element (in fact, the same commitment set K for every onion).) By a *basic building block* let us mean the fallbacks of O_* and P. In other words, the set of basic building blocks is $O_* \cup \{P\}$. Refining the suggestion of the previous paragraph, we stipulate that, in the case of revision, *the elements of any new onion* *should be constructed in terms of basic building blocks with the help of intersection and union.* Let us see where this programme leads. Supposing that O_* and P overlap (otherwise revision is trivial) let Z be the smallest element of O_* to intersect P. By a *fallback candidate* (with respect to revising O_* by P) we shall mean any set $V \cup PW$,where $V \in O_* \cup \{\emptyset\}$ and $W \in O_*$ and $V \subseteq W$ and $Z \subseteq W$. (We sometimes represent set theoretical intersection by juxtaposition; thus $V \cup PW = V \cup (P \cap W)$.) We define an onion as an *onion candidate* (with respect to revision by P) if all its elements are fallback candidates and, in particular, it contains PZ and K.We say that an onion candidate O^* is *maximal* (with respect to O and P) if it is not propertly included in any other onion candidate.Moreover, we say that $O^*$is *trusting* (with respect to O and P) if $PK \in O^*$, *skeptical* (with respect to O and P) if $Z \in O^*$. The strangeness of this terminology may be mitigated by replacing "onion" by "belief state". Including the agent in this terminology, we might say that an agent is *trusting in O_* with respect to P* if O^* is trusting in the sense just defined, and that the agent is *maximally trusting in O^* with respect to P* if O^* is maximal as well as trusting. Furthermore, we might say that the agent is *generally trusting in O^** if he is trusting with respect to all propositions, and that he is *maximally generally trusting in O^** if he is maximally trusting with respect to all propositions.And we might say that he is *in trusting mood* or in *maximally trusting mood* if in all of his possible belief states he is generally trusting or maximally trusting, respectively. In an analogous way we may define notions in which the term "trusting" is replaced by the term "skeptical". (This is more terminology than we need here, but the terminology is of some interest in its own right.)It is worth noting that the elements of a trusting (not necessarily maximal) onion candidate fall into one of two categories: an element is either of type PW where $Z \subseteq W \subseteq K$, or of type $V \cup PK$, where V is an element of the old onion. Similarly, the elements of a skeptical onion candidate fall into one of three categories: an element is either of type $V \cup PZ$ where $V \subseteq Z$ or $V = \emptyset$, or of type$V$ where $Z \subseteq V \subseteq K$,or of type $V \cup PW$, where V and W, in that order, are consecutive elements of the old onion such that $Z \subseteq V \subseteq K$. (By definition, two elements X and Y of O are consecutive, in that order, in O if $X \subset Y$ and, furthermore, for all $W \in O$, if $X \subseteq W \subseteq Y$ then $W = X$ or $W = Y$).

12.4 Two Examples

It is not surprising that revision in different moods generates different logics. In this section we give two prime examples. Consider the following two schemata:

(∗t) $\mathbf{k}(\varphi \wedge \psi) \rightarrow ([*\varphi][*\psi]\mathbf{B}\theta \leftrightarrow [*(\varphi \wedge \psi)]\mathbf{B}\theta)$,

(∗s) $(\mathbf{b}\top \wedge \mathbf{B}\psi \wedge [*\varphi]\mathbf{B}\neg\psi) \rightarrow (\mathbf{B}\theta \leftrightarrow [*\varphi][*\psi]\mathbf{B}\theta)$.

Lemma 12.1 *The schema (∗t) is valid in the maximally trusting mood.*

Proof. Let (U,T,H,R) be any onion frame with maximally trusting revision. Fix a valuation in (U,T). For any onions $O, O', O'', O^\star$ in H, assume

(1) $(O,O') \in R^{[\varphi]}$,

(2) $(O', O'') \in R^{[\psi]}$,
(3) $(O, O^\star) \in R^{[\varphi \wedge \psi]}$.

Furthermore assume, for any point u in the model,

(4) $(O, u) \vDash \mathbf{k}(\varphi \wedge \psi)$.

It will be enough to prove that O'' and $O^\star$ have the same belief set; that is, that $\bigcap O'' = \bigcap O^\star$.Let X be the smallest element of O that intersects $[\![\varphi]\!]$, let Y the smallest element of O' that intersects $[\![\psi]\!]$, and let Z be the smallest element of O that intersects $[\![\varphi \wedge \psi]\!]$. (Thanks to (4) and repleteness, we know that those elements exist.) Note:

(5) $\bigcap O' = X \cap [\![\varphi]\!]$,
(6) $\bigcap O'' = Y \cap [\![\psi]\!]$,
(7) $\bigcap O^\star = Z \cap [\![\varphi \wedge \psi]\!]$.

In these terms, it will be enough to prove that

$(*)$ $Y \cap [\![\psi]\!] = Z \cap [\![\varphi \wedge \psi]\!]$.

We divide the proof of $(*)$ into the two natural halves. Since the mood is maximally trusting and$Z$ is an element of O that intersects $[\![\varphi]\!]$, it follows from (1) that $Z \cap [\![\varphi]\!]$ is an element of O'. By (3), (4) and (7), $Z \cap [\![\varphi]\!]$ intersects $[\![\psi]\!]$.Hence $Y \subseteq Z \cap [\![\varphi]\!]$, by the minimality of Y. A fortiori, $Y \cap [\![\psi]\!] \subseteq Z \cap [\![\varphi]\!] \cap [\![\psi]\!] = Z \cap [\![\varphi \wedge \psi]\!]$, which yields the first half of $(*)$.For the other half, note that since Y in an element of $O'$and we are in trusting mood, either $Y = V \cap [\![\varphi]\!]$, for some element V in O, or else $Y = W \cup (K \cap [\![\varphi]\!])$, for some element W in O, The two subcases are treated separately. In the first subcase, note that Y intersects $[\![\psi]\!]$ by (2), (4) and (6), and therefore V intersects $[\![\varphi \wedge \psi]\!]$. Hence $Z \subseteq V$, by the minimality of Z. A fortiori, $Z \cap [\![\varphi \wedge \psi]\!] \subseteq V \cap [\![\varphi \wedge \psi]\!] = V \cap [\![\varphi[\!\cap\![\psi]\!] = Y \cap [\![\psi]\!]$, as we wanted. The second subcase is trivial: since $Z \subseteq K$, of course $Z \cap [\![\varphi \wedge \psi]\!] \subseteq K \cap [\![\varphi \wedge \psi]\!] \subseteq (W \cap [\![\psi]\!]) \cup (K \cap [\![\varphi \wedge \psi]\!]) = (W \cap [\![\psi]\!]) \cup (K \cap [\![\varphi]\!] \cap [\![\psi]\!]) = (W \cup (K \cap [\![\varphi]\!])) \cap [\![\psi]\!] = Y \cap [\![\psi]\!]$,again what we wanted. $\odot$

Lemma 12.2 *($*$t) is not valid in the maximally skeptical mood.*

Proof. The problem is to find a particular instance of ($*$t) that fails in a particular maximally skeptical model. We give an outline of the construction but not all details.What is needed is four sets A, B, P and Q answering to the following specifications: $A \subset B$, $PA \neq \emptyset$, $QA \neq \emptyset$, $PQA = \emptyset$ and $PQB \neq \emptyset$. Define $O = \{A, B\}$, obviously an onion. We wish to define onions O', O'' and $O^\star$ and a revision function R such that $(O, O') \in R^P, (O', O'') \in R^Q$ and $(O, O^\star) \in R^{PQ}$. If our model-to-be is to be maximally skeptical, we have no choice but to make the following definitions:

$O' = \{PA,\ A,\ A \cup PB,\ B\}$,
$O'' = \{QA,\ (P \cup Q)A,\ A,\ A \cup PQB,\ A \cup PB,\ A \cup (P \cup Q)B,\ B\}$,
$O^\star = \{PQB,\ A \cup PQB,\ B\}$.

Thus the belief sets QA of O'' and PQB or $O^\star$ are disjoint. To build an actual model, find a set of U of elements in which there are subsets A, B, P and Q as desired. For example, let $U = \{0,1,2\}$ and define $A = \{0,1\}, B = \{0,1,2\}, P = \{0,2\}$ and $Q = \{1,2\}$. Let T be the power set of U. Let H be the heap of onions and R the revision function generated by O and the condition of maximal skepticality (we already witnessed the generation of three new onions as well as one onion pair in each of R^P, R^Q and R^{PQ}; there are of course several onions and onion pairs to be added). With the onion frame (U,T,H,R) defined, let $\mathbb{P}, \mathbb{Q}$ and $\mathbb{R}$ be three propositional letters and consider a valuation in (U,T) mapping $\mathbb{P}$ to P, $\mathbb{Q}$ to Q, and $\mathbb{R}$ to either QA or PQB.You get your counterexample(s) by taking $\mathbb{P}, \mathbb{Q}$ and $\mathbb{R}$ as φ, ψ and θ, respectively. $\odot$

Lemma 12.3 *The schema (∗s) is valid in the maximally skeptical mood.*

Proof. Let O be any onion and u any point in any maximally skeptical onion model (U,T,H,R,V). Assume

(1) $(O,u) \vDash \mathbf{b}\top$,
(2) $(O,u) \vDash \mathbf{B}\psi$,
(3) $(O,u) \vDash [*\varphi]\mathbf{B}\neg\psi$.

Let O' and O'' be the onions such that $(O,O') \in R^{[\varphi]}$ and $(O',O'') \in R^{[\psi]}$. Write $X = \bigcap O$ and $Y = \bigcap O'$ and $Z = \bigcap O''$. Note that by (1)–(3)

(4) $X \neq \emptyset$,
(5) $X \subseteq [\![\psi]\!]$,
(6) $Y \cap [\![\psi]\!] = \emptyset$.

Since Y is the smallest element of O' and the modelling is maximally skeptical, $Y \cup X$ is the second smallest element of O'. By (6), Y does not intersect $[\![\psi]\!]$ but, by (4) and (5), X does.Consequently, $Z = (Y \cup X) \cap [\![\psi]\!] = X$ and hence$\bigcap O = \bigcap O''$. But if O and O'' have the same belief set, then they also support the same beliefs. $\odot$

Lemma 12.4 *The schema (∗s) is not valid in the maximally trusting mood.*

Proof. A finite countermodel can be built along the following lines. As in Lemma 12.2, we will not give all details.First one has to find nonempty sets A, B, C and P such that $A \subset B \subset C$ and $PA = \emptyset$ and $PB \neq \emptyset$. Define $O = \{A, B, C\}$. In a maximally trusting modelling with revision function R, the unique onion O' such that $(O,O') \in R^P$ is the set $\{PB, PC, PC \cup A, PC \cup B, C\}$. Let $Q = \overline{B} = U - B$. There is a unique onion O'' such that $(O',O'') \in R^Q$; in fact, $O'' = \{P\overline{B}C, \overline{B}C, PB \cup \overline{B}C, PB \cup A \cup \overline{B}C, C\}$. Note that the belief sets A of O and $P\overline{B}C$ of O'' are disjoint. Now determine a set U of elements and define A, B, C, and P as subsets of U in accordance with the preceding specifications. Most simply, let $U = \{0,1,2\}$ and define $A = \{0\}, B = \{0,1\}, C = \{0,1,2\}$ and $P = \{1,2\}$. Let T be the power set of U. Construct the heap H of onions and the revision relation R in accordance with maximal trust. Let V be a valuation in (U,T) such that $V(\mathbb{P}) = P$ and $V(\mathbb{Q}) = Q$, where $\mathbb{P}$

and $\mathbb{Q}$ are two different propositional letters. Put $\mathbb{P}$ for φ and $\mathbb{Q}$ for ψ. Let $\mathbb{R}$ be a third propositional letter, distinct from both $\mathbb{P}$ and $\mathbb{Q}$. The instances of (∗s) obtained by letting θ be either $\mathbb{R}$ or $\neg\mathbb{R}$ both fail in the resulting model. ⊙

The following result is an immediate consequence of the four lemmas:

Observation. *Of the two logics of maximally trusting revision and maximally skeptical revision, respectively, neither is included in the other.*

12.5 The Lattice of Fallback Candidates

It is clear that the fallback candidates form a lattice $\mathscr{L} = \mathscr{L}(O_*, P)$ with set theoretical union as the join and set theoretical intersection as the meet; furthermore, there is a unit element (K) and a null element (PZ). (See the Fig. 12.1 for an illustration.) Let us use the term *path* for any sequence of fallback candidates in the order of increasing set inclusion. There is a close connexion between paths and onion candidates. Clearly the sequence p_O of elements of an onion candidate O, ordered by set inclusion, is a path from PZ to K. Conversely, the set O_p of elements appearing in a path p from PZ to K is a set of closed sets satisfying NESTEDNESS; hence this set is an onion candidate if also the condition AINT is satisfied. Let us say that a path p is *meet-complete* if $\bigcap q$ is an element of p for every subpath q of p.

Observation. *If p is a meet-complete path from PZ to K, then O_p is a candidate onion.*

Proof. Assume that p is a meet-complete path from PZ to K. It is clear that O_p contains both PZ and K as elements and that NESTEDNESS is satisfied. Thus it will be enough to prove that O_p satisfies AINT. Let Sbe any nonempty subset of p. Then there must be a nonempty index set I such that $S = \{V_i \cup PW_i : i \in I\}$, where, according to the definition in the preceding section, for all $i \in I, V_i$ and W_i are elements of the old onion O_* such that $V_i \in O_* \cup \{\emptyset\}$ and $W_i \in O_*$ and $V_i \subseteq W_i$ and $Z \subseteq W_i$.It is readily seen (using the fact that O_* satisfies NESTEDNESS and AINT) that there are elements V' and W' of the old onion such that $V' \in O_* \cup \{\emptyset\}$ and $W' \in O_*$ and $V' \subseteq W'$ and $Z \subseteq W'$ and

$$\bigcap S = \bigcap_{i \in I}(V_i \cup PW_i) = V' \cup PW'.$$

Thus $\bigcap S$ is a fallback candidate. By the meet-completeness of p, therefore, $\bigcap S$ is an element of p. But then $\bigcap S$ is also an element of O_p, which is what we wanted to prove. ⊙

Corollary 12.1 *There is a one-one correspondence between meet-complete paths from PZ to K (inclusive) and onion candidates.*

The lattice $\mathscr{L}$ offers a picturesque representation of what, under our self-imposed limitation, can be the result of revising the onion O_* by the proposition P. By the

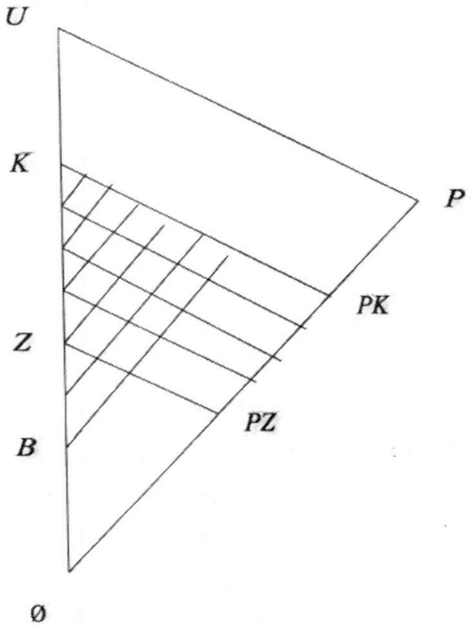

Fig. 12.1 Illustration of the "fallback candidates".

corollary, any meet-complete path from PZ to K yields an onion. Thus the problem of finding a method or mood for revision is formally the same as the problem of finding a strategy for choosing a meet-complete path with certain properties in the relevant lattice. In this paper, two such strategies have been examined, but obviously they are just the two extremes among infinitely many possible strategies. We may say of those two that they are *logically distinct*, meaning that, as remarked in the observation at the end of the preceding section, they give rise to different logics in the object language defined in this paper. It would be interesting to know how many logically distinct moods of revision there are.

A final note on maximality. If one uses the lattice $\mathscr{L}$ for the purpose of finding onion candidates, one must bear in mind that meet-complete paths need not be maximal: an onion candiate need not be a maximal onion candidate. This is an interesting fact and, from the point of view of application, important. The interest springs from the conflict between information and economy. Choosing anything less than a onion corresponding to a maximal path means that some memory will be lost. On the other hand, choosing an onion corresponding to maximal paths typically increases complexity. To illustrate how complexity grows, let us introduce the simplifying assumption that the elements of the onion O_0 is well-ordered by the inclusion relation. (The belief set of O_* is first in the well-ordering and the commitment set of O_* is last.)Let α and β be ordinals such that α is the order type of the set $\{X \in O_* : X \subset Z\}$ and β is the order type of the set $\{X \in O_* : Z \subset X \subset K\}$. The two extreme strategies are t, the maximally trusting one, and and s, the maximally skeptical one. Evidently the order type of the old onion is

$$1+\alpha+1+\beta+1$$

while that of the maximally trusting new onion is

$$1+\beta+1+\alpha+1+\beta+1$$

and that of the maximally skeptical new onion is

$$1+\alpha+2\cdot(\beta+1)+1,$$

where the dot stands for ordinal multiplication (and binding harder than the plus of ordinal addition). In the finite case, these entities are equal: if we write m for the finite ordinal α and n for the finite ordinal β, then the order type of the old onion is $m+n+2$ while the trusting order type and the skeptical order type both reduce to $m+2n+4$. (In the case depicted in the figure, $m=2$ and $n=3$. Thus both the maximally trusting onion candidate and the maximally skeptical one have twelve elements, as compared with the seven of the old onion.) But in the infinite case the two order types may differ. For example, if $\alpha=\omega$ and $\beta=\omega+\omega=\omega\cdot 2$ (and the old onion thus has $\omega\cdot 3+1$ elements), the maximally trusting onion candidate has $\omega\cdot 5+1$ elements, while the maximally skeptical onion candidate has$\omega\cdot 3+1$ elements. In the latter case, all onion candidates corresponding to maximal paths have ordinal type $\omega\cdot 3+1$ or $\omega\cdot 4+1$ or $\omega\cdot 5+1$.

12.6 Concluding Remarks

The treatment of belief revision in this paper should be seen as one representative instance of a general way of analysing belief change (in a static universe). There is a similar treatment of related kinds of belief revision, in particular contraction, an operation exhibiting even greater variation. In sum, the approach suggested in this paper is

- to treat belief change as a relation between belief states,
- to formalize belief states as hypertheories,
- to construct new hypertheories from material deriving from the old hypertheories and the new information with which the change is to be made.

In this paper, the only hypertheories to be studied are what we have called onions. But there are also more general alternatives.[1]

[1] The research reported in this paper was first presented on 13 September 2002 at a conference on formal epistemics at Villa Lanna, Prague and more recently on 13 November 2005 at a conference on the understanding the dynamics of knowledge at the Certosa di Pontignano, Siena. It is dedicated to the memory of Jasu, our beloved daughter-in-law.

References

1. Alchourrón C., Gärdenfors P., and Makinson D. On the logic of theory change, *J. Symbolic Log.*, 50: 510–530, 1985.
2. Grove A. Two modellings for theory choice, *J. philos. Log.*, 17: 157–170, 1988.
3. Lewis D. *Counterfactuals*, Blackwell, Oxford, 1973.
4. Segerberg K. The lattice of basic modal logics. In T. Childers and J. Palomäki, editor, *Between Words and Worlds: A Festschrift for Pavel Materna*, pages 170–183. Filosofia, Prague, 2000.
5. Segerberg K. The basic dynamic doxastic logic of AGM. In M.-A. Wiilliams and H. Rott, editor, *Frontiers of Belief Revision*, pages 57–84. Kluwer, Dordrecht, 2001.

Chapter 13
Towards a Logical Analysis of *Adjusted Winner*

Eric Pacuit

13.1 Introduction

It is often convenient to view a computational procedure, or a program, as a relation on a set of states, where a state can be thought of as a function that assigns a value to every possible variable and a truth value to all propositions. This idea was proposed by Pratt [19] and extends the work of of Floyd [4] and Hoare [6]. Harel, Kozen and Tiuryn [5] provide a very thorough discussion of computational procedures from this point of view. In [12], Rohit Parikh suggests that a similar framework can be developed for studying *social* procedures, such as fair division algorithms or voting protocols.

In fact, the first example of such an analysis can be found in an earlier paper of Parikh [11]. In [11], a **PDL**-style logic, called **game logic**, is developed for reasoning about multi-agent strategic situations. It is then used to show that the Banach-Knaster last diminisher procedure to fairly divide a cake is "correct". Here "correct" means that each agent has a *strategy* to ensure that it receives a piece of the cake that is fair *from its point of view*. See [11, 16] for a discussion of this result and the current state of affairs of game logic.

Recently, there have been a number of attempts to answer Parikh's challenge to develop logical methods for studying social procedures. The different approaches can roughly be divided into two categories. The first category makes use of model checking techniques to verify that a given game-theoretic mechanism satisfies a particular *specification* (written in some logical language). The idea is to draw an analogy with the formal verification of computer systems via model checking techniques where the desired specifications are expressed in some temporal logic. In the game-theoretic setting, modal strategy logics, such as alternating temporal logic (see [1]) or coalitional logic (see [14]), are used to express the desired specification.

Eric Pacuit
Tilburg University, Tilburg Institute for Logic and Philosophy of Science, Warandelaan 2, 5037 AB Tilburg, The Netherlands, e-mail: e.j.pacuit@uvt.nl

J. van Benthem et al. (eds.), *Proof, Computation and Agency*, Synthese Library 352, DOI 10.1007/978-94-007-0080-2_13, © Springer Science+Business Media B.V. 2011

See (Pauly M., and Wooldridge M. *Logic for Mechanisam Design – a Main festo.* Unpublished Manuscript.) [19], the recent dissertation [18] and references therein for details of this approach.

The second category of papers follow in the footsteps of Floyd [4], Hoare [6] and Pratt [19] and develops a formal calculus to reason about social procedures. This is the direction that Parikh took in [11] and more recently Pauly in [15]. A discussion of this approach can be found in Section 13.3.

In this paper, we apply the methods of [15] to fair division algorithms. In particular, we look at *Adjusted Winner (AW)* – an algorithm for "fairly " dividing n goods among two people invented by Steven Brams and Alan Taylor. See [2] for a discussion of *AW* and an excellent overview of other fair division algorithms. We will see that the techniques from [15] cannot be directly applied to the analysis of Adjusted Winner. The paper is organized as follows. The next section describes the Adjusted Winner procedure to "fairly" divide n divisible goods between two people. Section 13.3 outlines how to extend Pauly's framework to reason about Adjusted Winner and raises some difficulties.

13.2 The *Adjusted Winner* Procedure

We begin with an example that illustrates how *AW* works. Suppose there are two players, called Ann (A) and Bob (B), and n (divisible[1]) goods ($G_1, \ldots, G_n$) which must be distributed to Ann and Bob. The goal of the Adjusted Winner algorithm is to *fairly* distribute the n goods between Ann and Bob. We begin by discussing an example which illustrates the Adjusted Winner algorithm.

Suppose Ann and Bob are dividing three goods: G_1, G_2, and G_3. *Adjusted Winner* begins by giving both Ann and Bob 100 points to divide among the three goods. Suppose that Ann and Bob assign these points according to the following table.

Item	Ann	Bob
G_1	$\underline{10}$	7
G_2	$\underline{65}$	43
G_3	25	$\underline{50}$
Total	100	100

The first step of the procedure is to give G_1 and G_2 to Ann since she assigned more points to those items, and item G_3 to Bob. However this is not an equitable outcome since Ann has received 75 points while Bob only received 50 points (each according to their personal valuation). We must now transfer some of Ann's goods to Bob. In order to determine which goods should be transfered from Ann to Bob, we look at the ratios of Ann's valuations to Bob's valuations. For G_1 the ratio is $10/7 \approx 1.43$

[1] Actually all we need to assume is that *one* good is divisible. However, since we do not know before the algorithm begins *which* good will be divided, we assume all goods are divisible. See [2, 3, 9] for a discussion of this fact.

and for G_2 the ratio is $65/43 \approx 1.51$. Since 1.43 is less than 1.51, we transfer as much of G_1 as needed from Ann to Bob[2] to achieve equitability.

However, even giving all of item G_1 to Bob will not create an equitable division since Ann still has 65 points, while Bob has only 57 points. In order to create equitability, we must transfer part of item G_2 from Ann to Bob. Let p be the proportion of item G_2 that Ann will keep. p should then satisfy

$$65p = 100 - 43p$$

yielding $p = 100/108 = 0.9259$, so Ann will keep 92.59% of item G_2 and Bob will get 7.41% of item G_2. Thus both Ann and Bob receive 60.185 points. It turns out that this allocation (Ann receives 92.59% of item G_2 and Bob receives all of item G_1 and item G_3 plus 7.41% of item G_2) is *envy-free*, *equitable* and *efficient*, or *Pareto optimal*. In fact, Brams and Taylor show that Adjusted Winner *always* produces such an allocation [2]. We will discuss these properties in more detail below.

Suppose that $G_1, \ldots, G_n$ is a fixed set of goods, or items. A **valuation** of these goods is a vector of natural numbers $\langle a_1, \ldots, a_n \rangle$ whose sum is 100. Let $\alpha, \alpha', \alpha'', \ldots$ denote possible valuations for Ann and $\beta, \beta', \beta'', \ldots$ denote possible valuations for Bob. An **allocation** is a vector of n real numbers where each component is between 0 and 1 (inclusive). An allocation $\sigma = \langle s_1, \ldots, s_n \rangle$ is interpreted as follows. For each $i = 1, \ldots, n$, s_i is the proportion of G_i given to Ann. Thus if there are three goods, then $\langle 1, 0.5, 0 \rangle$ means, "Give all of item 1 and half of item 2 to Ann and all of item 3 and half of item 2 to Bob." Thus *AW* can be viewed as a function that accepts Ann's valuation α and Bob's valuation β and returns an allocation σ. It is not hard to see that every allocation produced by *AW* will have a special form: all components except one will be either 1 or 0.

We now give the details of the procedure. Suppose that Ann and Bob are each given 100 points to distribute among n goods as he/she sees fit. In other words, Ann and Bob each select a valuation, $\alpha = \langle a_1, \ldots, a_n \rangle$ and $\beta = \langle b_1, \ldots, b_n \rangle$ respectively. For convenience rename the goods so that

$$a_1/b_1 \geq a_2/b_2 \geq \cdots a_r/b_r \geq 1 > a_{r+1}/b_{r+1} \geq \cdots a_n/b_n$$

Let α/β be the above vector of real numbers (after renaming of the goods). Notice that this renaming of the goods ensures that Ann, based on her valuation α, values the goods $G_1, \ldots, G_r$ at least as much as Bob; and Bob, based on his valuation β, values the goods $G_{r+1}, \ldots, G_n$ more than Ann does. Then the *AW* algorithm proceeds as follows:

1. Give all the goods $G_1, \ldots, G_r$ to Ann and $G_{r+1}, \ldots, G_n$ to Bob. Let X, Y be the number of points received by Ann and Bob respectively. Assume for simplicity that $X \geq Y$.
2. If $X = Y$, then stop. Otherwise, transfer a portion of G_r from Ann to Bob which makes $X = Y$. If equitability is not achieved even with all of G_r going to Bob, transfer $G_{r-1}, G_{r-2}, \ldots, G_1$ in that order to Bob until equitability is achieved.

[2] When the ratio is closer to 1, a unit gain for Bob costs a smaller loss for Ann.

Thus the *AW* procedure is a function from pairs of valuations to allocations. Let $\mathsf{AW}(\alpha,\beta)=\sigma$ mean that σ is the allocation given by the procedure *AW* when Ann announces valuation α and Bob announces valuation β. In [2, 3], it is argued that *AW* is a "fair" procedure, where fairness is judged according to the following properties.

Let $\alpha=\langle a_1,\dots,a_n\rangle$ and $\beta=\langle b_1,\dots b_n\rangle$ be valuations for Ann and Bob respectively. An allocation $\sigma=\langle s_1,\dots,s_n\rangle$ is

- **Proportional** if both Ann and Bob receive at least 50% of their valuation. That is, $\sum_{i=1}^n s_i a_i \geq 50$ and $\sum_{i=1}^n (1-s_i)b_i \geq 50$.
- **Envy-Free** if no party is willing to give up its allocation in exchange for the other player's allocation. That is, $\sum_{i=1}^n s_1 a_i \geq \sum_{i=1}^n (1-s_i)a_i$ and $\sum_{i=1}^n (1-s_i)b_i \geq \sum_{i=1}^n s_i b_i$.
- **Equitable** if both players receive the same total number of points. That is $\sum_{i=1}^n s_i a_i = \sum_{i=1}^n (1-s_i)b_i$.
- **Efficient** if there is no other allocation that is strictly better for one party without being worse for another party. That is for each allocation $\sigma'=\langle s'_1,\dots,s'_n\rangle$ if $\sum_{i=1}^n a_i s'_i > \sum_{i=1}^n a_i s_i$, then $\sum_{i=1}^n (1-s'_i)b_i < \sum_{i=1}^n (1-s_i)b_i$. (Similarly for Bob).

In order to simplify notation, let $V_A(\alpha,\sigma)$ be the total number of points Ann receives according to valuation α and allocation σ and $V_B(\beta,\sigma)$ the total number of points Bob receives according to valuation β and allocation σ.

It is not hard to see that for two-party disputes, proportionality and envy-freeness are equivalent. For a proof, notice that

$$\sum_{i=1}^n a_i s_i + \sum_{i=1}^n a_i(1-s_i) = \sum_{i=1}^n a_i s_i + \sum_{i=1}^n a_i - \sum_{i=1}^n a_i s_i = 100$$

Then if σ is envy free for Ann, then $\sum_{i=1}^n a_i s_i \geq \sum_{i=1}^n a_i(1-s_i)$. Hence, $2\sum_{i=1}^n a_i s_i \geq \sum_{i=1}^n a_i = 100$. And so, $\sum_{i=1}^n a_i s_i \geq 50$. The argument is similar for Bob.

Conversely, suppose that σ is proportional. Then since $\sum_{i=1}^n a_i s_i \geq 50$, $\sum_{i=1}^n a_i s_i + \sum_{i=1}^n a_i s_i \geq 100 = \sum_{i=1}^n a_i$. Then $\sum_{i=1}^n a_i s_i + \sum_{i=1}^n a_i s_i - \sum_{i=1}^n a_i \geq 0$. Hence, $\sum_{i=1}^n a_i s_i - \sum_{i=1}^n a_i(1-s_i) \geq 0$. And so, $\sum_{i=1}^n a_i s_i \geq \sum_{i=1}^n a_i(1-s_i)$. The proof is similar for Bob.

Returning to *AW*, it is easy to see the *AW* only produces equitable allocations (equitability is essentially built in to the procedure). Brams and Taylor go on to show that *AW*, in fact, satisfies all of the above properties.

Theorem 13.1 (Brams and Taylor [2]) *AW produces an allocation of the goods based on the announced valuations that is efficient, equitable and envy-free.*

A formal proof of this Theorem is provided in [2]. See also [9] for a number of new results about the Adjusted Winner procedure.

13.3 Towards a Logical Analysis

Hoare logic contains expression of the form $\{P\}\alpha\{Q\}$ where α is intended to be a program and P and Q are intended to be pre- and post-conditions respectively. The intended interpretation is that if α starts in some state satisfying P, then α halts (*if* it does halt) in a state satisfying Q. Here the program α is a formal expression in some *programming language* (for example, the WHILE-language) and P and Q are formulas in some logical language (say first-order logic). Details are given below. Pauly's key idea in [16] is to extend the formal programming language with expressions $\mathsf{ch}_{\mathscr{A}}(\{x_i \mid i \in \mathscr{A}\})$ intended to mean

"*each agent $i \in \mathscr{A}$ independently chooses a value for the variable x_i.*"

It is assumed that the agents make their choice *simultaneously*. Thus, $\mathsf{ch}_{\mathscr{A}}(\{x_i \mid i \in \mathscr{A}\})$ is best thought of as a *strategic game form* (cf. [8]) where the choices for each agent $i \in \mathscr{A}$ is the set of possible values for the variable x_i.

13.3.1 Relevant Details

Pauly extends the work of Hoare to develop a formal calculus for reasoning about game theoretic mechanisms. In the interest of space, we will not go into full details of Pauly's framework, but rather sketch the main ideas (the interested reader can consult [15]).

The Mechanism Programming Language: The mechanism programming language (*MPL*) is a simple extension of the well-known WHILE-language (cf. [7]). Assume $\mathscr{A}$ is a non-empty set of agents, $\mathscr{V}$ a set of variables, $\mathbb{F}$ a set of function symbols and $\mathscr{R}$ a set of relation symbols. **Terms** t and **boolean expressions** φ are defined in the usual way:

$$t =_{\text{def}} x \mid f^k(t_1, \ldots, t_k)$$
$$\varphi =_{\text{def}} \top \mid R^k(t_1, \ldots, t_k) \mid \neg\varphi \mid \varphi_1 \wedge \varphi_2$$

where $f \in \mathbb{F}$ and $R \in \mathscr{R}$ have arity $k \in \mathbb{N}$. These boolean expressions and terms are used to create *game expressions*.

Definition 13.1 A **game expression** of *MPL* is generated according to the following grammar:

$$\gamma =_{\text{def}} x := t \mid \gamma_1 ; \gamma_2 \mid \mathsf{if}\ \varphi\ \mathsf{then}\ \gamma_1\ \mathsf{else}\ \gamma_2 \mid \mathsf{while}\ \varphi\ \mathsf{do}\ \gamma \mid \mathsf{ch}_{\mathscr{A}}(\{x_i \mid i \in \mathscr{A}\})$$

where t is a term and φ a boolean expression.

Pauly shows that this language is quite general in the sense that it can be used to describe a large number of game-theoretic mechanisms. For example, a solution to Solomon's Dilemma and various auctions can be described in this language (see [15] for details).

Semantics: As in Hoare logic, a **partial correctness assertion** has the form $\{P\}\gamma\{Q\}$, where P and Q are *predicates*. The intended interpretation is: *given an initial state* s_0, *there is a* subgame perfect equilibrium[3] *of the* game *described by* γ *consistent with the predicate* Q. We now make this precise.

Start with a standard first-order model containing a domain D and an interpretation $\mathscr{I}$ (which interprets elements of $\mathbb{F}$ and $\mathscr{R}$ as functions and relations over D respectively). A **state** is a function $s : \mathscr{V} \to D$. Given an interpretation $\mathscr{I}$ and an initial state s_0, game expressions are interpreted as *extensive game forms*,[4] denoted $G(s_0, \gamma, \mathscr{I})$. Instead of giving a formal definition, we work through a simple example. Suppose that the domains consists of two elements $D = \{a, b\}$, there are four variables $\mathscr{V} = \{x, x_i, x_j, x_k\}$ and three agents $\mathscr{A} = \{i, j, k\}$. Consider the game expression

$$(x := a); \mathsf{ch}_{\{i\}}(\{x_i\}); \mathsf{ch}_{\{j,k\}}(\{x_j, x_k\}).$$

Suppose the initial state is $s_0 : \mathscr{V} \to D$ is given by $s_0(v) = b$ for all $v \in \mathscr{V}$. The game form corresponding to this game expression is pictured below:

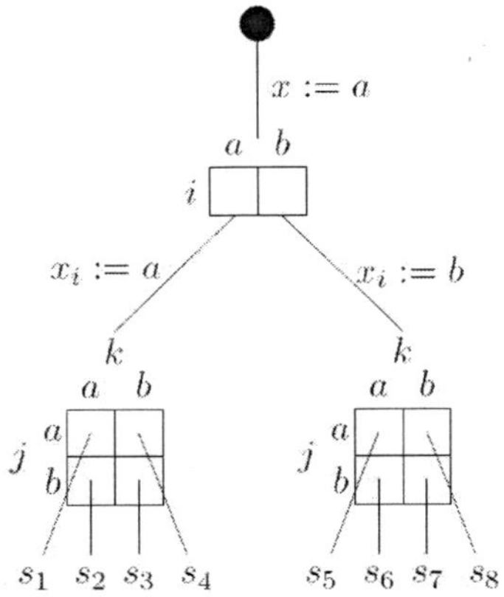

Note that the matrices in the above tree represent the fact that agent j and agent k must *simultaneously* choose values for the variables x_j and x_k respectively. In the above picture, each s_i is a state where, for example, s_2 is the following function: $x \mapsto a$; $x_i \mapsto a$; $x_j \mapsto b$; and $x_k \mapsto a$. The if and while expressions have the usual interpretation (branching and iteration). Formally, game expressions are interpreted as extensive game forms with perfect information and simultaneous moves (cf. [8], Section 6.3.2).

As we just noted, game expressions do not provide any information about the agents' preferences over the terminal histories. This information is provided by the

[3] A strategy profile in an extensive game is a **subgame perfect equilibrium** if no agent has a reason to deviate from the profile *and this is true for all subgames*. Consult [8] for details.

[4] That is, an extensive game (i.e., a game tree) without the preferences over the outcomes.

pre- and post-conditions (P and Q above). To that end, elements of the domain are interpreted as possible outcomes and it is assumed that each agent has a (reflexive, transitive and complete) preference over the domain. A **predicate** is any[5] collection of states. Let o be an element of the domain. An ***e*-state** is a pair (s,o) and a ***e*-predicate** is any set of e-states. An e-predicate P can be used to turn a semi-game into a game as follows. For each $(s,o) \in P$, let f_o be the function that assigns o to each terminal history whose last state is s. Given an e-predicate P, let $\hat{P}$ denote the set of such functions. In general, for a given semi-game and e-predicate P, $\hat{P}$ may be empty or contain more than one function. What is important is that each $f_o \in \hat{P}$ turns the semi-game under consideration into a game. Let P and Q be e-predicates. Given this machinery we can be more precise about what it means to say that $\{P\}\gamma\{Q\}$ is **valid in some first-order model**:

> For each $(s,o) \in P$ there is a outcome function $f \in \hat{Q}$ and a strategy profile σ such that in the game generated from γ and f, σ is a subgame perfect equilibrium and the last state on σ is mapped by f to o.

13.3.2 Formalizing *AW*

Describing *AW* in the above mechanism programming language is straightforward:

$$
\begin{array}{l}
\mathtt{ch}_{\mathscr{A}}(\{x_a,x_b\});\\
s := \mathrm{wta}(x_a,x_b);\\
\mathtt{while}\ \neg Eq(s,x_a,x_b)\ \mathtt{do}\\
\qquad s := \mathrm{t}(s,x_a,x_b);
\end{array}
$$

In the above program, $\mathscr{A} = \{a,b\}$ is the set of agents (Ann and Bob), the variables x_a and x_b are intended to represent Ann and Bob's valuations respectively and s represents the current allocation. To make this formal, it will be convenient to work in a two-sorted first-order structure with an allocation sort and a valuation sort. In what follows, we use $s,t,s_1,s_2,\ldots$ for allocation variables and $x,y,x_1,x_2,\ldots$ for valuation variables. The intended interpretation of the functions wta and t and the relation Eq is as expected:

- $\mathrm{wta}(x_a,x_b)$ is intended to represent the *winner take all* initial allocation given Ann's valuation x_a and Bob's valuation x_b. That is, wta will be interpreted as a function from pairs of valuations to allocations where the agent who assigns the most points to a particular item receives that item.
- $\mathrm{t}(s,x_1,x_2)$ is intended to represent the one-step transfer of goods as described in Section 13.2. The exact details of which good is transfered from which agent is described Section 13.2, so we will not repeat it here.

---

[5] Note that Pauly does *not* restrict to definable sets here. Thus he is provided an *extensional* semantics rather than an intensional semantics. Issues related to this, such as whether or not certain properties are expressible in a logical language (say first-order logic), is left for future work. We agree with Pauly that, although this raises some interesting questions, it will only complicate the current discussion.

- The relation Eq is intended to represent *equality* of each agents' valuation with respect to the current allocation. That is, $Eq(s, x_a, x_b)$ will hold whenever the allocation s is *equitable* with respect to the valuations x_a and x_b. Again, see Section 13.2 for details.

We also include in our language relation symbols Ef and *Pe* intended to mean envy-free and Pareto-efficient respectively. Formally, Ef will hold between an allocation σ and a pair of valuations α and β provided σ is envy-free with respect to α and β (see the definition in Section 13.2). Similarly for Pareto efficiency: *Pe* will hold[6] for σ, α and β provided the allocation σ is *Pareto efficient* with respect to α and β.

13.3.3 Discussion

After Ann an Bob make their choice, it is clear what we have to prove in order to show *AW* is "correct": we must show that after the `while`-loop the allocation contained in s is envy-free, equitable and Pareto efficient (with respect to the allocations chosen by Ann and Bob). This is exactly what Theorem 13.1 says.

The proof from [2] proceeds by first showing that *AW* produces a Pareto efficient allocation then noting that all Pareto efficient and equitable outcomes must be envy-free. Thus the real work is in showing that *AW* produces and Pareto efficient allocation. In this setting, this amounts to finding an appropriate *loop-invariant*. Here we can use the predicate $Pe(s, x_1, x_2)$ defined above. Thus the crucial part of the proof of Theorem 13.1 is showing that the *winner takes all* procedure produces a Pareto efficient allocation and that Pareto efficiency is preserved under repeated applications of the transfer function. Although not phrased this way, the relevant details can be found in [2] (Theorem 4.1, pp. 85–94).

Note that the discussion above is relevant *after* Ann and Bob have made their choices. However, the correctness assertions in both Parikh's framework and Pauly's framework are about outcomes that the agents *can achieve*. In [11], the Banach-Knaster last diminisher procedure[7] is proven correct by showing that a certain formula (in the language of game logic) is *derivable*. This formula essentially states that if certain preconditions are satisfied (i.e., the cake is big enough for the group) then each agent has a *strategy* to ensure it receives a piece of the cake that is fair from its point of view. Pauly's notion of correctness is similar except that there is

[6] Note that *Pe* is an atomic relation symbol. Alternatively, we may first define a relation $(\sigma, \alpha, \beta) \succ (\sigma', \alpha', \beta')$ iff either $\sum_i \sigma_i \alpha_i > \sum_i \sigma_i' \alpha_i$ and $\sum_i (1 - \sigma_i)\beta_i \geq \sum_i (1 - \sigma_i')\beta_i$ or $\sum_i \sigma_i \alpha_i \geq \sum_i \sigma_i' \alpha_i$ and $\sum_i (1 - \sigma_i)\beta_i > \sum_i (1 - \sigma_i')\beta_i$. Then the we can define the predicate *Pe* as follows: $Pe(\sigma, \alpha, \beta) := \forall s((\sigma, \alpha, \beta) \succ (s, \alpha, \beta))$. However, this formula is not in the language we described above as a quantifier is involved.

[7] In this cake-cutting algorithm, the first agent cuts a slice of the cake he considers fair. This piece is then inspected by each of the remaining agents in turn. Each agent can decide either leave the piece as is or diminish the piece and return the extra portion to the main part of the cake. The last person to diminish the piece receives that piece of the cake (or if no-one diminishes the piece, the cutter receives the piece). See [2] for details.

an additional requirement that the strategy profile must satisfy a certain equilibrium concept.

Formally verifying the correctness of Adjusted Winner in Pauly's framework provides us with an interesting challenge. The main issue is finding the correct notion of equilibrium for Ann and Bob. Verifying *AW* is the above framework can be broken down into two different tasks:

1. Given two valuations from Ann and Bob, prove that *AW* produces a equitable, envy-free and efficient allocation.
2. Argue that there is a joint strategy which is a subgame perfect equilibrium in the game generated by the *AW* program as described above. Of course, stating precisely what this means requires specifying a pre- and post-condition.

As argued above, the first step is relatively straightforward. However, solving the second problem raises some interesting issues. The central problem is making precise what it means to say that Ann and Bob have a *strategy* to ensure a fair outcome. A strategy for an agent amounts to choosing a particular valuation. Now if we assume that for each agent there is one valuation that is that agent's "true" valuation,[8] then Pauly's notion of correctness may not be applicable. The reason is because truthful announcement of valuations is not a Nash equilibrium.[9] This is illustrated from the following example from [2]. Suppose that Ann and Bob are dividing two paintings: one by Matisse and one by Picasso. Suppose that Ann and Bob's actual valuations are given by the following table.

Item	Ann	Bob
Matisse	75	25
Picasso	25	75
Total	100	100

Ann will get the Matisse and Bob will get the Picasso and each gets 75 of his or her points. However, this is *not* a Nash equilibrium. Suppose that Ann announces here valuation according to the following table.

Item	Ann	Bob
Matisse	26	25
Picasso	74	75
Total	100	100

So Ann will get the Matisse, receiving 26 of her announced (and insincere) points and Bob gets 75 of his announced points. Let p be the fraction of the Picasso that Ann will get, then we want

[8] This assumes that there are valuations which are, as a matter of *fact*, the agents' actual valuations. However, it may very well be that the players themselves cannot point to a valuation which they consider their "true" valuation. After all, valuations are simply a way to *represent* the players' preferences or utilities over the set of goods. See (Parikh R., and Pacit E. *Safe Votes, Sincere Votes, and Strategizing*. Unpublished Manuscript, 2005.) for a relevant discussion in the context of voting. Nonetheless, for this paper we assume that each player is associated with a unique "true valuation".

[9] Here we need only consider Nash equilibrium and not subgame perfect equilibrium as there is only one point in *AW* when the agents make a choice.

$$26 + 74p = 75 - 75p$$

Solving for p gives us $p = 0.33$ and each gets 50 of his or her announced preference. In terms of Ann's *true* preference, however, the situation is very different. She is getting from her true preference $75 + 0.33 * 25 = 83.33$.

There are two directions one can follow to extend Pauly's analysis to take into account the above issue. We do not go into details in this essay, but only sketch the main ideas.

Imperfect Information In order for Ann (or Bob) to take advantage of the fact that she is not playing her best response in the above example, she must *know* Bob's actual valuation *and that he will in fact play that valuation*. Note that this second point is important. For suppose that in the above example, that Bob has knowledge of Ann's valuation and decides to deceive Ann in the same way as Ann is deceiving him: Bob announces 74 for the Matisse and 26 for the Picasso. In this case, Ann will get the Picasso and Bob will get the Matisse each receiving 74 of their announced points. However, according to their actual valuations both Ann and Bob receive only 25 points! Thus, the information that each agent has about the other agent's valuation is relevant. This suggests two extensions to Pauly's framework. The first is to extend the basic model to include the information each agent has about the other agents' preferences.[10] The second is to allow agents to misrepresent their preferences and include "truthfulness" as a post-condition.

Safe Strategies As argued above, if Ann attempts to deceive Bob and is *wrong* about Bob's choice of valuation, then this can lead to devastating results for Ann (i.e., receiving less than 50 points). Thus while the honest strategy profile (each agent reports his/her "true" preference) is not necessarily a Nash equilibrium, it is a *safe strategy*, i.e., a strategy that guarantees at least 50 points. This is discussed in [2] and (Parikh R., and Pacuit E. *Safe Votes Sincere Votes, and Strategizing*. Unpublished Manuscript, 2005.). Using this notion of a safe strategy we can be more precise about what it means to say that a fair division algorithm is "correct". A fair division algorithm is correct if there is a *safe strategy* such that the outcome is envy-free, Pareto efficient and equitable.

13.4 Conclusion

Evidence of the usefulness of a rigorous analysis of social procedures can already be seen in many areas of economics and social choice theory. Classic results such as Arrow's Theorem have had profound effects on social choice theory and voting theory. A more concrete example is the analysis of the procedure used by King Solomon in the well-known biblical story. King Solomon is faced with two women each claiming that a baby is her own child. Solomon threatens to cut the baby in half causing one of the mothers to rescind her claim of motherhood; thus revealing

[10] Note that this does not mean moving to *imperfect information games*. Although this may be an interesting direction for future research, all that is needed here is a formal model of the agents information about the other agents' preferences.

herself as the true mother. However a formal analysis of this procedure reveals a mistake: it is possible for the false mother to misrepresent her actual preference and claim, as the true mother would, that the baby should be given to the other woman. What is often missing from the game-theoretic analyzes of social procedures is a clear and rigorous analysis of what it means for the social procedure to be *correct.*

The logics discussed in [15, 18, 19] (Pauly M., and Wooldridge M. *Logic for Mechanism Design – a Manifesto.* Unpublished Manuscript.) are all intended to be tools for providing such a rigorous analysis of social procedures. Focusing on the framework from [15], this essay highlights an important aspect of the analysis of many social procedures. This key component is the information, or knowledge, the agents have during the execution of the procedure (cf. [13] and [10] for an extended discussion of this point).

References

1. Alur R., Henzinger T., and Kupferman O. Alternating-time temporal logic, *JACM*, 49(5): 672–713, 2002.
2. Brams S. J., and Taylor A. D. *Fair Division: From Cake-Cutting to Dispute Resolution.* Cambridge University Press, Cambridge, 1996.
3. Brams S. J., and Taylor A. D. *The Win–Win Solution.* W. W. Norton and Company, New York, NY, 1999.
4. Floyd R. W. Assigning meanings to programs, *Proc. Symp. Appl. Math.*, 19: 19–31, 1967.
5. Harel D., Kozen D., and Tiuryn J. *Dynamic Logic.* MIT Press, Cambridge, MA, 2000.
6. Hoare C. A. R. An axiomatic basis for computer programming, *Commun. ACM,* 12: 576–583, 1969.
7. Nielson H. R., and Nielson F. *Semantics with Applications.* Wiley, Chichester, 1992.
8. Osborne M., and Rubinstein A. *A Course in Game Theory.* The MIT Press, Cambridge, 1994.
9. Pacuit E., Parikh R., and Salame S. Some recent results on adjusted winner. In U. Endriss and J. Lange, editors, *Proceedings of Computational Social Choice.* ILLC Technical, Amsterdam, The Netherlands, 2006.
10. Pacuit E., and Parikh R. *Introduction to Formal Epistemology.* Coursenotes for ESSLLI 2007. Available at staff.science.uva.nl/_epacuit/formep esslli.html
11. Parikh R. The logic of games and its applications. In M. Karpinski and J. van Leeuwen, editors, *Topics in the Theory of Computation,* (vol. 24 of *Annals of Discrete Mathematics*), Elsevier, Amsterdam, 1985.
12. Parikh R. Language as social software (abstract). In *International Congress on Logic, Methodology and Philosophy of Science,* page 415. 1995.
13. Parikh R. Knowledge and structure in social algorithms (extended abstract). Presented at Stony Brook Conference on Game Theory, 2007.
14. Pauly M. A modal logic for coalitional power in games, *J. Log. Comput.*, 12(1): 149–166, 2002.
15. Pauly M. Programming and verifying subgame perfect mechanisms, *J. Log. Comput.,* 15(3): 295–316, 2005.
16. Pauly M., and Parikh R. Game logic – An overview, *Stud. Log.*, 75(2): 165–182, 2003.
17. Pratt V. R. Semantical considerations on Floyd-Hoare logic. In *Proceedings of. 17th Symposium on Foundation of Computer Science.* IEEE, pages 109–121, 1976
18. van Otterloo S. *A Strategic Analysis of Multi-Agent Protocols.* PhD thesis, University of Liverpool, 2005.
19. Wooldridge M., Agotnes T., Dunne P., and van der Hoek, W. Logic for automated mechanism design — A progress report. In *Twenty-Second Conference on Artificial Intelligence* (AAAI-07), 2007.

Chapter 14
Temporal Logic with Preferences and Reasoning About Games

G. Venkatesh

14.1 Introduction

Game theory based semantics has provided good insights into the decidability and completeness of several modal logics [24]. On the other hand, we can formalise the language and reasoning used in game theory using appropriate types of modal logic [1, 3, 12–15, 23, 25]. This paper is concerned with the latter issue. The starting point in games is a set of players (agents) having certain strategies (decisions) and preferences on the game's outcomes. Hence we have to represent both the game structure and the agents' preference relations. When reasoning about the games, we wish to determine the properties that hold in the equilibrium to which the game naturally evolves.

Specifically, we will be interested in modeling three types of games, that have a wide range of applications – one stage games with simultaneous moves, finite games with sequential moves and infinite repeated games. Single stage games can be easily represented using propositional logic, and the agents' preference relations become orderings on the models of this logic. The ordering can be described using theories of propositional formulas, but we explore whether they can be captured using preference modal operators that represent the preference relations directly. For sequential and repeated games, a natural extension is to consider the use of propositional temporal logic (PTL), since PTL is able to capture both finite sequences and infinitely repeated sequences in a simple way. Using preference modal operators, we are also able to represent agents' preference relations for the finite sequential and infinitely repeated cases.

Once a game is fully represented, we are able to formulate the notion of game theoretic consequence which allows us to derive properties that hold in the equilibrium of the game. We thus introduce the temporal logic with preferences, whose models are obtained by ordering models of linear time temporal logic. We discuss

G. Venkatesh
Indian Institute of Management, Bangalore 560076, India

J. van Benthem et al. (eds.), *Proof, Computation and Agency*, Synthese Library 352,
DOI 10.1007/978-94-007-0080-2_14,

with examples the expressive power of such a logic to reason about both finite and repeated games and their outcomes.

14.1.1 Related Work

The idea of using a temporal logic (CTL) to model finite extensive games and to reason about backward induction was discussed in [3]. Intuitively, this should work since CTL frames have sufficient structure to represent the game trees of sequential games. Moreover, CTL also allows infinite behaviour to be represented, and hence could in principle represent some kinds of repeated games.

The harder part is to find an adequate representation of the agents' preference relations. What we need is a compact preference representation language. Such languages have been developed by the Knowledge Representation (KR) community. Most are based on propositional logic, or use utility networks [2, 16] or valued constraint satisfaction [27]. We follow the approach taken in [5] and [10], where the semantics is given by a total ordering of propositional interpretations, with appropriate syntax introduced to represent and reason about this.

Applications of such preference modeling to interactive situations, such as voting and collective choice has been studied in [17]. An attempt is made in [10] to find links between the interactive decision outcomes of Nash equilibria with those obtained from preference orderings. In [12], a game theoretic consequence mechanism is defined in which a propositional formula holds in the Nash equlibrium of a strategic game iff it is a consequence of theories representing the players' preference orderings. By using a modal logic, this idea is extended to extensive games and subgame pefect equilibria in [13].

In [14, 25], a logic programming setting is used to capture the game structure and players' preferences, so that properties of the game equilibria can be directly computed. Using a logic programming setting has the particular advantage that it allows infinite (repeated) games to be represented. The disadvantage is that it does not provide the insights that we can gain using model theory.

In [26], we motivate the use of a linear time temporal logic formulation to model and reason about strategic form, extensive form and infinite repeated games. This allows us to extend tableau based decision procedures for temporal logic [28] to compute game theoretic consequences as defined in [12]. The method used is to represent preferences as orderings on tableau nodes, and to convert the game into a negotiation game on tableau nodes.

14.1.2 Organisation

The paper is organised as follows: Section 14.1 defines games. Section 14.2 reviews preference orderings in the propositional logic setting. Section 14.3 extends this to propositional temporal logic. Section 14.4 introduces the temporal logic with preferences (TLP) based on this. Some examples are presented in Section 14.5. Section 14.6 discusses the expressive power of TLP. Section 14.7 summarises the paper and indicates the directions for future work.

14.2 Games

14.2.1 Strategic Form Games

Strategic game situations arise when a set of players make their moves simultaneously, and the outcome of the game is dependent on the combination of moves selected by the players. We follow the notation in [20], where the set of players is denoted by $\mathbf{N} = \{1, ..., n\}$, and for each player $i \in \mathbf{N}$, the (finite set of) actions (also called choices or strategies) available to her is denoted by A_i, with $A_i \cap A_j = \varphi$ for $i \neq j$. The set of consequences is denoted by $\mathscr{C}$. $\mathscr{A} = A_1 \times A_2 \times ... \times A_n$ is the set of *strategy profiles*. If $a = < a_1, ..., a_n > \in \mathscr{A}$ is a strategy profile, we write $< a_{-i}, a_i' >$ for the profile $< a_1, ..., a_i', ..., a_n >$.

A *strategic game form* is given by $\mathscr{G} = < \mathbf{N}, \mathscr{A}, \mathscr{C}, g >$ where $g : \mathscr{A} \rightarrow \mathscr{C}$ is a map that associates with each strategy profile a consequence. The player's preferences for the consequences is given by the complete, reflexive, transitive ordering $\succeq_i$ on $\mathscr{C}$. The preference ordering on the consequences $\mathscr{C}$ induces an ordering on the strategy profiles as follows: $a \succeq_i a'$ if $g(a) \succeq_i g(a')$.

A *strategic game* $< \mathscr{G}, \succeq_i >$ is a strategic game form together with a preference ordering for each player. A *Nash equilibrium of* the strategic game is any strategy profile $a = < a_1, ..., a_n >$ from which no player has the incentive to unilaterally deviate, i.e. $\forall i \in \mathbf{N} \forall a_i' \in A_i : g(a) \succeq_i g(< a_{-i}, a_i' >)$.

To model uncertainty between action and consequence, we consider a set of states Ω, and write the consequence relation $g : \mathscr{A} \times \Omega \rightarrow \mathscr{C}$. It is useful to recall [23], who considers consequences, acts and states as three ways of partitioning a space of possible worlds. *Consequences* partition the space by what matters to the agent, *Acts* by what the agent controls (or has the capacity to decide) and *States* by features of the world on which the consequences may depend, but over which the agent has no control.

14.2.2 Extensive Form and Repeated Games

When we have game situations where the players make their moves sequentially in turns, we use the extensive game form. We can also have repeated game situations in which the game enters a stage where a sequence of moves of the players is repeated again and again forever. Both sequential and repeated games can be fully represented by the runs of the games, which are usually modeled by the histories.

A *history* (see [20]) is a finite or infinite sequence of actions taken from the sets A_i. If $h = a_0a_1...a_k$ is a finite history, then $a_0a_1...a_l$, with $l < k$ is a called a *prefix* of h. Likewise, if $h = a_0a_1...a_l...$ is an infinite history, then all the finite histories $a_0a_1...a_l$ are prefixes of h. We consider finite or infinite sets H of histories that are prefix closed i.e. if h' is a prefix of $h \in H$, then $h' \in H$. The set of terminal histories $Z \subseteq H$ are those that are not prefixes of any history in H. In particular all infinite histories are terminal.

We will restrict our attention to the set of histories H for which we can define the function $\mathscr{P} : (H \backslash Z) \rightarrow \mathbf{N}$ that assigns to each prefix $h \in H$ a unique player $\mathscr{P}(h) = i \in \mathbf{N}$, and if $ha \in H$ then $a \in A_i$. This restriction ensures that we are only concerned with games in which only one player plays at a time, and there is a rule that determines who will play next. We also place an important restrictions on infinite histories in Z, that they are *eventually periodic*, i.e. each infinite history can be written in the form $a_0a_1...a_k(a'_0a'_1...a'_l)^\omega$, with $a'_r \neq a'_s$ for $r \neq s$ (x^ω denotes repetition of x infinitely many times). This restriction ensures that the infinite games we are dealing with are indeed repeated games, i.e. after some initial period in the game, there is a sequence of moves that will get repeated again and again.

We can now define an *extensive game form* as a tuple $\mathscr{G} =< \mathbf{N}, A_i, H, \mathscr{P}, \mathscr{C}, g >$ where $g : Z \rightarrow \mathscr{C}$ is a map that associates with each terminal history a consequence. An *extensive game* is such a tuple along with orderings $\succeq_i$ of $\mathscr{C}$ for each player i.

A strategy f_i for player i associates an action $f_i(h) \in A_i$ with each prefix $h \in H$ for which $\mathscr{P}(h) = i$. A strategy profile is a tuple $f =< f_1, ..., f_n >$. A strategy profile f leads to a unique terminal history $z_f = a_0a_1... \in Z$ in which $\forall k : a_{k+1} = f_{\mathscr{P}(a_0,...a_k)}(a_0, ..., a_k)$. We write $< f_{-i}, f'_i >$ for the strategy profile $< f_1, ..., f'_i, ..., f_n >$.

A *Nash equilibrium* of game $\mathscr{G}$ is a terminal history z_f from which no player has the incentive to unilaterally deviate, i.e. $\forall i \in N : g(z_f) \succeq_i g(z_{<f_{-i}, f'_i>})$. We define the subgame $\mathscr{G}_h$ rooted at h to consist of the histories $H_h = \{h' \,|\, hh' \in H\}$ i.e. the histories with prefix h from each of which we strip off the prefix. The functions $\mathscr{P}$, g and strategy profiles f can be extended in the obvious way to H_h. A *subgame perfect Nash equilibrium* is a terminal history z_f such that for all prefixes h of z_f, if $z_f = hz'_f$, then z'_f is a Nash equilibrium of G_h.

14.3 Preference Orderings in Propositional Logic

To capture uncertainty between action and consequence, we distinguish choices made by the decision maker and choices imposed by the environment. We will use the n disjoint sets of propositional variables $\mathscr{P} = P_1 \cup P_2 \cup ... \cup P_n$ with typical elements p_i^r (which denotes that the agent i has chosen an action $a_r \in A_i$). These are the agents' decision variables [18] (or controllable propositions [4]). We also use a set of propositional variables $\mathscr{W}$ with typical elements $b, c, ...$ (the world parameters or uncontrollable propositions) such that $\mathscr{P} \cap \mathscr{W} = \emptyset$. Out of these propositional variables, we distinguish the subset $\mathscr{C} \subset \mathscr{W}$ that represent *consequences*, which has typical elements c with subscripts. The propositional languages that are built up from P_i, $\mathscr{P}$, $\mathscr{W}$, $\mathscr{C}$, $\mathscr{P} \cup \mathscr{C}$ and $\mathscr{P} \cup \mathscr{W}$ variables using the usual propositional connectives $\wedge$, $\vee$, $\neg$ and the symbols $\top$, $\bot$ are denoted by $\mathscr{L}_{P_i}$, $\mathscr{L}_{\mathscr{P}}$, $\mathscr{L}_{\mathscr{C}}$, $\mathscr{L}_{\mathscr{W}}$, $\mathscr{L}_{\mathscr{P}\mathscr{C}}$ and $\mathscr{L}_{\mathscr{P}\mathscr{W}}$ respectively. Finally, we use variables φ, ψ,... to stand for any sentences of these languages.

14.3.1 Semantics

A state s is an assignment of truth values to the propositions taking into account the fact that each agent can choose at most one action, i.e.
$s(p_i^r) = True \Rightarrow \forall r' \neq r\, s(p_i^{r'}) = False$. Such a valuation can be extended to all propositional formulas φ in the usual way, so that $s(\varphi) = True$ or $s(\varphi) = False$. A model M consists of a single state s_M.

We say that M ***satisfies*** φ (written $M \models \varphi$) whenever $s_M(\varphi) = True$.
We write $Mod(\varphi)$ for the set of models satisfying φ, i.e. $Mod(\varphi) = \{M \mid M \models \varphi\}$.
If Γ is a set of formulas, $Mod(\Gamma) = \bigcap_{\varphi \in \Gamma} Mod(\varphi)$.

The ***logical consequence*** relation $\models$ between Γ and formula φ is defined by:
$\Gamma \models \varphi$ iff $Mod(\Gamma) \subseteq Mod(\varphi)$.

14.3.2 Preference Ordering of Models

We let $\mathscr{M}$ denote the set of all models. Agent i's preference relation $\succeq_i$ is given by a preorder, i.e., a reflexive and transitive binary relation on $\mathscr{M}$. $M \succeq_i M'$ means that situation M is at least as good (to the agent i) as the situation M'.

Such a relation is not necessarily complete, that is, it may be that neither $M \succeq_i M'$ nor $M' \succeq_i M$ holds for a pair M, M' in $\mathscr{M}$.
We write $M \succ_i M'$ for $M \succeq_i M'$ and $M' \not\succeq_i M$ (strict preference of M over M'), and $M \sim_i M'$ for $M \succeq_i M'$ and $M' \succeq_i M$ (indifference).
It is important to note that $M \sim_i M'$ means that the agent takes M and M' to be equally preferred, while the incomparability $M \not\succeq_i M'$ and $M' \not\succeq_i M$ simply means that no preference between them is expressed [9].

Each preference ordering $\succeq_i$ gives rise to a modal operator which we also write for convenience as $\succeq_i$. We thus use $\varphi \succeq_i \psi$ to compactly represent the ordering between the models satisfying φ and those satisfying ψ, i.e. $Mod(\varphi \succeq_i \psi) = \{M \succeq_i M' \mid M \models \varphi \text{ and } M' \models \psi\}$.

Example 14.1 (Prisoner's dilemma)
Game description: Two suspects in a crime are put into separate cells and interrogated. If they both confess, each will be sentenced to 3 years in prison. If only one

of them confesses, he will be freed and used as a witness against the other, who will receive 5 years. If neither confesses, they will both be convicted for a minor offense and spend 1 year in prison.

We use four action propositions $p_1^c, p_1^n, p_2^c, p_2^n$, where p_i^c means i chooses to confess and p_i^n means i chooses to not confess. The preference ordering of Agent 1 can be expressed using $p_1^c \wedge p_2^n \succeq_1 p_1^n \wedge p_2^n \succeq_1 p_1^c \wedge p_2^c \succeq_1 p_1^n \wedge p_2^c$ and Agent 2 using $p_2^c \wedge p_1^n \succeq_2 p_2^n \wedge p_1^n \succeq_2 p_2^c \wedge p_1^c \succeq_2 p_2^n \wedge p_1^c$.

14.3.3 Preferred Models and Non-monotonic Consequence

To provide a model theoretic framework for non-monotonic reasoning, several authors [19, 21] have introduced the notion of *preferred models*, which are maximal according to a preference ordering $\succeq$ of models. We write $PrefMod(\Gamma) = \{M \in Mod(\Gamma) \mid \not\exists\ M' \in Mod(\Gamma) : M' \succeq M\}$, the maximal models satisfying Γ.

Using this, a ***non-monotonic consequence relation*** $\models_N$ between Γ and a formula φ is defined by : $\Gamma \models_N \varphi$ iff $PrefMod(\Gamma) \subseteq Mod(\varphi)$.

The logical cosequence relation is monotonic because $Mod(\Gamma \cup \{\varphi\}) \subseteq Mod(\Gamma)$. However this may not hold with preferred models : $PrefMod(\Gamma \cup \{\varphi\})$ may not be included in $PrefMod(\Gamma)$. To see this, we can create a counter-example as follows: consider the case where $Mod(\Gamma) = \{M, M'\}$, with $M \succeq M'$, and $M' \models \varphi$ but $M \not\models \varphi$. This gives $PrefMod(\Gamma \cup \{\varphi\}) = \{M'\}$ and $PrefMod(\Gamma) = \{M\}$.

14.3.4 Modeling Preferences Using Theories

Let $[[\Gamma]] = \{Mod(\varphi) \mid \varphi \in \Gamma\}$.
Note that the statement $\Gamma \models \varphi$ in propositional logic depends only on $Mod(\Gamma)$ and $Mod(\varphi)$, where $Mod(\Gamma)$ is defined by the intersection $\bigcap[[\Gamma]]$.

In [12], it is observed that the intersection obliterates much of the structure contained in $[[\Gamma]]$. Since $[[\Gamma]]$ is a set of sets of models, instead of taking a simple intersection, we could actually try to define a partial preorder over these models.

For any set of sets $\mathscr{X} \subseteq 2^{\mathscr{M}}$, define the ordering relation $\succeq_{\mathscr{X}}$ on $\mathscr{M}$ such that for all $M, M' \in \mathscr{M} : M \succeq_{\mathscr{X}} M'\ iff\ \forall X \in \mathscr{X} : M \in X \Rightarrow M' \in X$.

Intuitively, elements outside a set $X \in \mathscr{X}$ are not ranked above elements in X in the order $\succeq_{\mathscr{X}}$. Thus, $\succeq_{[[\Gamma]]}$ gives a pre-order of models.

Example 1 (Prisoners' dilemma ...).
We use only two action propositions p_1, p_2, where p_i means i chooses to confess and $\neg p_i$ means i chooses to not confess. The preference ordering of Agent 1 can be expressed using the theory $\Gamma_1 = \{p_2 \Rightarrow p_1, \neg p_2, \neg p_2 \wedge p_1\}$ and Agent 2 using the theory $\Gamma_2 = \{p_1 \Rightarrow p_2, \neg p_1, \neg p_1 \wedge p_2\}$. To see this, note that:
$[[\Gamma_1]] = \{\{\{p_1\},\{p_1,p_2\},\emptyset\}, \{\{p_1\},\emptyset\}, \{\{p_1\}\}\}$ and

$[[\Gamma_2]] = \{\{\{p_2\},\{p_1,p_2\},\emptyset\}, \{\{p_2\},\emptyset\}, \{\{p_2\}\}\}$
where we let the model M be simply represented by the set of propositions that are satisfied by M. This gives the following order over models
$\{p_1\} \succ_1 \emptyset \succ_1 \{p_1,p_2\} \succ_1 \{p_2\}$ and $\{p_2\} \succ_2 \emptyset \succ_2 \{p_1,p_2\} \succ_2 \{p_1\}$.
which is exactly the same order of models that we saw earlier in Section 14.2.2.

14.3.5 Game Theoretic Consequence Relation

If we use Γ_i to represent the preference ordering of agent i, then the set of theories $\{\Gamma_1,...\Gamma_n\}$ represents a game. The ***game theoretic consequence relation*** $\{\Gamma_1,...\Gamma_n\} \models_G \varphi$ holds only if φ is true in all the Nash equilibria of the game. We directly state the characterisation derived in [12]. Let π be a partition of $\mathscr{M}$.

For each $X \subseteq \mathscr{M}$, we define $apr_\pi(X) = \bigcup\{Y \in \pi | Y \subseteq X\}$. Define the equivalence relation $\cong_{-i}$ by: $M \cong_{-i} M'$ iff M, M' differ only in the assignment of truth values to propositions in P_i. We write π_{-i} to denote the partition of $\mathscr{M}$ obtained through the equivalence relation $M \cong_{-i} M'$. Define $\mathscr{N}(\{[[\Gamma_1]],...,[[\Gamma_n]]\}) = \bigcap_{i=1}^{n} \bigcap_{X\in[[\Gamma_i]]}(X \cup apr_{\pi_{-i}}(-X))$.

Proposition 14.1 (Characterisation of game theoretic consequence)

$\{\Gamma_1,...\Gamma_n\} \models_G \varphi$ *iff* $\mathscr{N}(\{[[\Gamma_1]],...,[[\Gamma_n]]\}) \subseteq Mod(\varphi)$. *(See* [12]) □

14.4 Preference Orderings in Propositional Temporal Logic

14.4.1 Propositional Temporal Logic

As before, we use p_i^r to denote that player i has chosen an action $a_r \in A_i$. Consequences in $\mathscr{C}$ are described using propositional variables c_j, $j = 1,...,m$. We use an unlimited supply of propositional variables $b \in \mathscr{W}$. We will assume that the agents express all their preferences using only the consequence variables in $\mathscr{C}$. Define PTL by the BNF grammar:
$\varphi ::= p_i^r \,|\, c_j \,|\, b \,|\, \neg\varphi \,|\, \varphi_1 \wedge \varphi_2 \,|\, \bigcirc\varphi \,|\, \Box\varphi \,|\, \varphi_1 \,\mathcal{U}\, \varphi_2$

We abbreviate $\neg(\neg\varphi_1 \wedge \neg\varphi_2)$ by $\varphi_1 \vee \varphi_2$, $\neg\varphi_1 \vee \varphi_2$ by $\varphi_1 \Rightarrow \varphi_2$, $(\varphi_1 \Rightarrow \varphi_2) \wedge (\varphi_2 \Rightarrow \varphi_1)$ by $\varphi_1 \Leftrightarrow \varphi_2$, $\varphi_1 \Leftrightarrow \neg\varphi_2$ by $\varphi_1 \oplus \varphi_2$ and $\neg\Box\neg\varphi$ by $\Diamond\varphi$. $\mathscr{L}^*_{P_i}$, $\mathscr{L}^*_{\mathscr{P}}$, $\mathscr{L}^*_{\mathscr{C}}$, $\mathscr{L}^*_{\mathscr{W}}$, $\mathscr{L}^*_{\mathscr{P}\mathscr{C}}$ and $\mathscr{L}^*_{\mathscr{P}\mathscr{W}}$ refer to the PTL languages with propositional variables restricted to P_i, $\mathscr{P}$, $\mathscr{W}$, $\mathscr{C}$, $\mathscr{P} \cup \mathscr{C}$ and $\mathscr{P} \cup \mathscr{W}$ respectively.

14.4.2 Using PTL to Model Games

Using temporal logic, we can model the delay between action and consequence.

Example 1 (Prisoners' dilemma ...).
We use two action propositions p_1, p_2, where p_i means i chooses to confess and $\neg p_i$ means i chooses to not confess. The four consequences are: $c_{3y,3y}$, $c_{5y,free}$, $c_{free,5y}$, $c_{1y,1y}$, where $c_{p,q}$ represents sentence p for Agent 1 and q for Agent 2. The temporal formulas representing this game are:

$$(F1) \quad p_1 \wedge p_2 \Rightarrow \bigcirc c_{3y,3y} \qquad (F2) \quad p_1 \wedge \neg p_2 \Rightarrow \bigcirc c_{free,5y}$$
$$(F3) \quad \neg p_1 \wedge p_2 \Rightarrow \bigcirc c_{5y,free} \qquad (F4) \quad \neg p_1 \wedge \neg p_2 \Rightarrow \bigcirc c_{1y,1y}$$

The set $\Gamma = \{F1, F2, F3, F4\}$ represents the game.

Example 14.2 (Repeated prisoner's dilemma)

Game description: Two criminals form a partnership in crime. In each crime event that they commit together, there is a possibility that they are caught and interrogated, with the consequences as defined in the prisoners dilemma game.

We use the operator $\bigcirc$ to represent progression of stages of the game. Then the set $\Gamma = \{\Box F1, \Box F2, \Box F3, \Box F4\}$ represents the game.

The repeated game differs from the one-stage game because it provides the scope for long term agent strategies to emerge. More specifically, agents could form a partnership pact to co-operate (not confess), and may have individual strategies to respond appropriately in case their partner "defects" (confesses).

Consider for example Agent 1 who plays a *grim strategy* written as $\varphi_1 = \neg p_1 \wedge \Box(p_2 \Rightarrow \bigcirc\Box p_1) \wedge \Box(\neg p_2 \Rightarrow \bigcirc\neg p_1)$, i.e. Agent 1 will play according to the pact to not confess, but will forever punish Agent 2 if she defects from the pact. We can similarly examine the *Tit-For-Tat strategy* expressed by $\varphi_1 = \neg p_1 \wedge \Box(p_2 \Rightarrow \bigcirc p_1) \wedge \Box(\neg p_2 \Rightarrow \bigcirc\neg p_1)$ in which Agent 1 punishes Agent 2 only in the next round, and goes back to co-operating after that.

14.4.3 Semantics of PTL

As before, a state s is an assignment of truth values to the propositions taking into account the fact that each agent can choose at most one action – i.e. $s(p_i^r) = True \Rightarrow \forall r' \neq r\ s(p_i^{r'}) = False$.

We let S represent all such states, S^k represent state sequences of length $k+1$, and S^ω all the infinite sequences. A PTL model M is a sequence $s_0, s_1, ... \in S^\omega$. The satisfaction relation $\models$ between elements $M \in S^\omega$ and PTL formulas is defined as follows, for all $k \geq 0$:

$$s_k, s_{k+1} \ldots \models p_i^r \qquad \text{if } s_k(p_i^r) = True$$
$$s_k, s_{k+1} \ldots \models c_j \qquad \text{if } s_k(c_j) = True$$

$s_k, s_{k+1} \ldots \models b$ if $s_k(b) = True$
$s_k, s_{k+1} \ldots \models \varphi \wedge \psi$ if $s_k, s_{k+1} \ldots \models \varphi$ and $s_k, s_{k+1} \ldots \models \psi$
$s_k, s_{k+1} \ldots \models \neg\varphi$ if not $s_k, s_{k+1} \ldots \models \varphi$
$s_k, s_{k+1} \ldots \models \bigcirc\varphi$ if $s_{k+1}, s_{k+2} \ldots \models \varphi$
$s_k, s_{k+1} \ldots \models \Box\varphi$ if $\forall l \geq k\ s_l, s_{l+1} \ldots \models \varphi$
$s_k, s_{k+1} \ldots \models \varphi \bigcup \psi$ if $\exists l \geq k\ s_l, s_{l+1} \ldots \models \psi$ and $\forall u : k \leq u < l\ s_u, \ldots \models \varphi$.

14.4.4 Eventually Periodic Models

We let $EP = \bigcup_{k\geq 0, l>0} S^k(S^l)^\omega$ denote the set of all eventually periodic sequences $M = (s_0, s_1, \ldots s_k)(s'_0, \ldots, s'_l)^\omega$ with $s'_i \neq s'_j$ for $i \neq j$. M_r denotes s_r if $0 \leq r \leq k$, and $s'_{(r-k-1)\, mod\, l}$ when $r > k$.

If $\Gamma \subseteq$PTL, then $Mod(\Gamma) = \{M \in S^\omega | \forall \varphi \in \Gamma : M \models \varphi\}$. Γ is *PTL-satisfiable* if $Mod(\Gamma) \neq \emptyset$, and *EP-satisfiable* if $Mod(\Gamma) \cap EP \neq \emptyset$.

Proposition 14.2 (Equivalence)

For any set Γ of PTL formulas, Γ is PTL-satisfiable iff it is EP-satisfiable.

Proof. TLP-satisfiability obviously implies PTL-satisfiability. If $M \in S^\omega$ satisfies Γ, then we can construct an equivalent eventually periodic sequence from M that also satisfies Γ by using the tableau for Γ (see [28]).

We write $\Gamma \models_{PTL} \varphi$ if $\forall M \in S^\omega : M \models \Gamma \Rightarrow M \models \varphi$.
Clearly $\Gamma \models_{PTL} \varphi$ iff $\Gamma \cup \{\neg\varphi\}$ is not PTL-satisfiable. Similarly, we write $\Gamma \models_{EP} \varphi$ if $\forall M \in EP : M \models \Gamma \Rightarrow M \models \varphi$. Then $\Gamma \models_{EP} \varphi$ iff $\Gamma \cup \{\neg\varphi\}$ is not EP-satisfiable.

Corollary 14.1 $\Gamma \models_{PTL} \varphi$ *iff* $\Gamma \models_{EP} \varphi$ □

In the sequel, we will consider the universe of models to be *EP*, and write $Mod(\Gamma)$ for $Mod(\Gamma) \cap EP$.

14.4.5 Preference Ordering

Representing agents' preference ordering over models in *EP* can be quite tricky as it requires us to represent how agents compare infinite sequence of situations. In particular, we have to account for agents' preference for *short term* outcomes, and how they compare it with *long term* outcomes. In [20] (pp. 136–139), three orderings are suggested for infinite repeated games – Discounting (in which the value of time is taken into account), Limit of means (in which finite periods of bad circumstances are ignored), and Overtaking (which favours the long term to the short term). All of these are based on cumulating utility values for each period over the infinite time horizon.

On the other hand, it is not reasonable to assume that agents have the mental capabilities to compare infinitely many futures of infinite time horizon. This issue has been raised and addressed by bounded rationality proponents [8, 22]. However the alternative of defining choice through search procedures could make analysis difficult or even intractable.

In the rest of this paper, we take a pragmatic approach combining these two viewpoints. We make three critical assumptions:

- *Assumption 1:* Agents care only about the consequences, and hence each agent can always express a preference between two states based only on the consequences in $\mathscr{C}$ that are true in the state. The preference order for each agent on the set of states is thus complete.
- *Assumption 2:* Agents value for time is modest compared to their value for a consequence. Thus a time delay in the occurrence of a consequence will be ignored.
- *Assumption 3:* Agents are interested more in the long term outcomes than in the short term outcomes. We thus adopt the overtaking criterion, and ignore any initial finite sequence of outcomes.

As in Section 14.2.2, we can introduce a modal operator $\succeq_i$ to compactly represent ordering between models satisfying two formulas – i.e $\varphi \succeq_i \psi$ compactly represents $\{M \succeq_i M' \,|\, M \models \varphi \; and \; M' \models \psi\}$.

Example 1 (Prisoners' dilemma ...).
The preference ordering in a single stage game is given by $c_{free,5y} \succeq_1 c_{1y,1y} \succeq_1 c_{3y,3y} \succeq_1 c_{5y,free}$, and $c_{5y,free} \succeq_2 c_{1y,1y} \succeq_2 c_{3y,3y} \succeq_2 c_{free,5y}$.

Example 2 (Repeated Prisoners' dilemma ...).
The preference ordering $\Diamond\Box c_{free,5y} \succeq_1 \Diamond\Box c_{1y,1y} \succeq_1 \Diamond\Box c_{3y,3y} \succeq_1 \Diamond\Box c_{5y,free}$ for Agent 1 and $\Diamond\Box c_{5y,free} \succeq_2 \Diamond\Box c_{1y,1y} \succeq_2 \Diamond\Box c_{3y,3y} \succeq_2 \Diamond\Box c_{free,5y}$ describes the long term interests of both agents – that they are interested only in the asymptotic sustained values of consequences, and no preferences are expressed for histories which do not lead to such sustained consequences.

14.4.6 Ordering EP Models

Let $S_{\mathscr{C}}$ denote the states in which at least one proposition of $\mathscr{C}$ is assigned *True*. Let $EP_{\mathscr{C}}$ be the set of eventually periodic models with states drawn from $S_{\mathscr{C}}$.

Given a reflexive, transitive and complete ordering $\succeq_i$ on $S_{\mathscr{C}}$, we can extend $\succeq_i$ to sequences in $S_{\mathscr{C}}^{\omega}$: $s_1 s_2 \ldots \succ_i^* s' s'_2 \ldots$ iff $\exists k \geq 1 : s_k \succ_i s'_k$ and $\forall j \geq 1 : s'_j \succeq_i s_j \Rightarrow s_j \succeq_i s'_j$, i.e. at each instant of time, either i prefers the state from the former sequence, or else i is indifferent between the corresponding states.

For $M = (s_0, s_1, ...s_k)(s'_0, ..., s'_l)^\omega \in EP_{\mathscr{C}}$, we use M_r to denote the rth state in the sequence, i.e. $M_r = s_r$ if $0 \leq r \leq k$, and $M_r = s'_{(r-k-1) mod\, l}$ when $r > k$. We also use M^{r+} to denote the tail starting at r, i.e. $M^{r+} = M_r, M_{r+1}, ...$

Note that by Assumption 1, i has to express a preference between corresponding states. This ensures that the ordering $\succeq_i$ of $S^\omega_{\mathscr{C}}$ is transitive. When we take into account Assumption 3 that the agent is more interested in the longer term outcome, the agent may wish to express the asymptotic relation $M \succ_i M'$ if the relation holds in the eventually periodic portion of the two models, i.e. when $M_u \succ_i M'_u$ occurs infinitely often.

Thus, for $M, M' \in EP_{\mathscr{C}}$, we take the asymptotic case: $M \succ_i M'$ iff there exists $t > 0$ such that $\forall w \geq t \, \exists r \geq w : M^{r+} \succ^*_i M'^{r+}$, i.e. we can ignore the initial portion (upto t) of the two models M, M' after which there are infinitely many occasions at which M is preferred to M'. Note that in particular we have $M^{t+} \succ^*_i M'^{t+}$. The ordering will be transitive since if $M^{t_1+} \succ^*_i M'^{t_1+}$ and $M'^{t_2+} \succ^*_i M''^{t_2+}$, then $M^{t+} \succ^*_i M''^{t+}$ where $t = max(t_1, t_2)$.

Note that we have defined orderings only for models in $EP_{\mathscr{C}}$, i.e. for models in which at least one consequence is true at any point in time. In order to extend it to models in EP, we will need to allow comparison of states across different points in time (since for any k, the kth consequence may occur at different times in two different models). Thus for any model M, we can define the model $M_{\mathscr{C}}$ inductively as follows: If $M_{\mathscr{C}} = s_0 s_1 ...$ and $s_0 \in S_{\mathscr{C}}$ then $M_{\mathscr{C}} = s_0 M^+_{\mathscr{C}}$, else $M_{\mathscr{C}} = M^+_{\mathscr{C}}$, where $M^+_{\mathscr{C}} = s_1 ....$ Then $M \succ_i M'$ iff $M_{\mathscr{C}} \succ_i M'_{\mathscr{C}}$.

14.4.7 Modeling Preferences Using Theories

As in Section 14.2.4, we can ask whether the preference ordering of $EP_{\mathscr{C}}$ models can be captured using subsets of $\mathscr{L}^*_{\mathscr{C}}$. Firstly note that any ordering $\succeq_i$ of $S_{\mathscr{C}}$ can be captured by a theory $\mathscr{T}_{\succeq_i} \subseteq \mathscr{L}_{\mathscr{C}}$ which is also a theory of $\mathscr{L}^*_{\mathscr{C}}$. Now consider the theory $\{\bigcirc^l \varphi \,|\, \varphi \in \mathscr{T}_{\succeq_i}, \, l \geq 0\}$. This clearly represents the ordering $\succ^*_i$ on sequences of states.

Let $EP^r_{\mathscr{C}} = \bigcup_{k \leq r, l \leq r} S^k (S^l)^\omega$ be the eventual periodic sequences each containing at most $2r$ distinct states. Note that $EP^0_{\mathscr{C}} \subseteq EP^1_{\mathscr{C}} \subseteq EP^2_{\mathscr{C}}...$, and $EP_{\mathscr{C}} = Lim_{r \to \infty} EP^r_{\mathscr{C}}$, and so we can get arbitrarily close approximations to $EP_{\mathscr{C}}$ through the sets $EP^r_{\mathscr{C}}$.

It is easy to check that the theory $\Gamma^r_i = \{\bigcirc^l \varphi \,|\, \varphi \in \mathscr{T}_{\succeq_i}, \, l \geq 2r\}$ should represent the order $M \succ_i M'$ for $M, M' \in EP^r_{\mathscr{C}}$ and hence we can capture the ordering for arbitrarily close approximations of $EP_{\mathscr{C}}$ through theories of $\mathscr{L}_{\mathscr{C}}{}^*$.

The question of whether the ordering on $EP_{\mathscr{C}}$ or EP can be fully captured through a theory remains an open problem.

14.4.8 Nash Equilibrium for Repeated Games

Define the equivalence relation $\cong^r_{-i}$ by: $M \cong^r_{-i} M'$ iff $M_k = M'_k$ for $k \neq r$ and M_r, M'_r differs only in the assignment of truth values to propositions in P_i. We write π^r_{-i} to denote the partition of $EP_{\mathscr{C}}$ obtained through the equivalence relation $M \cong^r_{-i} M'$.

Let $\mathscr{N}(\{[[\Gamma_1^r]],...,[[\Gamma_n^r]]\}) = \bigcap_{i=1}^n \bigcap_{X \in [[\Gamma_i^r]]} (X \cup apr_{\pi^r_{-i}}(-X))$ be the set of models from which no agent has the incentive to deviate at a certain time point r. Then the *Nash equilibrium models* are given by $\bigcap_{r\geq 0} \mathscr{N}(\{[[\Gamma_1^r]],...,[[\Gamma_n^r]]\})$.

Given a formula $\varphi \in \mathscr{L}^*_{\mathscr{C}}$, we define the *length of the formula* $|\varphi|$ to be twice the number of subformulas occuring in φ. Let $r(\varphi) = 2^{|\varphi|}$.

The tableau for φ will have at most $r(\varphi)$ states (see [28]), and hence it is enough to consider models from $EP_{\mathscr{C}}^{r(\varphi)}$. Then $\{\Gamma_1^{r(\varphi)},...\Gamma_n^{r(\varphi)}\}$ gives an adequate representation of the ordering when we restrict our attention to models of φ.

The game theoretic consequence defined on formulas in $\mathscr{L}^*_{\mathscr{C}}$ is written as $\{\Gamma_1^{r(\varphi)},...\Gamma_n^{r(\varphi)}\} \models_G \varphi$ and holds iff φ is true in all the Nash equilibrium models.

Proposition 14.3 (Characterisation of game theoretic consequence)

$$\{\Gamma_1^{r(\varphi)},...\Gamma_n^{r(\varphi)}\} \models_G \varphi \ iff \ \bigcap_{r\geq r(\varphi)} \mathscr{N}(\{[[\Gamma_1^{r(\varphi)}]],...,[[\Gamma_n^{r(\varphi)}]]\}) \subseteq Mod(\varphi). \qquad \square$$

14.5 Temporal Logic with Preferences (TLP)

We are now ready to introduce the temporal logic with preferences, which is built from PTL formulas using three types of modal operators to represent the agent's preferences and selections based on these preferences.

14.5.1 The Language

The set of TLP formulas is defined through the following BNF grammar:
$\psi ::= \ \varphi \,|\, \varphi_1 \succ_i \varphi_2 \,|\, \varphi_1 \bowtie_i \varphi_2 \,|\, \varphi_1 \rhd_i \varphi_2$.

We read $\varphi_1 \succ_i \varphi_2$ as "i prefers φ_1 to φ_2". This modal operator, as has already been explained earlier has to be interpreted using a pair of PTL models. Intuitively, it means that if we take any pair of models M and M', with M satisfying φ_1 and M' satisfying φ_2, then player i will prefer M to M'.

We read $\varphi_1 \bowtie_i \varphi_2$ as "i can select between φ_1 and φ_2". Again, this modal operator is to be interpreted over a pair of PTL models. Intuitively, it means that if we take any pair of models M and M', with M satisfying φ_1 and M' satisfying φ_2, then M and M' coincide upto a time point at which it is the turn of player i to move, at which point, player i can choose an action that either takes the path of M or that of M'.

Finally, we read $\varphi_1 \rhd_i \varphi_2$ as "i can pick φ_1 over φ_2". This operator has to be interpreted over a relevant set of PTL models, keeping a specific pair of models M, M' in focus. To understand this operator, we consider the situation where player i prefers M to M', and can select between M and M' as above. Then player i can choose M and discard M'. However, before discarding M', i must be reasonably sure that her choice of M is not going to be veto-ed by future players. The operator $\rhd_i$ confirms this condition.

Note that this operator gives us a model elimination tool (see [26]) that can lead to a negation inference rule.

14.5.2 Models of TLP

As we saw in the previous section, we interpret TLP formulas over sets of PTL models keeping in mind the preferences of the players. Thus, a **TLP model** is a tuple $\mathscr{M}_{\succeq} = <\mathscr{M}, \succeq_i>$ where $\mathscr{M} \subseteq EP$, and $\succeq_i$ is a preference ordering of $S_{\mathscr{C}}$ for each agent i. We extend $\succeq_i$ to $\mathscr{M}$ as in Section 14.3.6 above. We can now order sets of models: $\mathscr{M} \succeq_i \mathscr{M}'$ iff $\forall M' \in \mathscr{M}' \ \exists M \in \mathscr{M} : M \succeq_i M'$, i.e. the maximal models of $\mathscr{M}'$ lie below the maximal models of $\mathscr{M}$.

We write $s \equiv_i s'$ if $s(p_i^r) = s'(p_i^r) \forall a_r \in A_i$, and $M \equiv_i^k M'$ if $\forall l < k M_l \equiv_i M_l'$. We write $M \bowtie_i M'$ when $\exists k : \neg(M \equiv_i^k M')$ and $\forall j \neq i \ M \equiv_j^k M'$ i.e. M and M' co-incide in choices of all agents till i makes a choice.

Let $\mathscr{M}_\varphi = Mod(\varphi) \cap \mathscr{M}$.
The relation $\models_{TLP}$ between TLP models $\mathscr{M}_{\succeq}$ and formulas ψ is given by:
$\mathscr{M}_{\succeq} \models_{TLP} \varphi$ iff $\mathscr{M}_\varphi = \mathscr{M}$.
$\mathscr{M}_{\succeq} \models_{TLP} \varphi \succ_i \varphi'$ iff $\mathscr{M}_\varphi \succeq_i \mathscr{M}_{\varphi'}$ but not $\mathscr{M}_{\varphi'} \succeq_i \mathscr{M}_\varphi$.
$\mathscr{M}_{\succeq} \models_{TLP} \varphi \bowtie_i \varphi'$ iff $\forall M \in \mathscr{M}_\varphi, M' \in \mathscr{M}_{\varphi'} : M \bowtie_i M'$
$\mathscr{M}_{\succeq} \models_{TLP} \varphi \rhd_i \varphi'$ iff $\mathscr{M}_{\succeq} \models_{TLP} \varphi \bowtie_i \varphi'$, $\mathscr{M}_{\succeq} \models_{TLP} \varphi \succ_i \varphi'$ and
$\forall M \in \mathscr{M}_\varphi, M' \in \mathscr{M}_{\varphi'}, M'' \in \mathscr{M}, j \neq i : M' \bowtie_i M'', M \bowtie_j M'' \Rightarrow M \succeq_j M''$.

i.e. $\varphi \rhd_i \varphi'$ holds if i's preference φ is assured because agents j who choose after i also prefer it.
Note that $M \bowtie_i M'$, $M' \bowtie_i M''$ and $M \bowtie_j M''$ necessarily implies that between the three models, i chooses first between M' and the remaining two, following which j chooses between M and M''.

Note also that the satisfaction relation is **not monotonic**:
It is possible that we have $\mathscr{M}' \subset \mathscr{M}$: $\mathscr{M}_{\succeq} \models_{TLP} \varphi \succ_i \varphi'$, but $\mathscr{M}'_{\succeq} \not\models_{TLP} \varphi \succ_i \varphi'$. As an example, consider $\mathscr{M}' = \mathscr{M}_{\varphi'} \subset \mathscr{M}$.

Let Γ, Γ' be sets of TLP formulas. We write $\Gamma \models_{TLP} \Gamma'$ if
$\forall \mathscr{M}_{\succeq} : (\forall \psi \in \Gamma \ \mathscr{M}_{\succeq} \models_{TLP} \psi) \Rightarrow (\forall \psi' \in \Gamma' \ \mathscr{M}_{\succeq} \models_{TLP} \psi')$.

When $\Gamma' = \{\varphi\}$ is a singleton set, we simply write $\Gamma \models_{TLP} \varphi$ instead of $\Gamma \models_{TLP} \{\varphi\}$

14.5.3 Game Theoretic Consequence

We now define game theoretic consequence recursively through a set of rules.

The rules of game theoretic consequence $\models_G$ between Γ and TLP formulas are:
If $\Gamma \models_{TLP} \varphi$ then $\Gamma \models_G \varphi$ (Temporal logic inference)
If $\Gamma \models_G \varphi \rhd_i \varphi'$ then $\Gamma \models_G \neg\varphi'$ (Game theoretic inference)
If $\Gamma \models_{TLP} \Gamma'$ and $\Gamma' \models_G \varphi$ then $\Gamma \models_G \varphi$ (Logical Inference)
If $\Gamma \cup \{\varphi\} \models_G \psi$, $\Gamma \cup \{\varphi'\} \models_G \psi'$ and
$\Gamma \models_G \psi \succeq_i \psi'$ then $\Gamma \models_G \varphi \succeq_i \varphi'$ (Order Inference)

While TLP consequence is monotonic, **game theoretic consequence is not monotonic**. To see this, note that $\Gamma \models_G \varphi \rhd_i \varphi'$ but $\Gamma \cup \{\neg\varphi\} \not\models_G \varphi \rhd_i \varphi'$.

The first rule makes all TLP consequences of Γ also game theoretic consequences of Γ (the starting set). The second rule is the negation inference rule arising from model elimination that we discussed in Section 14.4.1. The third rule fully utilises the monotonicity of TLP consequence. The last rule allows preferences between models to be deduced from partial preference specifications.

For any set Γ of PTL formulas, we let $\bigcirc\Gamma = \{\bigcirc\varphi | \varphi \in \Gamma\}$.

Proposition 14.4 (Time invariance of game theoretic consequence)

If $\Gamma \models_G \varphi$ *then* $\bigcirc\Gamma \models_G \bigcirc\varphi$. □

14.6 Modeling Games in TLP

14.6.1 Strategic Form Games

To model the statement "If Agent 1 chooses a_1 and Agent 2 chooses a_2 then the consequence is c_1" we use $p_1^1 \wedge p_2^2 \Rightarrow \bigcirc c_1$. To express that each agent has to choose exactly one of her actions, we use choice formulas $p_i^r \Rightarrow \bigwedge_{r' \neq r} \neg p_i^{r'}$. Let $\mathscr{G}$ be a strategic game form. $\Gamma(\mathscr{G})$ denotes all such action-consequence and choice formulas arising from the game form $\mathscr{G}$. We abbreviate $\Gamma(\mathscr{G})$ to Γ whenever $\mathscr{G}$ is clear from the context. To show that a certain formula φ holds in the equilibrium of the game $\mathscr{G}$, we attempt to use the game theoretic consequence relation $\Gamma(\mathscr{G}) \models_G \varphi$.

Example 1 (Prisoners' dilemma ...).
The game is represented by the set $\Gamma = \{F1, F2, F3, F4\}$ (see section 14.3.2).
We wish to show that both agents will serve 3 years in prison, i.e. $\Gamma \models_G \bigcirc c_{3y,3y}$.
First, note that $\Gamma \not\models \bigcirc c_{3y,3y}$. So, we will need to use preference orderings.

Let $\Gamma' = \Gamma \cup \{\bigcirc c_{free,5y} \succeq_1 \bigcirc c_{1y,1y}, \bigcirc c_{1y,1y} \succeq_1 \bigcirc c_{3y,3y}, \bigcirc c_{3y,3y} \succeq_1 \bigcirc c_{5y,free}\}$
$\cup \{\bigcirc c_{5y,free} \succeq_2 \bigcirc c_{1y,1y}, \bigcirc c_{1y,1y} \succeq_2 \bigcirc c_{3y,3y}, \bigcirc c_{3y,3y} \succeq_2 \bigcirc c_{free,5y}\}$

We can prove the following consequences:
$\Gamma' \models_G \neg\bigcirc c_{1y,1y}$ from $\Gamma' \models_G \bigcirc c_{free,5y} \rhd_1 \bigcirc c_{1y,1y}$
$\Gamma' \models_G \neg\bigcirc c_{5y,free}$ from $\Gamma' \models_G \bigcirc c_{3y,3y} \rhd_1 \bigcirc c_{5y,free}$

$\Gamma' \models_G \neg \bigcirc c_{free,5y}$ from $\Gamma' \models_G \bigcirc c_{3y,3y} \rhd_2 \bigcirc c_{free,5y}$
$\Gamma' \models \bigcirc c_{1y,1y} \vee \bigcirc c_{3y,3y} \vee \bigcirc c_{free,5y} \vee \bigcirc c_{5y,free}$

From these we can deduce that $\Gamma' \models_G \bigcirc c_{3y,3y}$.

Example 2 (Repeated Prisoners' dilemma ...).
The game is represented by $\Gamma = \{\Box F1, \Box F2, \Box F3, \Box F4\}$.

From Proposition 14.4, we can deduce that:
$\Gamma' \models_G \bigcirc^l c_{3y,3y}\ \forall l > 0$, which means both agents will confess in each round, i.e. their equilibrium strategies are described by $\Box p_2^c$. This is not entirely satisfactory as agents usually establish their reputation and use it in the game. To remedy this, we consider the long term preferences of the players. Let $\Gamma' = \Gamma \cup$
$\{\Diamond\Box c_{free,5y} \succeq_1 \Diamond\Box c_{1y,1y}, \Diamond\Box c_{1y,1y} \succeq_1 \Diamond\Box c_{3y,3y}, \Diamond\Box c_{3y,3y} \succeq_1 \Diamond\Box c_{5y,free}\} \cup$
$\{\Diamond\Box c_{5y,free} \succeq_2 \Diamond\Box c_{1y,1y}, \Diamond\Box c_{1y,1y} \succeq_2 \Diamond\Box c_{3y,3y}, \Diamond\Box c_{3y,3y} \succeq_2 \Diamond\Box c_{free,5y}\}$

Consider for example Agent 1 who plays a *grim strategy* written as:
$\varphi_1 = \neg p_1 \wedge \Box(p_2 \Rightarrow \bigcirc\Box p_1) \wedge \Box(\neg p_2 \Rightarrow \bigcirc\neg p_1)$. Agent 1's strategy now causes interaction between the stages of the game. With the grim strategy, $\Box p_2$ results in the sequence $\{c_{5y,free}\}(\{c_{3y,3y}\})^\omega$, i.e. $\Gamma' \cup \{\varphi_1, \Box p_2\} \models_G c_{5y,free} \wedge \bigcirc\Box c_{3y,3y}$. Similarly, $\Box\neg p_2$ results in $(\{c_{1y,1y}\})^\omega$, i.e. $\Gamma' \cup \{\varphi_1, \Box\neg p_2\} \models_{TLP} \Box c_{1y,1y}$.

Also, asymptotically $(\{c_{1y,1y}\})^\omega$ is preferred by Agent 2 to $\{c_{5y,free}\}(\{c_{3y,3y}\})^\omega$, i.e. $\Gamma' \models_G \Diamond\Box c_{1y,1y} \succ_2 \Diamond\Box c_{3y,3y}$. By order inference, $\Gamma' \cup \{\varphi_1\} \models_G \Box\neg p_2 \succ_2 \Box p_2$.

From the game theoretic consequence this leads to $\Gamma' \cup \{\varphi_1\} \models_G \neg\Box p_2$.
We are thus able to conclude that Agent 1 is effective in using a grim strategy to dissuade Agent 2 from playing p_2.

14.6.2 Extensive Form Games

Consider an extensive form game $\mathscr{G}$ with histories H and $Z \subseteq H$ being the set of terminal histories. Let $a_0a_1...a_k \in Z$ be a finite terminal history, and let $\mathscr{P}(a_0a_1...a_l) = i_l$, for all $l \leq k$. To state that a consequence c_r holds in this history, we can use the TLP formula $p_{i_0}^{a_0} \wedge \bigcirc p_{i_1}^{a_1} \wedge ... \wedge \bigcirc^k p_{i_k}^{a_k} \Rightarrow \bigcirc^{(k+1)} c_r$. Let $a_0a_1...a_k(a'_0a'_1...a'_l)^\omega \in Z$ be an eventually periodic (infinite) terminal history, and let $\mathscr{P}(a_0a_1...a_j) = i_j$, for all $j \leq k$ and $\mathscr{P}(a'_0a'_1...a'_j) = i'_j$, for all $j \leq l$. To state that a consequence c_r holds in this history, we use the TLP formula: $p_{i_0}^{a_0} \wedge \bigcirc p_{i_1}^{a_1} \wedge ... \wedge \bigcirc^k p_{i_k}^{a_k} \wedge \bigcirc^{(k+1)} p_{i'_0}^{a'_0} \wedge$
$\bigcirc^{(k+1)} \Box((p_{i'_0}^{a'_0} \Rightarrow \bigcirc p_{i'_1}^{a'_1}) \wedge ... \wedge (p_{i'_{l-1}}^{a'_{l-1}} \Rightarrow \bigcirc p_{i'_l}^{a'_l})) \Rightarrow \bigcirc^{(k+1)} \Diamond c_r$.
Let $\Gamma(\mathscr{G})$ be the collection of all such temporal formulas derived from H.

Example 14.3 (Chain store game)

Game description: A single local competitor in each city decides whether to start a store competing with the chain store. If it does, the chain store could start a price war (when both make losses) or could accomodate the competitor (in which case they share the profits). The best situation for the chain store is if the competitor stays out, when it enjoys monopoly profits.

We use propositions $p_1^E, p_1^O, p_2^F, p_2^A$ where Agent 1 is the local competitor whose actions are to enter or stay out, and Agent 2 is the chain store who can fight or accomodate. We use $c_{loss}, c_{mpoly}, c_{share}$ to model consequences with the ordering being $c_{share} \succeq_1 c_{mpoly} \succeq_1 c_{loss}$ and $c_{mpoly} \succeq_2 c_{share} \succeq_2 c_{loss}$.

We first consider the one-city case. The consequence formulas are given by:
$\Gamma = \{p_1^O \Rightarrow \bigcirc c_{mpoly}, p_1^E \wedge \bigcirc p_2^F \Rightarrow \bigcirc\bigcirc c_{loss}, p_1^E \wedge \bigcirc p_2^A \Rightarrow \bigcirc\bigcirc c_{share}\}$.
Let $\Gamma' = \Gamma \cup \{\bigcirc\bigcirc c_{share} \succeq_1 \bigcirc c_{mpoly}, \bigcirc c_{mpoly} \succeq_1 \bigcirc\bigcirc c_{loss}\} \cup$
$\{\bigcirc c_{mpoly} \succeq_2 \bigcirc\bigcirc c_{share}, \bigcirc\bigcirc c_{share} \succeq_2 \bigcirc\bigcirc c_{loss}\}$

We can prove the following consequences:
$\Gamma' \models_G \neg\bigcirc\bigcirc c_{loss}$ from $\Gamma' \models \bigcirc\bigcirc c_{share} \rhd_2 \bigcirc\bigcirc c_{loss}$
$\Gamma' \models_G \neg\bigcirc c_{mpoly}$ from $\Gamma' \models \bigcirc\bigcirc c_{share} \rhd_1 \bigcirc c_{mpoly}$
$\Gamma' \models \bigcirc c_{mpoly} \vee \bigcirc\bigcirc c_{share} \vee \bigcirc\bigcirc c_{loss}$
From these we can deduce that $\Gamma' \models_G \bigcirc\bigcirc c_{share}$.

Consider a repeated game in which the chain store competes in several cities. Without loss of generality, let Agent 1 represent all the competitors.
$\Gamma'' = \{\Box\varphi \mid \varphi \in \Gamma\} \cup \{\bigcirc^l \varphi \succeq_i \bigcirc^l \varphi' \mid \varphi \succeq_i \varphi' \in \Gamma', l \geq 0\}$.
Using Proposition 4, $\Gamma'' \models_G \bigcirc\bigcirc\bigcirc^l \neg c_{share}$, i.e. the competitor always enters ($\Box p_1^E$) and the chain store always shares profits ($\Box p_2^A$).

The **chain store paradox** occurs because it is not credible that a chain store will share profits in all cities. It may choose to create a reputation as a fighter to deter competition. We can model this through the formula $\varphi_2 = \Box(p_2^F \bigcup p_1^O)$. Then $\Gamma'' \cup \{\varphi_2\} \models \Box p_1^O \succ_1 \Box p_1^E$, which yields $\Gamma'' \cup \{\varphi_2\} \models_G \neg\Box p_1^E$, i.e. the reputation is indeed effective in deterring Agent 1 from always entering.

Note that if Agent 1 chooses the strategy $\Box p_1^O$ of staying out, then φ_2 does not require Agent 2 to fight at all. φ_2 is thus a credible threat.

14.7 Expressive Power of TLP

Proposition 14.5 (Expressive completeness)
(1) For any finite strategic form game $\mathscr{G}$, there is an equivalent set of formulas Γ such that elements of $Mod(\Gamma)$ directly correspond to strategy profiles of $\mathscr{G}$.
(2) For any extensive form game $\mathscr{G}$, there is an equivalent set of formulas Γ such that elements of $Mod(\Gamma)$ directly correspond to histories of $\mathscr{G}$.

Proof. (1) is obvious. (2) follows from the definition of history set through eventually periodic sequences, and from the fact that we can express any eventually periodic sequence in propositional temporal logic using additional propositions in $\mathscr{W}$.

Proposition 14.6 (Characterising Nash equilibria)

If Γ is the set of temporal formulas representing the strategic game $\mathscr{G}$ and Γ' the set of TLP formulas representing the orderings on consequences, then:

(1) The game has a Nash equilibrium iff $\Gamma \cup \Gamma' \models_G True$.
(2) φ *is true in the Nash equilibria iff* $\Gamma \cup \Gamma' \models_G \varphi$

Proof. Follows from the equivalence between the tableau method in [26] and the game theoretic consequence rules above.

Let Γ be the set of temporal formulas representing a finite extensive form game and Γ' the set of TLP formulas representing the orderings on consequences. The operator $Cl : 2^{PTL} \rightarrow 2^{PTL}$ is defined by $Cl(\Gamma) = \{\varphi \,|\, \Gamma \cup \Gamma' \models_G \varphi\}$. The **closure** $Cl^*(\Gamma)$ of Γ is the fixed point of Cl. Note that $\Gamma \subseteq Cl(\Gamma)$, which means that the closure $Cl^*(\Gamma)$ exists and is unique.

Proposition 14.7 (Characterising Sub-game perfect equilibria)
(1) φ *holds in the sub-game perfect equilibrium iff* $Cl^*(\Gamma) \cup \Gamma' \models_G \varphi$.
(2) $Cl^*(\Gamma) \cup \Gamma' \models_G True$ *i.e. the game has a subgame perfect equilibrium.*

Proof. Follows from the equivalence between the tableau method in [26] and the game theoretic consequence rules above.

14.8 Conclusions and Possible Extensions

We showed that propositional temporal logic has the expressive power to capture the structure of finite games and repeated games in which the history sets are eventually periodic. We extended the preference ordering relationships defined for propositional logic models to eventual periodic models. We showed how arbitrary approximations of this preference ordering can be represented using theories.

We then introduced the language and semantics of the temporal logic of preferences – TLP. TLP allows representation of agents choices and preferences. Using TLP, we could find an expression for game theoretic consequence.

Agents may not know their preferences. To help agents express their preferences, interactive elicitation procedures work by finding relevant questions to ask, until the agents preferences are consistent and complete [6, 7, 11]. In a game situation, agents may know their own preferences, but may not have knowledge of the other agents' preferences. The choices made by an agent in a game reveals her preferences to the other agents. By allowing preferences to be explicitly modeled, we provide a path for dealing with such game evolutions.

References

1. Asheim G. B., and Dufwenberg M. Deductive reasoning in extensive games, *Econ. J.*, 113: 305–325, April 2003. Blackwell Publishing.
2. Bacchus F., and Grove A. Graphical models for preference and utility. In *Proceedings of the 11th Conference on Uncertainty and AI (UAI '95)*. pages 3–10, 1995.
3. Bonanno G. Branching time logic, perfect information games and backward induction. In *3rd Conference on Logic and Foundations of Game and Decision Theory*, International Centre for Economic Research (ICER), Torino, Italy, Dec 1998.

4. Boutilier C. Toward a logic for qualitative decision theory. In *Proceedings of the KR94*, pages 75–86, 1994.
5. Boutilier C. Conditional logics of normality: A modal approach, *Artif. Intell.*, 68: 87–154, 1994.
6. Boutilier C. A POMDP formulation of preference elicitation problems. In *Proceedings of AAAI-02*, AAAI Press, 2002.
7. Boutilier C., Brafman R. I., Domshlak C., Hoos H. H., and Poole D. Preference based constrained optimisation based on CP-Nets. *Comput. Intell.*, 20(2): 137–157, 2004.
8. Conlisk J. Why bounded rationality? *J. Econ. Lit.*, XXXIV: 669–700, June 1996.
9. Coste-Marquis S., Lang J., Liberatore P., and Marquis P. Expressivity and succinctness of preference representation languages. In *Proceedings of the 9th International Conference on Knowledge Representation and Reasoning (KR-2004)*, AAAI Press, pages 203–212, 2004.
10. Dastani M., and van der Torre L. Decisions and games for BD agents. In *Proceedings of AAAI-02*, AAAI Press, 2002.
11. Gonzales C., and Perny P. Graphical models for utility elicitation. In *Proceedings of 9th Knowledge Representation and Reasoning Conference (KR-2004)*, AAAI Press, pages 224–233, 2004.
12. Harrenstein P. A game-theoretical notion of consequence. In *5th Conference on Logic and Foundations of Game and Decision Theory*, International Centre for Economic Research (ICER), Torino, June 2002.
13. Harrenstein P., van der Hoek W., Meyer J.-J., and Witteven C. A modal characterization of nash equilibrium, *Fundam. Informaticae*, 57: 281–321, 2003.
14. De Vos M., and Vermeir D. Choice logic programs and Nash equilibria in strategic games. In J. Flum and M. Rodriguez-Artalejo, editors, *Computer Science Logic (CSL '99), LNCS-1683*, pages 266–276. Springer, 1999.
15. De Vos M., and Vermeir D. *Dynamically Ordered Probabilistic Choice Logic Programming. (FST & TCS -20, LNCS*-1974), Springer, 2000.
16. La Mura P., and Shoham Y. Expected utility networks. In *Proceedings of UAI '99.* pages 366–373, 1999.
17. Lang J. Logical preference representation and combinatorial vote. *Ann. Math. Artifl. Intell.*, 42(1–3): 37–71, 2004.
18. Lang J. Conditional desires and utilities – An alternative approach to qualitative decision theory. In *Proceedings of the European Conference on Artificial Intelligence (ECAI96)*, pages 318–322, 1996.
19. Makinson D. General theory of cumulative inference, In M. Reinfrank et al., editors, *Nonmonotonic Reasoning*. Springer, Berlin, 1989.
20. Osborne M. J., and Rubinstein A. *A Course in Game Theory*, (3rd edition). The MIT Press, Cambridge, MA; London, , 1996.
21. Shoham Y. *Reasoning About Change: Time and Causation from the Standpoint of Artificial Intelligence*. MIT Press, Cambridge, 1988.
22. Simon, H A. A behavioral model of rational choice, *Q. J. Econ.*, 69(1): 99–118, Feb. 1955.
23. Stalnaker R. Extensive and strategic forms: Games and models for games, *Res. Econ.*, 53: 293–319, 1999. Academic Press.
24. van Benthem J. *Logic and Games. Lecture Notes*. ILLC Amsterdam & Stanford University, 1999.
25. van Otterloo S., van der Hoek W., and Woolridge M. preferences in game logics. In AAMAS 2004. New York, NY, http://www.aamas2004.org/proceedings/021_otterloos_preferences.pdf
26. Venkatesh G. Reasoning about game equilibria using temporal logic. In *Proceedings of the FST & TCS Conference 2004.* LNCS-3328, pages 503–514, Springer, 2004.
27. Wilson N. Extending CP-nets with stronger conditional preference statements. In *Proceedings of AAAI-04*. 2004.
28. Wolper P. The tableau method for temporal logic – An overview. *Logique Anal.*, 28: 119–152, 1985.

Part V
Logic, Language and Cognition

Chapter 15
From Sentence Meanings to Full Semantics

Wilfrid Hodges

Many questions about meaning are straightforwardly empirical, like the question whether "Gift" means the same in German as it does in English.

Other questions about meaning are more abstract: for example the question whether in metaphors a word acquires a new meaning, or simply contributes its meaning in a different way to a phrase containing it. This is a question about the architecture of our preferred semantic theory, and perhaps not a question about facts at all. For some theories of language the notion of meaning, like the notion of cause in physics, vanishes altogether when we reach fundamentals.

One task of any semantic theory is to draw a line between semantics and grammar/syntax. Noam Chomsky hints at this when he says: ([6] p. 144):

> … at every point in the stream of discourse the speaker must choose a particular single word, and it makes sense to ask to what extent his choice of a particular word was governed by the grammatical structure of the language, and to what extent it was governed by other factors.

Chomsky's phrasing suggests an equation:

$$\text{Language} = \text{Syntax} + (\text{Other factors}), \tag{15.1}$$

where the other factors presumably include meaning. If (15.1) was an equation in an abelian group, we could solve for (Other factors) by subtracting Syntax from Language. Unfortunately the mathematical structure underlying the equation not an abelian group, but something quite different that I describe in Section 15.2.4 below. (If mathematical readers detect some resemblance to Galois theory, that's good.)

In broad outline, if we have a language with a given or presumed grammatical structure, and an account of how sentences of the language are used, then together these items induce a formal structure on the rest of the language. Somewhere inside this formal structure there lies a shadow of meanings. In Section 15.1 I describe the formal structure, and Section 15.2 is about how one extracts the meanings.

Wilfrid Hodges
Queen Mary, University of London, London, UK, e-mail: w.hodges@qmw.ac.uk

J. van Benthem et al. (eds.), *Proof, Computation and Agency*, Synthese Library 352,
DOI 10.1007/978-94-007-0080-2_15,

This paper is an extract from talks given during the last couple of years in meetings at the University of Düsseldorf, the Sorbonne, Stanford University and the Indian Institute of Technology, Bombay. I warmly thank my hosts and audiences in all those places.

15.1 Mathematical Theory

We consider a language L. For the present our grammatical assumptions about L will be thin. We assume that we can recognise expressions of L, and the grammatical constituents of these expressions. A constituent of an expression is again an expression in its own right. We know what it is to replace a constituent in an expression by another expression. To make these notions formal, we adopt the following definition.

Definition 15.1

(a) By a *constituent structure* we mean an ordered pair of sets $(\mathbb{E}, \mathbb{F})$, where the elements of $\mathbb{E}$ are called the *expressions* and the elements of $\mathbb{F}$ are called the *frames*, such that the four conditions below hold. (Here and below, e, f etc. are expressions. F, $G(\xi)$ etc. are frames.)

1. $\mathbb{F}$ is a set of nonempty partial functions on $\mathbb{E}$. ("Nonempty" means their domains are not empty.)
2. (Nonempty Composition) If $F(\xi_1, \dots, \xi_n)$ and $G(\eta_1, \dots, \eta_m)$ are frames, $1 \leqslant i \leqslant n$ and there is an expression

$$F(e_1, \dots, e_{i-1}, G(f_1, \dots, f_m), e_{i+1}, \dots, e_n),$$

then

$$F(\xi_1, \dots, \xi_{i-1}, G(\eta_1, \dots, \eta_m), \xi_{i+1}, \dots, \xi_n)$$

is a frame.

Note that if $H(\xi)$ is $F(G(\xi))$ then the existence of an expression $H(f)$ implies the existence of an expression $G(f)$.

3. (Nonempty Substitution) If $F(e_1, \dots, e_n)$ is an expression, $n > 1$ and $1 \leqslant i \leqslant n$, then

$$F(\xi_1, \dots, \xi_{i-1}, e_i, \xi_{i+1}, \dots, \xi_n)$$

is a frame.
4. (Identity) There is a frame $1(\xi)$ such that for each expression e, $1(e) = e$.

(b) We say that an expression e is a *constituent* of an expression f if f is $G(e)$ for some frame G; e is a *proper constituent* of f if e is a constituent of f and $e \neq f$.

(c) We refer to $F(e_1, f, e_3)$ as *the result of replacing the occurrence of* e_2 *in second place in* $F(e_1, e_2, e_3)$ *by* f. (This notion depends on F, of course.)

(d) We say that a set Y of expressions is *cofinal* if every expression of L is a constituent of an expression in Y.

Our only syntactic assumption on L (for the moment) is that L has a constituent structure. By the *expressions* of L we mean the expressions of its constituent structure.

Definition 15.2 By a *partial semantics* for L we mean a function $\mu : X \to Y$ where X is a set of expressions of L. (There are no further requirements on Y or μ.) A *total semantics* for L is a partial semantics $\mu : X \to Y$ where X is the set $\mathbb{E}$ of all expressions.

The constituent structure and the partial semantics μ together induce two equivalence relations on the set of expressions as follows.

Definition 15.3(a) Let e, f be expressions of L and μ a partial semantics for L with domain X. We write $e \equiv_\mu f$ if for every 1-ary frame $G(\xi)$,

(i) $G(e)$ is in X if and only $G(f)$ is in X;
(ii) if $G(e)$ is in X then $\mu(G(e)) = \mu(G(f))$.

We say e, f have *the same* $\equiv_\mu$*-value*, or for short *the same fregean value*, if $e \equiv_\mu f$.

(b) The relation $\sim_\mu$ is defined exactly as $\equiv_\mu$ but with clause (ii) deleted.

We assume some choice of labels for the equivalence classes of $\equiv_\mu$, and we write $|e|_\mu$ for the label of the equivalence class of the expression e. We call $|e|_\mu$ the *fregean value* of e. The function $|.|_\mu$ is a total semantics for L. I call it the *fregean semantics* (or among computer scientists, the *fully abstract* semantics).

Lemma 15.1 (Lifting Lemma) *Suppose $F(e_1, \ldots, e_n)$ is a constituent of some expression in X, and for each i, $e_i \equiv_\mu f_i$. Then:*

(a) $F(f_1, \ldots, f_n)$ is an expression.
(b) $F(e_1, \ldots, e_n) \equiv_\mu F(f_1, \ldots, f_n)$.

Proof. By Nonempty Substitution we can make the replacements one expression at a time. So it suffices to prove the lemma when $n = 1$.

Assume $F(e)$ is an expression, $H(F(e))$ is in X and $e \equiv_\mu f$.

Claim (a): $F(f)$ is an expression.

By Nonempty Composition $H(F(\xi))$ is a frame $G(\xi)$. Since $e \equiv_\mu f$ and $G(e)$ is in X, $G(f)$ is in X. But $G(f)$ is $H(F(f))$, proving (a).

Claim (b): $F(e) \equiv_\mu F(f)$.

Let $G(\xi)$ be any 1-ary frame such that $G(F(e))$ is an expression in X. By Nonempty Composition $G(F(\xi))$ is a frame $J(\xi)$. Since $e \equiv_\mu f$ and $J(e)$ is in X, $J(f)$ is in X and $\mu(J(e)) = \mu(J(f))$. So $\mu(G(F(e)) = \mu(G(F(f))$ as required, proving (b). □

Note that the Lifting Lemma holds also for $\sim_\mu$ in place of $\equiv_\mu$; in fact the proof of the lemma already proves this.

Lemma 15.2 *Suppose the domain X of μ is cofinal (Definition 15.1(d)). Then there is, for each n-ary frame F, a n-ary map* $h_F : V^n \to V$*, where V is the class of* $\equiv_\mu$*-values, such that whenever* $F(e_1,\dots,e_n)$ *is an expression,*

$$|F(e_1,\dots,e_n)|_\mu = h_F(|e_1|_\mu,\dots,|e_n|_\mu).$$

Proof. By (b) of the Lifting Lemma, if $e_i \equiv_\mu f_i$ for each i then

$$F(e_1,\dots,e_n) \equiv_\mu F(f_1,\dots,f_n)$$

provided these expressions exist. So F and the fregean values of the e_i determine the fregean value of $F(e_1,\dots,e_n)$. □

The map h_F of the lemma is essentially unique; its values are determined on all n-tuples that actually occur as values of the constituents $e_1, \dots, e_n$ of an expression $F(e_1,\dots,e_n)$. We call this map the *Hayyan function* of F.

The name is from the distinguished Andalusian linguist Abu Ḥayyān al-Gharnāṭī, who was born in Granada in 1256 and worked in Egypt until his death in 1345. His claims to fame include a textbook of Turkish for Arabs [7] and a monograph on Mongolian. See [11] for details of the following quotation from him.

> I find it astonishing that people allow a sentence construction in a language, even when they have never heard a construction like it [before]. Are Arabic constructions different from the words in the dictionary? Just as one can't use newly-invented single words, so one can't use [newly-invented] constructions. Hence all these matters are subject to convention (*waḍ*ᶜ), and matters of convention require one to follow the practice of the speakers of the relevant language. The difference between syntax and lexicography is that syntax studies universal [rules], whereas lexicography studies items one at a time. These two sciences interlock in [describing] the conventions [on which language is based].

The word *waḍ*ᶜ is crucial for understanding this passage. It is the term used for the assumed imposition of words on their meanings in the origins of language. Abu Ḥayyān is arguing that syntactic constructions have meanings just as words do, but the meaning of a construction is "universal"—in modern terminology, it's a function defined for all possible inputs to the construction.

Definition 15.4

(a) Let $\simeq$ be an equivalence relation on expressions. We say that $\simeq$ is *compositional* if for every pair of expressions $F(e_1,\dots,e_n)$ and $F(f_1,\dots,f_n)$,

$$e_1 \simeq f_1 \text{ and } \dots \text{ and } e_n \simeq f_n \Rightarrow F(e_1,\dots,e_n) \simeq F(f_1,\dots,f_n).$$

(b) Let φ be a function defined on expressions. We say that φ is *compositional* if for each expression $F(e_1,\dots,e_n)$, the value

$$\varphi(F(e_1,\dots,e_n))$$

is determined by F and the values $\varphi(e_i)$.

(Cf. Partee et al. [15] p. 318.) Parts (a) and (b) match; the fregean semantics $|.|_\mu$ is compositional if and only if $\equiv_\mu$ is compositional. In this terminology, Lemma 15.2 tells us that $\equiv_\mu$ is in fact compositional. The same argument shows that $\sim_\mu$ is compositional too.

Lemma 15.3 *Suppose* $e \equiv_\mu f$ *and* e *is an expression in* X. *Then* f *is in* X *and* $\mu(e) = \mu(f)$.

Proof. This is immediate from the definition, by applying the identity frame $1(\xi)$. □

It follows from Lemma 15.3 that there is a function p_μ such that for every expression e in X, $\mu(e) = p_\mu(|e|_\mu)$. This function p_μ is uniquely determined in a similar sense to the Hayyan functions. We will call it the *read-out function* of μ.

Lemma 15.4 *The following are equivalent:*

(a) For all e, f *in* X, $e \equiv_\mu f$ *if and only if* $\mu(e) = \mu(f)$.
(b) For all e, f *in* X *and every frame* $F(\eta)$,

$$\begin{aligned} &\mu(e) = \mu(f) \text{ and } F(e) \in X \\ \Rightarrow\ &F(f) \in X \text{ and } \mu(F(e)) = \mu(F(f)). \end{aligned}$$

Proof. Again this is immediate from the definition. □

When the conditions of Lemma 15.4 hold, we can assume that the representatives $|e|_\mu$ with e in X were chosen so that $|e|_\mu = \mu(e)$. The read-out function p_μ is then the identity function.

Our next result needs a further assumption on the constituent structure.

Definition 15.5

(a) We say that the constituent structure is *well-founded* if there are no infinite sequences of expressions

$$e_0, e_1, e_2, \ldots$$

where for every n, e_{n+1} is a proper constituent of e_n.

(b) Assuming that the constituent structure is well-founded, we define the *complexity* $c(e)$ of an expression e to be the least ordinal strictly greater than $c(f)$ for each proper constituent f of e. A standard set-theoretic argument shows that c is a well-defined function from the set of expressions to the ordinal numbers. (If L is any natural language or a finitary formal language, then the complexity of each expression will be a natural number.) We say that an expression is an *atom* if its complexity is 0; otherwise the expression is *complex*.

Theorem 15.1 (Abstract Tarski theorem) *Let* L *be a language with a well-founded constituent structure, and* μ *a function whose domain is a cofinal set* X *of expressions of* L. *Then* μ *has a definition of the following form. A function* ν *is defined on all expressions of* L *by recursion on complexity. The basis clause is*

- *$\nu(e) = |e|_\mu$ for each atom e.*

The recursion clause is

- *$\nu(F(e_1,\dots,e_n)) = h_F(\nu(e_1),\dots,\nu(e_n))$ for each complex expression $F(e_1,\dots,e_n)$.*

Then for each expression e in X,

$$\mu(e) = p_\mu(\nu(e)).$$

This concludes our main mathematical development. Two further directions are worth mention.

15.1.1 Equivalence of Frames

In analogy to the definition of $\equiv_\mu$ on expressions, we can define $\equiv_\mu$ on frames as follows. For simplicity I give the definition for binary frames, but it extends straightforwardly. Again we suppose we have a partial semantics $\mu : X \to Y$.

Definition 15.6 Suppose $F(\xi_1,\xi_2)$ and $F'(\xi_1,\xi_2)$ are frames. Then $F(\xi_1,\xi_2) \equiv_\mu F'(\xi_1,\xi_2)$ if and only if for all expressions e, f and frames $G(\eta)$,

(a) $G(F(e,f))$ is in X if and only if $G(F'(e,f))$ is in X, and
(b) if $G(F(e,f))$ is in X then $\mu(G(F(e,f)) = \mu(G(F'(e,f))$.

Then $\equiv_\mu$ on frames is an equivalence relation, and one can prove analogues of the results for $\equiv_\mu$ on expressions.

15.1.2 The Case Where X Is Not Cofinal

Lemmas 15.1 and 15.4 extend μ (under the conditions in Lemma 15.4) to a compositional semantics ν on the set of all expressions that occur as constituents of expressions in X. A dual problem is to take a compositional semantics μ on a set X of expressions that is closed under constituents, and extend μ to a total compositional semantics. There is a trivial solution if we assume that $\mu(e) = \mu(f)$ implies $e \sim_\mu f$ (a condition called *husserlian* in [10]). When the husserlian condition fails, the problem is much harder. Dag Westerståhl [22] gave a positive answer under the assumption that L is a subset of a term algebra.

15.2 Commentary

15.2.1 The Freedom to Choose a Grammar

Broadly speaking, the results of Section 15.1 are independent of any choice of grammatical *theory* (at least among the better known theories), but the data delivered ($\equiv_\mu, h_F$ etc.) is highly dependent on the choice of a *constituent structure* for a given language.

Lemma 13 of my [10] was a version of the Lifting Lemma under the stronger assumption that the language is a subset of a term algebra. (More precisely I assumed that it's a homomorphic image of a subset of a term algebra, but in fact the work all takes place on the term algebra itself.) In Chapter 2 of his book [18] Mark Steedman argues for a more flexible notion of constituents. For example they might overlap. Not being a linguist, I can't comment on the force of Steedman's arguments. But they did convince me that the Lifting Lemma in [10] was not at the right level of generality. The version in this paper is the result, and it is certainly simpler than the earlier version (which is now a special case).

Are there other approaches to grammar that won't accommodate the Lifting Lemma? Fortunately I didn't need to hunt around on this, because at exactly the right time Edward Keenan and Edward Stabler published a notion of "bare grammar" [12] that is designed to pick out a common core from "otherwise rather different specific theories of grammar: Relational Grammar, Arc-Pair Grammar, Categorial Grammar, Lexical Functional Grammar, Head Driven Phrase Structure Grammar, Government Binding Theory/Minimalism, and others" ([12] p. 1). They build on the two notions of constituent and category.

Every bare grammar *G* yields a constituent structure in either of two ways. The first way is to take as the set of expressions the "language generated by *G*", which is the closure of the lexicon of *G* under the syntactic rules of *G*, and as the set of frames the closure under substitution of the "generating or structure building functions" of *G* ([12] p. 14f). The resulting expressions are ordered pairs consisting of (at first approximation) a phrase and the category of the phrase; for example at word level we might find two expressions

$$\text{(win,N)}, \quad \text{(win,VT)} \tag{15.2}$$

for the noun "win" and the transitive verb "win". The second way of extracting a constituent structure from the bare grammar is to ignore the categories and take the expressions to be the first terms of these pairs.

Bare grammars are general enough to handle non-configurational languages, where the word order within clauses is not significant. One of the examples in [12] (p. 59ff) is "free order Korean", a version of Korean that the authors have deliberately souped up to make it even less configurational than Korean normally is. I mention this because I was asked – for example about Sanskrit in Mumbai – whether the results of Section 15.1 make sense for non-configurational languages.

For these languages one would expect a large number of equivalences of the form $F(\xi_1, \xi_2) \equiv_\mu F(\xi_2, \xi_1)$.

It's true that Section 15.1 applies to non-configurational languages. But it's not the whole truth. Non-configurational languages tend to decorate each word stem with a cluster of prefixes, suffixes and infixes that carry the information given in other languages by the order. In fact most languages go some way down this road; thus Gothic *handus* is subject and *handu* is object. If you want to separate out the meanings of the stem *hand* and the suffixes, it makes sense to adopt a grammar where *hand*, *-us* and *-u* are all atoms. (Then probably you need a frame for combining stem and suffix into a single word.)

English has separated out the Old Germanic stem *hand* by dropping the suffixes. But you may want to think of the old *-us* and *-u* as implied constituents of a modern English sentence. One way of doing this is to have atoms NOM and ACC for nominative and accusative. In the resulting grammar "hand-NOM" and "hand-ACC" will be distinct expressions, even though "NOM" and "ACC" aren't pronounced. In the same spirit you may want to regard the English word "me" as how we spell and pronounce what is really a compound word "I-ACC". Or as (15.2) illustrates, you might want to count one inscription as two different words (for syntactic reasons in the case of (15.2), but reasons might come from anywhere).

And so on, really. A grammar for a language doesn't arrive ready-made. To a great extent it's a theoretical construct from the empirical data.

In that spirit, one way of looking at the results of Section 15.1 is as follows. We have a language L, and we propose a grammar for it. The grammar, together with a function μ describing the usage of sentences, induces fregean values, and we try to make sense of these values. If they are hopelessly obscure or perversely difficult to describe, then we look for a different grammar and/or a different way of describing the usage of sentences, and we try again. This is a familiar kind of heuristic.

15.2.2 Freedom in the Choice of $|e|_\mu$

The notion $e \equiv_\mu f$ conveys that e and f make exactly the same contribution to the μ-values of expressions in X that have them as constituents. Is this the only possible formalisation of this informal notion?

It seems that it is. The key point to note is that the informal notion is an *equivalence relation*; so any formalisation of it must be an equivalence relation too. This immediately knocks out some alternatives that one might propose. Take for example the relation $\equiv^\star$ defined by:

> For any expressions e and f, $e \equiv^\star f$ if and only if for all frames $F(\xi)$, if $F(e)$ and $F(f)$ are both in X then $\mu(F(e)) = \mu(F(f))$.

If e and f are so unrelated that there is no frame $F(\xi)$ for which both $F(e)$ and $F(f)$ are expressions, then $e \equiv^\star f$. That should already make us uneasy. What truly knocks this definition on the head is that there is no reason at all to expect the defined relation $\equiv^\star$ to be transitive. I think if you try out some further possibilities, you will soon convince yourself that our definition of $\equiv_\mu$ is the only one that makes sense in all cases.

When we know that two expressions have the same fregean value, this information on its own doesn't tell us what that fregean value actually is. This might seem a

devastating gap in the theory of Section 15.1, but in practice it isn't. In concrete cases, when we find out how to tell whether two expressions have different fregean values, that information normally gives us all we need for assigning sensible labels to the equivalence classes. We will see some examples below.

15.2.3 The Centrality of Sentence Meanings

The meanings of sentences of a language are decisive for the meanings of all phrases of the language. There are many arguments to support this view; here are three.

(1) We express ourselves in sentences.

(2) In practice the normal method for distinguishing the meanings of two closely related words is to give a sentence using one of them, that would lose its meaning or become inappropriate if the other was used. As a typical example, Webster's Dictionary of English Usage [8] distinguishes between "less" and "lesser" by calling on sentences from its corpus, including this from Virgil Thompson:

...lesser composers analyze music better than they used to. (15.3)

I'm not sure whether the result of putting "less" in place of "lesser" in (15.3) means anything, but it certainly doesn't mean the same as (15.3).

(3) Slightly more technical: the meaning of a word includes the semantic argument structure of the word. For example you don't know the meaning of the word "mother" until you know that "my mother" normally means a person who bears a certain relation to me, not just someone who is both mine and a mother. (This is a semantic property of "mother"; to the best of my knowledge there are no purely syntactic differences between these two uses of the phrase "my mother".) But words taken on their own don't show their argument structure. To discover that structure you have to understand how the words fit into phrases. Granted not all phrases are sentences; but at least the sentences containing a word give enough evidence to place its semantic argument structure.

We can almost read off the following

NECESSARY CONDITION FOR DIFFERENCE OF MEANING. Let L be a language, X the set of sentences of L and for each sentence e let $\mu(e)$ be the meaning of e. If two expressions have different meanings then they have different fregean values.

This condition is in danger of lapsing into triviality. Arguably if e and f are any two distinct expressions of a natural language, then there will be at least one sentence e – even a sentence not using quotation – whose meaning changes if we put one expression in place of the other in e.

But there are degrees of synonymy. Phrases can have broadly similar meanings and differ in nuances; otherwise we'd have no use for the notion of a synonym. To investigate a weak kind of synonymy by means of fregean values, there are two possible moves to make. The first is to restrict to a limited part of the language, either by shedding vocabulary or by tightening the requirements of grammaticality. Dropping adverbs of degree removes a distinction between "angry" and "enraged" (you can be mildly angry but not mildly enraged, cf. Apresjan [1] p. 125). And so on. The second possible move is to weaken the notion of synonymy on sentences. For example if we take note only of when sentences are true or false, and ignore their social status, we cut away the difference in meaning between "intoxicated" and "pissed". And so on.

In short, the apparatus of Section 15.1 gets its bite by being applied to limited parts of language – just as the physicist normally applies physical theory to bounded systems. Sections 15.2.6 and 15.2.7 will discuss some languages where Section 15.1 applies without any limitation. But these are austere formal languages of logic.

As a consequence, the Necessary Condition above will sometimes fail, but benignly. If there really is a difference between the meanings of e and f, it's safe to assume that some difference between meanings of sentences will serve to show this, though we might have to strengthen the language or the function μ to show this. At worst we might have to accept interjections like "Ouch!" as sentences.

15.2.4 Splitting Syntax From Semantics

Assume a language L is given, together with a constituent structure and a partial semantics μ on sentences. We saw how this data yields two equivalence relations $\sim_\mu$ and $\equiv_\mu$ on the class of expressions. Also $\equiv_\mu$ is a refinement of $\sim_\mu$: if $e \not\sim_\mu f$ then $e \not\equiv_\mu f$ too.

Now granting that the class of sentences is syntactically distinguishable, the relation $\sim_\mu$ is purely syntactic. Recall Chomsky's Other Factors and the question of distinguishing them from syntax. Imagine that we can define a third equivalence relation $\approx_\mu$, where $e \approx_\mu f$ holds if and only if e and f agree in their their meanings, i.e. those Other Factors that have a detectable effect on μ. Then we will have

$$e \equiv_\mu f \;\Leftrightarrow\; e \sim_\mu f \text{ and } e \approx_\mu f. \tag{15.4}$$

Equation (15.4) solves equation (15.1). There is always at least one solution of equation (15.4), namely where $\approx_\mu$ is $\equiv_\mu$. It should be obvious that in general there are bound to be many other solutions. This is where semantic theory takes off.

The common claim that meaning is compositional is a claim about the proper form of $\approx_\mu$. By our general theory, $\equiv_\mu$ and $\sim_\mu$ are both compositional; but it doesn't follows that $\approx_\mu$ is. Here let me make a comment in passing. Some of the arguments in favour of the compositionality of meanings, in particular those which refer to processes by which an interpreter of sentences discovers their meanings, are fully met by the observation that $\equiv_\mu$ is compositional; they provide no evidence that $\approx_\mu$ *as well as* $\equiv_\mu$ must be compositional. But as we've seen, the compositionality of $\equiv_\mu$ is a purely mathematical artefact. Unlike the compositionality of $\approx_\mu$, it has no empirical content at all. I believe these facts go a long way towards accounting for the feeling that compositionality of meaning is both necessary and elusive. (There are other reasons of a methodological kind for wanting meanings to be compositional.)

My own belief is that the choice of $\approx_\mu$ within $\equiv_\mu$ is *never* purely empirical; it always depends on a mixture of a priori theory and plain subjective taste. This is not the sort of thing one can hope to prove. But let me discuss four types of example.

Example 15.1 The chief difference between the English phrases "in spite of" and "notwithstanding" is that "notwithstanding" is usable as a postposition, whereas "in spite of" is always a preposition:

> She looked stunning, her tight schedule notwithstanding.
> ⋆She looked stunning, her tight schedule in spite of.

Probably most people would say at once that this is a purely syntactic difference between "in spite of" and "notwithstanding".

We can analyse this example as follows. Take $F(\xi_1, \xi_2)$ to be the construction putting a preposition at ξ_1 in front of a noun at ξ_2, and $F'(\xi_1, \xi_2)$ to be the construction putting a postposition at ξ_1 after a noun at ξ_2. Then

$$F(\text{in spite of}, \xi_2) \equiv_\mu F'(\text{notwithstanding}, \xi_2) \tag{15.5}$$

in the sense of Definition 15.6, and hence these two 1-ary frames have the same meaning. But F and F' themselves have no semantic content; they are pure constructors. Hence the meanings of the 1-ary frames in (15.5) come from their first arguments. Furthermore these arguments never occur in any other context. So we should reckon that the two arguments have the same meaning.

Before we rest easy with this argument, take another example. It's artificial but I hope it makes its point. Imagine a linguist coming on English for the first time, and recognising the morphemes "dis-" and "em-". She notes that "disgruntled" and "embittered" mean near enough the same, and she infers that the difference between "gruntled" and "bittered" is the purely syntactic one that the first takes "dis-" and the second takes "em-". This is essentially the same argument as above. But here it doesn't work, because the prefix "dis-" carries a semantic content not in "em-", namely that it negates what follows it.

As this second example emphasises, the argument for Example 15.1 rests on our already having decided that in English there is no semantic difference between prefixing and postfixing. Technically, it means we have already made some decisions about $\approx_\mu$ on frames as well as on expressions. (Not that this particular decision is controversial.)

Example 15.2 Another example, already referred to, is the difference between "I" and "me". You might want to say that these words have the same meaning, in spite of the fact that there are almost no sentences in which we can change between "I" and "me" without altering or losing the sense.

There are several possible arguments to support this view. One crude argument is that in any context where they are uttered, the two terms have the same reference. Since it's hard to maintain that reference is the whole of meaning, this argument is weak. A stronger argument would point to the practical advantages of analysing "I" and "me" as "I-NOM" and "I-ACC", and then observe that the difference between NOM and ACC is purely syntactic. The use of semantic theory in this argument is clear. (Also I suspect not everybody would accept that NOM and ACC are semantically neutral. They have more than a hint of agent and patient.)

My guess is that in Examples 15.1 and 15.2 most people will find it quite easy to separate the syntax from the semantics. The purpose of the examples was to show the involvement of background semantic theory even in uncontroversial cases. In the next two examples it seems to me genuinely puzzling where or how one should draw the line. Perhaps the distinction in these cases is purely within one's semantic theory, without any empirical content to it at all.

Example 15.3 Compare "murderer" and "person who has killed someone":

> A person who has killed someone most often already knew him or her.
> A murderer most often already knew him or her.

The second sentence doesn't carry the straightforward meaning of the first.

The sentences show that the phrase "person who has killed someone" includes a term for the object, which anaphoric pronouns can latch onto, whereas "murderer" has no such term. Is this a semantic difference or a purely syntactic one? How would one decide?

This example is complicated by the fact that there is another difference between the two expressions, as shown in

> That man is the murderer of Caroline.
> That man is the person who has killed someone of Caroline.

The second sentence is syntactically ugly, but the main point is that it can't mean the same as the first. Together the sentences show that "murderer" has a semantic argument place that's not available in "person who has killed someone". I think this is a clear difference of meaning, but I could name leading linguists who seem to disagree.

Example 15.4 Compare "liked" and "enjoyed". The general pattern seems to be that in frames where both words are acceptable, they carry the same sense. But "enjoyed" is limited to frames where it refers to an activity. For example "He enjoyed Bach" forces us to interpret as "He enjoyed listening to Bach", or in some contexts "He enjoyed playing Bach". "He enjoyed aluminium" is uninterpretable out of context. "He enjoyed Susan" strongly suggests an activity. (Pustejovsky [16] p. 135.)

With examples of this kind, the evidence for a semantic difference is that certain sentences don't normally occur. But exactly the same evidence could be used to make a case that English syntax recognises certain distinctions (here, between activity verbs and other verbs). So which wins, syntax or semantics?

A formal description of Example 15.4 would say that there are expressions e and f such that $e \not\equiv_{\mu} f$ but $F(e) \equiv_{\mu} F(f)$ in the many cases where $F(e)$ and $F(f)$ both exist. This seems to be a common phenomenon. An example with adjectives is "cold" and "cool". Apresjan ([1] p. 52) notes that we tend not to use "cool" for felt sensations in particular parts of our bodies: "My fingers feel cool/cold". In this case he opts for syntax, but he has an unusually articulate semantic theory in the background.

15.2.5 The Indeterminacy of Translation

Quine ([17] chapter ii) famously argues that the notion of a correct translation from one language to another is underdetermined. He allows that languages contain "observation sentences" which "wear their meanings on their sleeves". Between sentences of this kind, translations are objectively right or wrong, up to the "normal inductive" uncertainties. But to extend the translation downwards to constituents of observation sentences is to make "analytical hypotheses", and many different and incompatible analytical hypotheses might do the job.

Markus Werning [21], working as closely as possible from Quine's own assumptions, argues that the Lifting Lemma (Lemma 15.1) puts serious constraints on the possible analytical hypotheses, and so very much reduces the indeterminacy of translation, at least for words and phrases that occur as constituents of observation sentences. Hannes Leitgeb [13] replies to Werning.

15.2.6 Tarski's Definition of Truth

In 1933 Tarski [19] undertook to give a definition of the predicate "φ is a true sentence of the language L" where L is a fully interpreted formal language. One of his requirements was that the definition should use only notions from set theory and syntax, together with the notions expressible in L.

For the languages L that Tarski had in mind, the class of sentences is cofinal. Hence Theorem 15.1 applies, where X is the set of sentences and for each sentence φ, $\mu(\varphi)$ is the truth value of φ. That theorem gives us the format of a truth definition. For Tarski's purposes we need only check that syntax, set theory and "the notions expressible in L" suffice to define $|.|_\mu$ on atoms, the Hayyan functions and the read-out function p_μ.

The condition of Lemma 15.4 holds for Tarski's languages L: If a sentence φ is a constituent of a sentence ψ, and we replace this constituent by a sentence φ' with the same truth value as φ, the result is again a sentence, and it has the same truth value as ψ. So we can ignore the read-out function. With any reasonable notion of constituents for these languages, one can show that two formulas e, f have the same fregean value if and only if (1) e and f have the same free variables and (2) if α is an assignment to the free variables of e, then α satisfies e if and only if it satisfies f. Hence we can take $|e|_\mu$ to be the ordered pair $(FV(e), |e|)$ where $FV(e)$ is the set of variables free in e and $|e|$ is the set of assignments to $FV(e)$ which satisfy e. (Except when e is false under all assignments, $|e|$ determines $FV(e)$ anyway.) The set $FV(e)$ is defined syntactically, and for atomic formulas $R(x_1, \ldots, x_n)$ (the atoms of Tarski's paper) it's plausible that $|R(x_1, \ldots, x_n)|$ is a "notion expressible in L". It remains to define the Hayyan functions; Tarski's set-theoretic definitions of them are familiar.

In short, Tarski's truth definition in [19] is in all essentials the definition of truth given by the Abstract Tarski Theorem 15.1.

Actually Tarski doesn't determine for each formula e the set $|e|$ consisting of those assignments *to the free variables of* e that satisfy e; he determines the set consisting of those assignments *to all variables* that satisfy e. Some model theorists have preferred to rewrite his definition using $|e|$, but it's not what he wrote.

We can account for this discrepancy as follows. For Tarski's languages, $e \sim_\mu f$ if and only if $FV(e) = FV(f)$. Write $e \approx_\mu f$ if and only if among the assignments to all variables, those that satisfy e are those that satisfy f. Then Equation (15.4) holds. As it happens, $\approx_\mu$ is compositional too.

Later Tarski and Vaught [20] gave a truth definition for uninterpreted model-theoretic first-order languages. In this setting one can ask for truth definitions at several different levels of generality:

(a) Given a particular structure A and corresponding language L, define which sentences of L are true in A.

(b) Given a particular signature σ of relation symbols and corresponding language L, define for each sentence of φ the class of σ-structures in which A is true.

(c) The same as (b), but for all signatures simultaneously.

The Abstract Tarski Theorem applies as before. Case (a) is a special case of the 1933 definition. For case (b) one defines $\mu(\varphi)$ to be the class of all σ-structures in which φ is true. It turns out that two formulas with the same signature have the same fregean value if and only if they have the same set of free variables and are logically equivalent. So the Abstract Tarski Theorem delivers the usual model-theoretic definition of satisfaction.

Case (c) can be an exercise, as can the question how to deal with constant and function symbols if they are included as atoms. The Abstract Tarski Theorem brings home the bacon again. In cases (b) and (c) there is a set-theoretical problem about the definition of μ and the definition of the resulting fregean value function $|.|_\mu$. But if you know enough set theory to see the problem, you almost certainly know enough set theory to fix it.

15.2.7 Tarski-Style Semantics for Other Languages

People have sometimes asked whether this or that logical language "has a Tarski-style truth definition". In cases where the language has infinitary relations, or its constituent structure is not well-founded, the Abstract Tarski Theorem doesn't apply (at least in its present form). But in the vast majority of cases the Abstract Tarski Theorem does guarantee that there is a Tarski-style truth definition. That is, unless you have your own notion of what Tarski's style is, in which case you should tell us.

One well-known case is the branching quantifier logic of Leon Henkin. Jon Barwise asserts on the first page of [2] that for this logic the meaning of a certain formula "cannot be defined inductively in terms of simpler formulas, by explaining away one quantifier at a time". As it happens, the formula is tree-shaped. Our notion of constituent structure can cope with that, but there is no need, since Jaakko Hintikka translated Henkin's formulas into linear form within his IF logic. The Abstract Tarski Theorem applies straightforwardly to IF logic, so we know that Barwise's claim must be wrong. (We can write down the fregean values for Hintikka's IF logic explicitly, as in [9].)

At the end of his paper Barwise proves his claim. What he actually proves (p. 75f with some small changes of notation) is that

> The relation "φ is true in A" is not an inductive verifiability relation in the sense of Barwise-Moschovakis [3].

I don't think anybody with any experience of Henkin quantifiers would have expected otherwise. In fact it's possible to show [5] that fregean values for Henkin's language can't be given in terms of any notion of satisfaction by assignments of elements to variables.

Nevertheless the fregean semantics for IF and similar logics is a close analogue of Tarski's; the Hayyan functions have an obvious resemblance to his. The semantics is natural enough that Rohit Parikh and Jouko Väänänen [14] can use it as a basis for a semantics for a "finite information logic". The semantics uses sets of assignments rather than single assignments, but it's similar enough to standard model-theoretic semantics to serve for defining fixed point operators in IF and similar logics, as Julian Bradfield [4] showed.

The existence of a tractable fregean semantics for IF logic also has a consequence for natural language semantics: A discrepancy between syntactic scope and semantic scope of quantifiers is no longer automatically a barrier to a Tarski-style formal semantics.

•

References

1. Apresjan J. *Systematic Lexicography*. Oxford University Press, Oxford, 2000.
2. Barwise J. On branching quantifiers in English, *J. Philos. Log.,* 8: 47–80, 1979.
3. Barwise J., and Moschovakis Y. Global inductive definability, *J. Symbolic Log.*, 43: 521–534, 1978.
4. Bradfield J. Parity of imperfection, or fixing independence. In *CSL '03*, (Lecture Notes in Computer Science 2803), Springer, New York, NY, 2003.
5. Cameron P., and Hodges W. Some combinatorics of imperfect information, *J. Symbolic Log.*, 66: 673–684, 2001.
6. Chomsky N. *The Logical Structure of Linguistic Theory.* Plenum Press, New York, NY, 1975.
7. Ermers R., *Arabic Grammars of Turkic: Arabic Linguistic Model Applied to Foreign Languages and Translation of 'Abu Hayyan Al-Andalusi's "Kitab Al-Idrak Li-Lisan Al-Atrak"*. Brill, Leiden, 1999.
8. E. Ward Gilman, editor. *Webster's Dictionary of English Usage*. Merriam-Webster, Springfield, MA, 1989.
9. Hodges W. Some strange quantifiers. In J. Mycielski, et al., editor, *Structures in Logic and Computer Science*, (Lecture Notes in Computer Science 1261), pages 51–65. Springer, Berlin, 1997.
10. Hodges W. Formal features of compositionality, *J. Log. Lang. Inf.,* 10: 7–28, 2007.
11. Hodges W. Two doors to open. In D. M. Gabbay, S. S. Goncharov, and M. Zakharyaschev, editors, *Mathematical Problems from Applied Logic I, New Logics for the XXIst Century.* Springer, New York, NY, 2005.
12. Keenan E. L., and Stabler E. P. *Bare Grammar*. CSLI, Stanford, CA, 2003.
13. Leitgeb H. Hodges' theorem does not account for determinacy of translation: a reply to Werning, *Erkentis*, 62(3): 411–425, 2005.
14. Parikh R., and Väänänen J. Finite information logic, *Ann. Pure Appl. Log.,* 134: 83–93, 2005.
15. Partee B. H., ter Meulen A., and Wall R. E. *Mathematical Methods in Linguistics*, Kluwer, Dordrecht, 1990.
16. Pustejovsky J. *The Generative Lexicon*, MIT Press, Cambridge, MA, 1998.
17. Van Orman Quine W. *Word and Object*, MIT Press, Cambridge, MA, 1960.
18. Steedman M. *The Syntactic Process*, MIT Press, Cambridge, MA, 2000.
19. Tarski A. The concept of truth in formalized languages. English translation In J. Corcoran, editor, *Logic, Semantics, Metamathematics*, pages 152–278. Hackett Publishing Co, Indianapolis, IN, 1983.
20. Tarski A., and Vaught R. Arithmetical extensions of relational systems, *Compositio Math.,* 13: 81–102, 1957.
21. Werning M. Compositionality, context, categories and the indeterminacy of translation, *Erkenntnis,* 60: 145–178, 2004.
22. Westerståhl D. On the compositional extension problem, *J. Philos. Log.*, 33: 549–582, 2004.

Chapter 16

Some Reflections on Discrete Mathematical Models in Behavioral, Cognitive and Social Sciences

B. D. Acharya and Shalini Joshi*

16.1 Cybernetics and Systems

From times immemorial, man has wondered about nature and has tried hard to understand it in order to be able to use the knowledge gained through his incessant endeavours to his advantage. This very spirit underlies his progress, right up to what we see all around us today. Of course, his ability to conceive of alternative ways to look at things has led him to diversify fields of his knowledge; principally, while on hand he *analysed* an observed phenomenon he *synthesized* various pieces of information about the phenomenon towards developing an *integrated* view of the phenomenon. These two approaches are in some sense complementary to each other in accordance with modern *system theory*, a central theme of *cybernetics* the essential feature of which is control of *communication* processes taking place amongst various *modules* that constitute the system. As Stafford Beer [18] put it,

> ... cybernetics is the science of control and communication whenever these occur in whichever kinds of systems. The core of cybernetics research is the discovery that there is unity of natural law in the way control must operate, whether the system controlled is animate or inanimate, physical or biological, social or economic.

Hence, a *cybernetic approach* to solve a problem involves the following steps: (i) the overall system relevant to the problem is defined in terms of its *elements* and

B. D. Acharya
Department of Science and Technology, Government of India,"Technology Bhawan", New Delhi - 110 016, India, e-mail: bdacharya@yahoo.com

Shalini Joshi
Department of Psychology, University of Allahabad, Allahabad - 211 002, India, e-mail: shalinijoshi2000@yahoo.com

* The second author is currently visiting the Department of Studies in Mathematics, University of Mysore, Manasagangotri, Mysore - 570 005

J. van Benthem et al. (eds.), *Proof, Computation and Agency*, Synthese Library 352,
DOI 10.1007/978-94-007-0080-2_16,

their *interrelations*; (ii) the problem is defined quantitatively (as far as possible) in terms of elements of the system; (iii) the communication and control mechanisms of the system are studied by taking into effect the *deterministic* or *stochastic* variables involved in the description of the system; and finally, (iv) a *decision mechanism* of how to act under various situations is prescribed. Thus, studying the changes in the system and adopting decisions accordingly form an essential feature of a cybernetic approach.

For identifying a system, we have to know its *behavioural characteristic* which is represented by (i) the manner in which various elements within the system are interrelated; (ii) the manner in which the entire system or its elements react to any *external influence*, called alternately as the system's *external environment*, and (iii) evolutionary aspects of its overall buildup including existence of functional impressions of its past forms in its presently active or ongoing functions, if any [34]. Effects of the environment on the system are called *stimuli* while effects of the system on the environment are called *responses* of the system. The response of a system to any stimulus is dictated to a great extent upon the way the elements are organized to function within the system.

A system is said to be *closed* if there is negligible effect of the environment on the system; it is *open* if there is considerable effect of the environment on the system; and it is *partially closed* if the environment has considerable effect only on a proper subset or a few such subsets of elements in the system. The role of logic is central in prescribing the *thresholds* underlying these definitions.

Formally, a *cybernetic system* is thus a triple $S=(X,R,f)$ where $X=\{x_1,x_2,\ldots,x_n\}$ is the set of elements called *inputs*, R denotes the *behavioural pattern* (or, the "characteristic set") of the system and $f:(X,R)\rightarrow Y$ is the mapping relation which indicates that $Y=(y_1,y_2,\ldots,y_m)$ (the *output vector*) is determinable from the knowledge of (X,R) if the map f is known.

For example, if the system is defined only in terms of the set X and the knowledge as to whether there is a communication from x_i to $x_j (i,j\in\eta=\{1,2,\ldots,n\})$, then we can represent it by a simple directed graph (or, a *digraph* as treated in Harary et al [33]) D in which $R^d=(r^d_{ij})$ where $r^d_{ij}=1$ if there is a communication from x_i to x_j and $r^d_{ij}=0$ otherwise; here, by the existence of a communication from x_i to x_j we mean existence of a *directed path* $(x_i=y_0,y_1,\ldots,y_d=x_j)$ *from* x_i *to* x_j of some fixed *length* d in D. If $d=1$ in this definition, then "communication from x_i to x_j" means "direct communication from x_i to x_j" whence $R:=R^1$ is called the *adjacency matrix* of D.

In the foregoing, we have prepared a background to present some of the thoughts that are basic to lay the foundation of a cybernetic approach to deal with keeping in view the standpoints of social and behavioral scientists on the use of mathematical methods in general, and those of combinatorial theory in particular, towards gaining better insights into some of the intriguing areas of behavioral/social sciences. In fact, it is not uncommon that in almost all the fields of inquiry in humanities mathematical methods are used. One of the most famous examples is in the field of population studies where well known statistical models have been successfully used.

16.2 Combinatorial Models

In recent times, there has been an unprecedented surge in the application of very special techniques of network theory in the field of social sciences. Nevertheless, *any modeling effort should keep in view a note of caution that any quantitative description such as mathematical/statistical/ computer representation of a natural phenomenon in its entirety is fraught with limitations on its validity in comparison with the phenomenon as observed.* Even in the face of such admonitions, increasing popularity of network analytical methods in behavioural, cognitive and social sciences appears to be primarily due to the following reasons:

1. Representation of *attributes* and *relationships* amongst them distinctly by means of *points* on the euclidean plane and *line segments* interconnecting pairs of them those are interrelated in a specified way provides accurate structural description of a social/cognitive system;
2. Possibility of gaining a *global view* of (usually large scale) *social systems* endowed with such structures, called *networks*, thereby enabling identification, and possibly solution, of problems concerning different categories of attributional character of specified or specifiable types associated not only with individual entities but also with groups of them constituting a functional module;
3. Any dynamical aspect may then be brought into the logical framework of mathematical modeling by means of specifying functions that describe various *states* attained by the *nodes* at the intervals of observation taken up for investigation and quantifying the *strengths* of various *links* interrelating them, and then subject the resulting "state matrix" to mathematical analysis by applying the rules typical to the problem at hand.
4. The "logical framework" referred to in 3 has to have the flexibility to accommodate constraints possibly affecting the system from its "hitherto external environment".

Here, by an "attribute" one means any generic entity which has an intrinsically distinctive individual character [40] and the character of one attribute might influence that of another, depending on their "proximity" in the social network or "frequency" of their contact (or interaction) in the course of evolution of the social system.

The jump from physical and chemical systems to social systems, which consist of living and cognitive individuals endowed with ability to make decisions, has always seemed to involve concepts, which are beyond simple physics and chemistry. At one time this "extra ingredient" was explained by the *postulate of spirits and vital forces*. Gradually, as more and more of experimental science replaced philosophy, some of the biochemical components were isolated and identified, fractionation techniques improved, the question of the existence of vital forces became a joke. Eventually, the "reductionist view" became dominant. Yet, something was missing. Hints began to appear that breaking down structure to get to pieces to study in isolation also broke down function, often those functions which were under investigation. That "missing something" was named *organization*. Prominent biologists pointed out that the reason biology was higher in the hierarchical structure of science than physics and chemistry was that it dealt with *levels of organization* not found among objects of study in those disciplines. Indeed, the molecules and the physicalchemical laws are the same, but things are put together in different ways when they are alive! Rene Thom [66] labeled these two philosophical trends as *reductionism* and *vitalism* or *wholism* in more concretized terminology; he further stated that if one of these approaches is metaphysical, it is reductionism and not vitalism. One main question of interest in this debate relevant to behavioural, cognitive and social sciences is whether the network approach is of any help in resolving the philosophical dilemma. The conviction is that to the extent that it provides a method for dealing with more complicated organizational patterns, it must. On the other hand, reducing a social or a cognitive system is not far from reducing it to a collection of attributes. The networks, as models, are more models of our theories and hypotheses about how the social or cognitive system works than of the actual system itself. By creating the appropriate network, various notions we have about the workings of a social or cognitive organization can be quantitatively tested and lot of handwaving and speculation can be done away with. This, it seems, will be the role of network models in the next phase of understanding the complex nature of social or cognitive or social systems. Before the network approach, the task of analyzing highly organized systems looked hopeless. Now, that part of the problem seems almost trivial, a situation in tune with the wellknown view of the late Aharon Katzir-Katchalsky that "The goal of all science is to reduce all its problems to triviality".

To appreciate the above philosophical points, let us consider the following example: If the set of attributes is taken as the set of people in a town and if we categorize them by their prominent orientations toward various political parties active in the region then each of the subgroups in such a categorization would specify the set of people having prominent orientation toward one particular political party; such a categorization can be modeled invoking an abstract combinatorial object called a *hypergraph* in discrete mathematics literature [19]. Further, if we consider people as *points* and pairs of them belonging to a given subgroup as *lines*, one obtains a network, which is called a *2-section* of the hypergraph that represents the categorization (e.g., see Fig. 16.1).

What unusual inferences can one draw out of such a configuration? There is a fine point we would like to bring to notice, that would become naturally evident from this model and otherwise difficult to comprehend: While people in each of the sub-

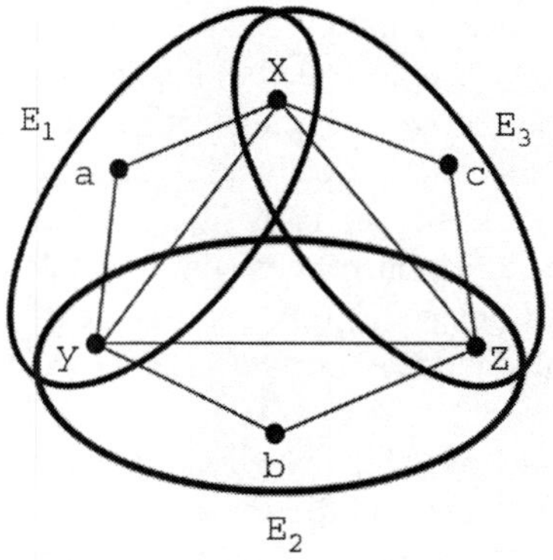

Fig. 16.1 Network *2-section* embedded in a hypergraph.

groups would form a *clique* in the network, not all members of an arbitrary clique chosen from the network structure may belong to the same subgroup. For example, in the categorization of the social group $X = \{a,b,c,x,y,z\}$ represented by the hypergraph $\mathscr{H} = (X, E = \{E1, E2, E3\})$ of Fig. 16.1, the clique $\{x,y,z\}$ does not represent a subgroup of people with uniform political orientations while every dyad in it is such a subgroup. While such a possibility led to define one of the most important classes of hypergraphs, called *conformal hypergraphs* in discrete mathematics literature, implications of this possibility to political psychoanalysts could be quite perplexing with regard to complexities it poses for planning propaganda campaigns as it would involve development or implementation of methods to distinguish "right cliques" (i.e., those which are constituted by people with predominantly similar political orientations) from "wrong cliques" (i.e., those in which not all the people have predominantly similar political orientations, possibly some of them might be under influence of some other political party).

16.3 Opinion Influence Processes

In scientific psychology, one of the fundamental questions arising in situations like the above is to describe what is meant by *behaviour*. It has been argued that this description should start with the *axiom:* ***Behaviour is a controlled consequence of output*** where the last step is a causal chain that begins in the cognitive environment created by the brain in response to given external forcing. In this axiom, the term "control" is meant to be a real objective phenomenon that involves the production of consistent results under varying environmental constraints. *Mathematical control theory* was developed to explain the phenomenon of control [50]. Even though there have been efforts to quantify behaviour [63], comparison of a quantitative description of control with a physical analysis of behaviour has shown that the events referred to as behaviour always involve control. The appearance of behaviour as programmed output or response to situation has been shown to be a consequence of ignoring two fundamental characteristics – of *control reference states and disturbance resistance*, a view that allows treating a controlled event as a physical variable that remains stable in the face of factors that should produce variability. There have been mathematical models in control theory that enable describing the nature and dynamics of evolution of behaviour (taken in the sense of the above axiom) to a reasonable degree of satisfaction. The network that represents correlation matrix specified by the relative variations in the variables (attributes) of such a description in any control theory model of social behaviour has been found to be a fundamental tool to extract important features in the dynamics of evolution of social networks since many physical and social phenomena are embedded within networks of interdependencies, the so-called "context", of these phenomena. In determining their opinions and behaviour in accordance with the constraints and possibilities imposed by the network, actors are assumed to be responsive to the contextual cues provided by the opinions and behaviour of significant others. By appropriately taking into account the opinions and behaviours displayed by their significant others, actors thus establish their own behaviour. Of course, opinions and behaviours are not solely determined by those of others (interaction), but also by reaction to various other constraints and opportunities granted by the social system (local effects). In sociology, this type of processes is typically represented as an *auto-correlation model* of the form

$$y = \rho \mathbf{W} y + x\beta + \varepsilon$$

or

$$y = x\beta + \varepsilon, \ \varepsilon = \rho \mathbf{W} \varepsilon + u$$

("Ego's opinion is a weighted version of the opinions of his alters".)

Parameter estimates and inferences based on such autocorrelation models hinge upon the chosen specification of *weight matrix* **W**. This matrix represents the influence process assumed to be present in the network and can be operationalized in many different ways. **W** is supposed to represent the theory a researcher has about the structure of the influence processes in the network. Since any conclusion drawn on the basis of auto-correlation matrices is conditional upon the specification of **W**, the scarcity of attention and justification researchers pay to the chosen operationalisation of **W** would inhibit the conclusion they would draw with inconsistencies.This is especially so, since different specifications of **W** typically lead to different empirical results [48]. In view of the seminal work of Shepard & Arabie [61], it would be of interest to extend this model in which weight matrix W represents the influence processes that are assumed to be present in the hypergraph whose edges represent similarity classes of opinions of individuals in the social group.

16.4 Simmelian Triads

Recently, researchers have begun to draw upon sociological theory and cognitive anthropology to forge a social network approach to culture that emphasizes underlying structures of relations rather than the content of ceremonies and rituals. A focus on the ways in which patterns of informal relations affect patterns of understanding can help pose and answer "a novel set of researchable questions" concerning organizational culture [53]. The genesis of this thought lies in the fact that individuals who interact with each other are likely to have a higher agreement concerning the culture than non-interacting individuals. Further, some relations (strong ties, for example) are likely to produce more cultural agreements than others. In fact, Simmel [62] had moved beyond the distinction between *strong* and *weak* ties by examining the special nature of *dyadic ties* embedded within triads. He suggested that ***relations embedded in a triad are stronger, more durable and in particular more able to produce agreement between actors than relations not so embedded***. Research confirms that dyadic relations embedded in triads (relative to dyadic relations in general) are more stable over time and exert more pressure on people to conform to clique norms and behaviour [44, 45]. Due to Simmel's pioneering work in this area, dyadic ties embedded in three-person cliques are called *simmelian ties*. Extensive questionnaire-based survey has led to a hypergraph model. The questionnaire response data may be used to form the matrix of raw dyadic ties, say $\mathbf{R} = (\mathbf{R}_{ij})$. To generate the $\mathbf{S}$ matrix of simmelian ties from $\mathbf{R}$, the hypergraph $\mathscr{H}$ matrix is first created by recording every instance in which an actor belonged to a *complete triad* (defined as a set of three individuals each member of which is tied with every other in it – in general, a set of any number of such individuals is called a *clique*); specifically, if there are n actors then one defines $\mathscr{H} = (\mathscr{H}_{ij})$ by letting $\mathscr{H}_{ij} = 1$ if and only if actor a_i is a member of the triad T_j and $\mathscr{H}_{ij} = 0$ otherwise. We can use this representation to uncover simmelian triads by multiplying the matrix form of the hypergraph $\mathscr{H}$ by its transpose $\mathscr{H}'$ and then taking the boolean of that matrix, i.e., $\mathbf{S} = \mathbf{bool}\ [\mathscr{H}'\mathscr{H}]$. $\mathbf{S}$ will then be an $n \times n$ (read, "n by n") $(0,1)$ matrix so that $\mathbf{S}_{ij} = 1$ if actors a_i and a_j are simmelian tied to each other and $\mathbf{S}_{ij} = 0$ otherwise.

One implication of this matrix $\mathbf{S}$ of simmelian ties is that it not only reveals who are in the same strongly connected informal group (clique) in the organization [45] but also that two actors being simmelian tied implies that they are co-members of the same clique and *vice-versa*, or in other words, $\mathbf{S}$ is a dichotomized clique co-membership matrix.

Using the above hypergraph model, Krackhardt and Kilduff [47] have confirmed the validity of the following two interesting hypotheses in the theory of organizational dynamics:

Hypothesis 16.1 *Relative to dyads in general, dyads embedded in simmelian triads are likely to have higher agreement concerning who is tied to whom in the organization.*

Hypothesis 16.2 *Relative to dyads in general, dyads embedded in simmelian triads are likely to have higher agreement concerning who is embedded together in triads in the organization.*

In general, the role of specific structural type of networks in studying socio-psychological phenomena can hardly be underemphasized. For instance, in the very foregoing example, the degree of agreement concerning the informal social structure of organizations amongst simmelian tied dyads (relative to dyads in general) has been observed to vary depending on the type of network and the particular organization. Structure of an organization consists of the relationships among the actors of that organization, observed as invariant during a given epoch of time of its evolution. In network terms, the structure is a set of dyadic statements describing who is related to whom on particular dimensions.

Cultural agreements are far from uniform across the organization but rather occur between pairs of actors. Romney et al [60] discovered that the *agreement takes on a group form, and different groups can create their own cultural definitions* – this was an important step. One step further is the suggestion above that *dyadic process of agreement formation becomes particularly powerful in the context of simmelian triad* [47], since disagreements within the clique are more likely to be mediated by a third party positively oriented toward the two antagonists in a three-person clique than in a dyad wherein any process of appeasement must be confined within the dyad itself acting as an autonomous system; further, *sense-making processes* within such cliques are likely to be particularly effective in providing individuals with opportunities to compare beliefs with similar others, thus facilitating the process of clarifications concerning many aspects of organizational culture including information about who the informal leaders are and who is connected to whom [22]; it is at such contexts that normal laws of common-sense logic might often fail, yet seemingly transient processes responsible for stability of the system, which are commonly referred to as *nonlinear processes*, might be at work. For instance, in view of the remark about the "right" and the "wrong" triads made above, we see that there is a way to identify them as follows: A dyad uv is regarded *positive* if all the common "neighbours" of u and v are in just one cultural subgroup E_i containing u and v and is regarded *negative* otherwise. For instance, this rule applied to the 2-section $\mathscr{H}_{(2)}$ of the hypergraph $\mathscr{H}$ depicted in Fig. 16.1 above would render all the dyads in its central triad $\{x,y,z\}$ negative and all the other dyads positive; thus, in this particular example, all the triads are "wrong" ones because every triad would then contain an antagonistic dyad, appeasement in which is inhibited by the presence of people from different cultural subgroups.

Krackhardt & Kilduff [46] argue that *cultural beliefs emerge through a negotiated dyadic process*. Nevertheless, though cultural agreements are far from being uniform across the organization it appears more reasonable to expect that they rather occur between actors in dyads. Subcultures that have evolved as one group E_i within the system forge agreement on one set of beliefs, say $\mathscr{B}_i$. while other groups emphasize different cultural truths; i.e., we may assume that the *set-labeling* $\mathscr{B}_i \rightarrow E_i$

of the edges E_i of the hypergraph $\mathscr{H}$ is a bijection (loosely speaking, different subgroups receive different sets of beliefs as labels). Thus, "culture" itself then becomes structured to the extent that different actors agree with other specific actors within the system specified by the hypergraph $\mathscr{H} = (X, E)$. Hence, each dyad can be characterized by the extent to which the two individuals in the dyad agree on a particular cultural domain; this level of agreement between the actors constitutes a belief relationship between those two actors. "Agreement" here may be represented mathematically as follows: Let $\mathscr{B}(u)$ denote the set-theoretical union ($\bigcup$) of $\mathscr{B}_i$s associated with the edges of $\mathscr{H}$ that contain u. Then, "agreement" between u and v is naturally the set-theoretical intersection of $\mathscr{B}(u)$ and $\mathscr{B}(v)$, *viz.*, $\mathscr{B}(u) \cap \mathscr{B}(v)$. In other words, each dyad uv may be thought of being associated with the set $\mathscr{B}(u) \cap \mathscr{B}(v)$ of beliefs shared in common by the individual actors u and v in the dyad (or, the so-called "dyadic partners"). The *aggregate set of dyadic belief relations among the actors of an organization can be considered one aspect of the structure of culture*; that is,

$$\bigcup_{uv \in \mathscr{H}_{(2)}} \mathscr{B}(u) \cap \mathscr{B}(v)$$

is the set of all beliefs which are shared in common by the dyads in the social system specified by $\mathscr{H}$. This elaboration supports the theory that *the social structure influences cultural understandings.*

One may go on to consider the notion of *strength of ties* by associating with each tie uv the cardinality $|\mathscr{B}(u) \cap \mathscr{B}(v)|$ of the set $\mathscr{B}(u) \cap \mathscr{B}(v)$, of beliefs that are common between u and v, as an instance since Simmel would perhaps have argued that *cocliqued relations will be stronger relations following his notion of dyads embedded in triads being stronger than those which are not.*

16.5 Roberts' Energy Use Sidigraph Model

We shall next see how to construct a dynamic **signed digraph model** for describing the behavioural strategies of a social system involved in its structural evolution through time. For this purpose, the following definition of human behaviour is used: "*Behaviour* is the way in which an individual or a group responds to a stimulus received from the immediate environment in which it is found imbedded". The following "Energy Use" dynamical model of Roberts [58] is found to fit the bill!

A *signed digraph* (or, simply "sidigraph") consists of a digraph together with an assignment of a *sign* "+" or "$-$" to each of its arcs. In many real-life applications, the vertices of a digraph are taken to be variables relevant to the problem being studied (e.g., population, energy capacity in a given region, energy price, level of public appreciation of a policy, etc.). There is an *arc* from a vertex x to a vertex y if change in x has a significant effect on y. This arc (x, y) is assigned a "plus"(+) sign if there is an effect of augmenting (or, reinforcement), i.e., all other things being equal, an increase (decrease) in x leads to an increase (decrease) in y; and a "minus"($-$) sign if the effect is inhibiting (or, weakening), i.e., all other things being equal, an increase (decrease) in x leads to a decrease (increase) in y.

Figure 16.2 shows a sample sidigraph model for energy demand in electrical power usage in a given area, constructed for illustration. One can, for example, interpret the network as follows: There is an arc from *population* **P** to *energy use* **U** with a plus sign because as population goes up, energy use goes up. There is a negative arc from energy use **U** to *quality of (physical) environment* **Q** because as energy use goes up, the quality of the (physical) environment goes down as the result of increased smog, thermal pollution, carbon dioxide, etc.

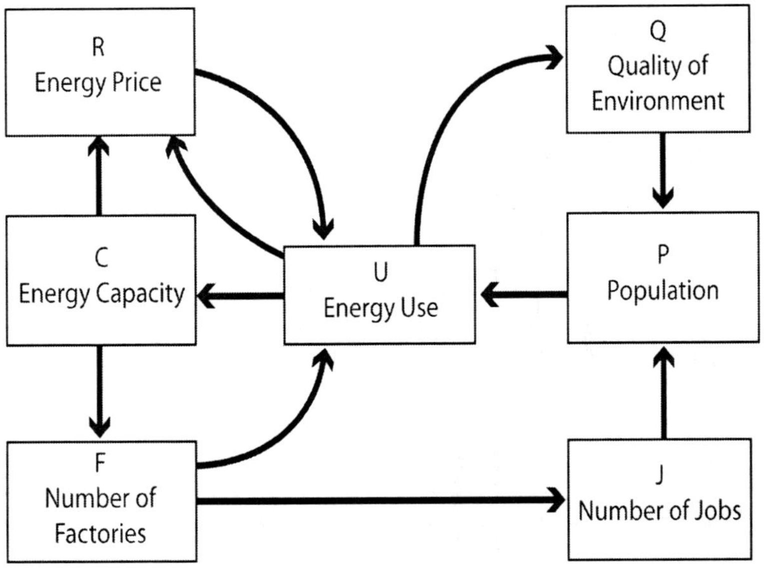

Fig. 16.2 Signed digraph for the energy demand in electrical power (Adapted from Roberts, [59]).

The usage of sign of an arc has an interesting interpretation as noted by Roberts [58]: According to the present system, for example, the more you use energy (**U**) the less you pay (**R**; per kilowatt hour) so that the arc (**U, R**) is negative. It has now been suggested, say, that the rate structure should be *inverted*, and that large users should pay more rather than less. This *strategy*, known as *inverting rate structure*, corresponds to *changing the sign* of the arc (**U, R**) from $-$ to $+$. In the same way, other changes in the sidigraph, in particular other changes of sign, might correspond to potential strategies for modifying the **energy use system**. One is hence interested in evaluating these strategies.

The sidigraph model has in it many oversimplifications. For example, some effects of variables on others are stronger than other effects. The sidigraph model, however, assumes that all effects are equally strong, by placing unit (+1 or -1) weights on each arc. It might be more reasonable to place a different weight $w(x,y)$ on each arc (x,y) of a given digraph, thus yielding a weighted digraph (or, what is commonly known as a *directed network*). The weight is interpreted as the relative strength of the effect, and can be still positive or negative in character. Even more realistic than assigning a weight to each arc is to assume that the *strength of an effect corresponding to the arc (x,y) changes depending on the levels of the variables x and y*. But even weights are hard to estimate in practice, especially if the variables themselves are hard to measure or even to define.

Most important omission in the sidigraph model is the *time lag* involved before a *change* in x has an effect on y. For example, an increase in population **P** will lead almost immediately to an increase in energy use **U**, while there is a time lag after an increase in the number of jobs **J** before that attracts more population **P** to that area. The sidigraph model assumes that all effects take place in one unit time. Thus, a more realistic model would introduce a time lag corresponding to each effect. As with weights, time lags are hard to estimate, and there is a tradeoff between the generality of the model and the possibility of estimating its parameters (weights, time lags, etc.) in a realistic way.

Some rather interesting conclusions can be reached if we introduce a simple *dynamic model* for the propagation of changes through the vertices of a signed or weighted digraph. Let us for simplicity begin with a signed digraph, and let its vertices be denoted $x_1, x_2, \ldots, x_n$. We suppose that each vertex x_i attains a value $v_i(t)$ at each discrete time $t = 1, 2, \ldots$. The succeeding value $v_i(t+1)$ is determined from $v_i(t)$, from an outside *pulse* $p_i^0(t+1)$ introduced at vertex x_i at time $t+1$ and from information about whether other vertices x_j adjacent to x_i went up or down at the last time period. Specifically, it is assumed that if there is an arc x_j to x_i and x_j goes up by $\dashv$ units at time t, then as a result x_i goes up at time $t+1$ by an amount equal to $\dashv$ times the sign of the arc (x_j, x_i). Moreover, x_i must increase by an amount equal to any external change $p_i^0(t+1)$ introduced at x_i at time $t+1$. To make all this precise, one defines

$$v_i(t+1) = v_i(t) + p_i^0(t+1) + \Sigma_j sgn(x_j, x_i) p_i(t) \tag{16.1}$$

where $sgn(x_j, x_i)$ is +1, -1 or 0 according to whether the arc (x_j, x_i) is positive, negative or nonexistent in the sidigraph, and $p_j(t)$ equals $v_j(t) - v_j(t-1)$ or $p_i^0(0)$ according to whether $t > 0$ or $t = 0$. The quantity $p_j(t)$ is called the pulse at vertex x_j at time t. A *pulse process* (Roberts, 1971) on a sidigraph D is then defined by the rule (1), by an *initial vector* of values $V(0) = (v_1(0), v_2(0), \ldots v_n(0))$ and by vectors giving the outside pulse introduced at each vertex at each time period. These vectors are denoted by $\mathbf{P}^0(t) = (p_1^0(t), p_2^0(t), \ldots, p_n^0(t))$. One can also use the *pulse vector* $\mathbf{P}(t) = (p_1(t), p_2(t), \ldots, p_n(t))$.

In applications, one usually determines $V(0)$ as follows: Suppose we know the *starting value* $v_i(start)$ at each vertex x_i. Then, $v_i(0)$ is defined by $v_i(0) = v_i(start) + p_i(0)$; that is, $v_i(0)$ is the starting value at the vertex x_i plus the initial pulse introduced at vertex x_i. Thus, usually the pulse process is defined by giving the vector $V(start) = (v_1(start), v_2(start), \ldots, v_n(start))$ rather than the vector $V(0)$.

Using the above notion of a pulse process, Roberts [58] described ways to analyse and deduce various viable strategies for deriving *stability conditions* for a complex system to attain.

As indicated in the beginning of the section, one can apply this very model for deriving various consistent behavioural patterns of individuals in a social system given the initial "pulse vector" and the initial "state vector" for the effect of external environment on the social system and the "cognitive states" of the individuals in the social system, respectively. The concept of "consistent" or "balanced" states of a social or cognitive system is itself a highly studied notion of structural stability of the system (e.g., see [3]). We will present salient features of such a model in the succeeding sections.

16.6 Structural Balance

Towards understanding the role of network structure in ongoing social processes, the members of the social group are represented as *nodes* (often called *vertices*) of the network in which any two nodes A and B are interconnected by at most all of the four *arcs* $(A,B)^+, (A,B), (B,A)^+$ and (B,A) where a "plus" (" + ") or a "minus" (" − ") sign in the super fix of an ordered pair represents the fact that the corresponding (oriented) relation is respectively "*positive*" or "*negative*" in character; such a network has been called an ***ambivalent signed digraph*** or ***ambisidigraph*** in short.

For instance, Fig. 16.3 depicts an ambisidigraph "state" of a "small" social group consisting of five persons A, B, C, D and E; in this "social network", (A,C) is a negative arc, $(C,E)^+$ is a positive, (B,A) is an *ambivalent arc* in the sense that B has positive as well as negative evaluations of A each represented by the corresponding type (positive or negative) of arc, so on. Its basic structure is a digraph in which every pair of vertices is *joined* by at most two arcs in each direction.

In fact, ambisidigraph model of a social group takes into account the state of ***ambivalence***, viz., and simultaneous existence of "positivity" and "negativity" of evaluative opinions of an individual exhibited in respect of another, in the analyses of the social group. It also facilitates treatment of the notion of "indifference" by which we mean a "*state of neutrality*" that may exist between two individuals in the social group who exert on each other some sort of "*balance of attitudes/feelings*"

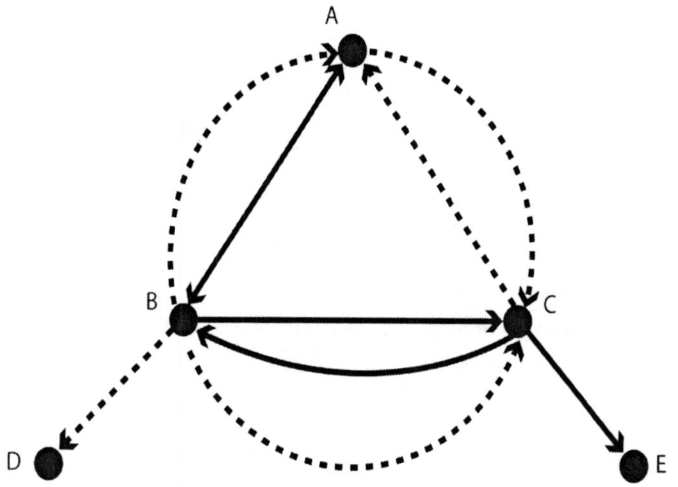

Fig. 16.3 An ambisidigraph state of a "small" social group.

that represents a "*tacit agreement*" to maintain "*status quo*" in respect of ongoing processes of interaction between them.

One may further assign a positive real number to each arc to represent the *intensity* of the interpersonal relation measured on some reasonable scale. However, for simplicity to begin with, we initiate a study of these networks by disregarding such assignment of *weights* (equivalently, by treating all the weights to be unity) to their arcs.

A concept of fundamental theoretical importance related to the structure of a network is that of *traversal* of its arcs, like we *walk* on a street. In fact, the term *walk* has been used in literature to mean a sequence $W_k = (v_0, a_1, v_1, a_2, v_2, a_3, \ldots, a_k, v_k)$ of vertices $v_0, v_1, v_2, v_3, \ldots, v_k$ and arcs $a_1, a_2, a_3, \ldots, a_k$ of S such that $a_i = (v_{i-1}, v_i)$ for each $i, 1 \leq i \leq k$; here, the positive integer k is called the *length* of the walk and the walk is said to *join* the *initial* vertex v_0 to the *terminal* vertex v_k, or simply that W_k is a $v_0 - v_k$ *walk*. For example, in the ambisidigraph of Fig. 16.3, the "alternating" sequence $(A, (A,C)^-, C, (C,B)^+, B, (B,C)^-, C, (C,B)^+, B, (B,C)^-, C, (C,E)^+, E)$ is an A-E walk of length six. If, in a walk, all the arcs are distinct then the walk is called a *trail* and if all the vertices are distinct (then observe that all the arcs must also be distinct) then the walk is called a *path*. A walk in which the initial and the terminal vertices are identical (i.e., one and the same) then the walk is said to be *closed*; it is said to be *open* otherwise. For example, in the ambisidigraph of Fig. 16.3, $(C, (C,B)^+, B, (B,C)^-, C, (C,B)^+, B, (B,C)^-, C)$ is a closed walk of length four. A closed walk in which all the vertices are distinct is called a *cycle*; for example, in Fig. 16.3, $(C, (C,B)^+, B, (B,C)^-, C)$ is a cycle of length two and $(A, (A,C)^-, C, (C,B)^+, B, (B,A)^-, A)$ is a cycle of length three (such a cycle is often called a *triangle*).

In the ambisidigraph S of Fig. 16.3, consider the two triangles
$T = (A,(A,C)^-,C,(C,B)^+,B,(B,A)^-,A)$ and
$T' = (A,(A,C)^-,C,(C,B)^+,B,(B,A)^+,A)$.
We wish to point out an interesting difference between these two triangles. In T, A has a negative opinion about C, C has a positive opinion about B and B has a negative opinion about A; or, by a simpler interpretation, A dislikes C, C likes B and B dislikes A. In T', with similar interpretation, we have the situation in which A dislikes C, C likes B and B likes A; ***what can one infer from this situation about C's disposition toward A***?

There could be many practically appealing interpretations of the situation. First of all, considered on their own right as triads, one may easily identify some straightforward features in which they differ: (i) in T the number of negative arcs is even whereas in T' this number is odd; (ii) there is exactly one vertex in T at which both "incoming" arc and the "outgoing" arc are negative whereas in T' there is exactly one vertex at which both incoming and the outgoing arcs are positive.

On an interpretative basis, at the very outset, one can sense some inherent differences between T and T'. Let us try to find out. Towards this end, some *realities in life* must be considered! One of the principal concepts in social psychology in analyzing such situations is that of ***reciprocity***. The behavioural phenomenon of *reciprocity* is well imbedded in the so-called "Dyadic Interaction Paradigm" (DIP) according to which

> The defining hallmark of interaction is *influence*; each partner's behaviour influences the other partner's subsequent behaviour ... within a single interaction episode and each interaction episode influences future episodes ... dyadic relationships are maintained by two individuals ... the nature of their interaction is determined not by one partner's properties but by the interaction of the properties of both partners, by the social and physical environments in which they interact, and by how all these causal conditions interact with each other. [56].

In this light, one can define *reciprocity* as the *behavioural tendency* of an individual (or even a "group") to respond to another in a manner consistent with the nature of the latter's displayed behaviour or actions; in literature, this definition appears to have been taken as granted (e.g., *see* [54]; p. 5). However, in *real-life*, it is not *difficult* to notice situations in which *response* of an individual (or group) need not be "consistent with" *stimulus* in general, but it could be even "inconsistent with" the nature of the stimulus; for example, *aiding a habitual cheater may not bring any benefit*. In effect, we need to modify the above definition of the term *reciprocity* to mean just a behavioural tendency of an individual (or group) to respond to another whenever the latter displays certain behaviour or action toward the former.

From the broad points of view mentioned above, we now get back to analyse the situations presented in the question raised about the inherent differences in the two triads, T and T'. Hence, as read from the relational structure of T, A dislikes C, C likes B and B dislikes A whereas in T', A dislikes C, C likes B and B likes A. Applying the notion of reciprocity to the dyadic relations imbedded in these two triads, we may deduce the following analysis assuming that in each case expressed attitudes or behaviours of one are noticed by the other in each dyadic interaction and may or may not cause a response by the other.

Situation presented by the structure of T : A dislikes C whence C may or may not respond to A's dislike for him. If, however, C decides to respond, his response to A may either be positive or negative or both ("mixed" or ambivalent reaction). If it is positive or even mixed, then it would represent some what unnatural situation in which C's liking A even to a limited extent despite the fact that C knows A's dislike for him would create *tension* not only between A and C but also between B and C since B who may have responded positively to C due to C's liking for B would not like C's liking A whom B doesn't like. Since so much tension in the triad would render the triad $\{A,B,C\}$ *unstable* and since the stability of a larger social group is perceived to be a commonly felt need for accommodating well being of its individuals and subgroups C would find it beneficial (or safe!) to maintain status quo with regard to his being disliked by A. Similar arguments in respect of other dyadic interaction existing in T would lead to the same conclusion. Thus, *there is no motivation to change the pattern of existing interrelationships among the members of* T. Such a structural state of a social group is termed ***structurally stable state*** of the group and hence the triad is said to be ***balanced.***

Situation presented by the structure of T′ : Here, A likes C whence C may or may not respond to A's like for him. If, however, C decides to respond, his response to A may either be positive or negative or both ("mixed" or ambivalent reaction). Negative response to a positive stimulus is unnatural and hence unlikely; however, it may be mixed sometimes, as for instance in the initial or in a fleeting episode of encounter, expression of A's liking for C may be met with a response by C inhibited by suspicion or distrust! But, in any case, some amount of positivity in the reaction of C toward A in response to A's expression of liking for C must exist, and this may be *full* in the sense that C's response to A may be fully positive too. However, since C likes B and B does not like A, C would not tend to be fully positive with A for otherwise it may alienate him from B who may show his displeasure about C's noticeable positivity with A whom B does not like. Thus, C would have to keep his positive inclinations, if any, toward A to an extent not noticeable by B! This would cause tension in C which would force him to alienate from A eventually. On noticing such a subdued and negatively inhibited response from C would then prompt A to switch his liking for C to dislike whence the triad $\{A,B,C\}$ would eventually become identical to T which represents a stable state of the triad. Since in T' there is such a *tendency toward transition in to a stable state* as described above, this ***structurally unstable state*** presented by T' qualifies T' to be called an ***unbalanced*** triad.

The kind of sociopsychological arguments to decide whether a given social group is structurally stable or unstable would become extremely complex and formidable when the number of individual members in the group, referred to as ***order*** of the corresponding social network, increases and generally the ground situation being so in practice, we need to appeal to mathematical representation of the essential ideas involved in the analysis for then we could apply appropriate mathematical methods to draw meaningful inferences about the sociopsychological phenomena being studied. Hence, in what follows, we will provide a glimpse into some elementary mathematical ideas toward this end.

Before we embark on presenting general theoretical framework to treat problems on structural stability in social systems, it may be rewarding to understand another subtle difference in the triads T and T' considered above. This will be presented in terms of *Matrix Theory*, a well-established field of mathematics.

By an $m \times n$ ***matrix***, written briefly as $\boldsymbol{A} = (a_{ij})_{m \times n}$ or simply as (a_{ij}) one simply means an arrangement of given entities in an array consisting of some m number of *rows* and some n number of *columns*, and looks like

$$A = \begin{array}{ccccc} a_{11} & a_{12} & a_{13} & \cdots & a_{1n} \\ a_{21} & a_{22} & a_{23} & \cdots & a_{2n} \\ \vdots & \ddots & \vdots & & \\ a_{m1} & a_{m2} & a_{m3} & \cdots & a_{mn} \end{array}$$

For example, a 3×2 and a 3×3 matrices are shown below.

$$M = \begin{array}{cc} A & B \\ A & A \\ B & A \end{array} \qquad N = \begin{array}{ccc} 1 & 3 & 7 \\ 2 & 6 & 5 \\ 4 & 2 & 1 \end{array}$$

Next, given any signed digraph S (that is, an ambisidigraph in which every pair of distinct vertices (*representing individuals in the social group*) is interconnected by at most one arc (*representing evaluative relation existing between the individuals*) which may be either positive or negative; in sociopsychological literature, such configurations are called *sociograms*. Given a sociogram S of order p one can associate with S a $p \times p$ matrix whose elements are from the set $\{1, 0, 1\}$, called the ***adjacency matrix*** of S and denoted $\boldsymbol{A}(S) = (a_{ij})$, defined as follows: Let the vertices of S be labeled as $v_1, v_2, v_3, \ldots, v_p$. Then, let $a_{ij} = 1, 0$ or 1 according to whether the arc (v_i, v_j) is negative, (v_i, v_j) is not an arc in S or (v_i, v_j) is a positive arc in S. For example, $\boldsymbol{A}(T)$ and $\boldsymbol{A}(T')$ obtained in this manner are given below.

$$M = \begin{array}{ccc} 0 & 0 & -1 \\ -1 & 0 & 0 \\ 0 & 1 & 0 \end{array} \qquad N = \begin{array}{ccc} 0 & 0 & 1 \\ -1 & 0 & 0 \\ 0 & 1 & 0 \end{array}$$

Conversely, given any *square* $(-1,0,1)$*-matrix* $\boldsymbol{A} = (a_{ij})$ of order p one can construct a signed digraph $S(\boldsymbol{A})$ uniquely as follows: Take p vertices $v_1, v_2, v_3, \ldots, v_p$ and join the vertex v_i to the vertex v_j by a positive (negative or no) arc whenever $a_{ij} = 1$ (respectively, 1 or 0). Thus, it is easily seen that *signed digraphs are uniquely represented by their adjacency matrices.* This is mathematical representation of a signed digraph. In particular, the "three-by-three" $(-1,0,1)$-matrices $\boldsymbol{A}(T)$ and $\boldsymbol{A}(T')$ shown above uniquely represent the triads T and T', respectively. So, what next? Using this representation, can we show some how show that T and T' are *mathematically different in some nontrivial sense*? Such a conclusion, as mentioned already, is essential if we want to derive meaningful inferences about larger social systems using available mathematical methods. In this particular case, we expect to use methods of matrix theory to analyse states of *socio-psychological stability* of a social group. This, obviously, would need a proper mathematical representation of the aforesaid socio-psychological notion of structural stability of a social group of an arbitrarily given order as, in practice, we may encounter social groups of any order. Hence, in what follows, we shall introduce this notion.

16.7 Theory of Balance in Social Systems

It is now well established that *affect* (attitude) has a direct influence on behaviour. Further, the different *attitudes* that an individual holds are *consistent* in his or her mind when they do not cause a state of *psychological tension* to the person. *Psychological consistency* of a person, therefore, depends on holding of attitudes, which are not contradictory with one another. Based on this fundamental notion in *personal psychology*, Heider [36] defined *interpersonal interaction* between two individuals as being in a state of *balance* whenever there is no conflicting attitude noticed between them, expressed by one and felt by the other, and described the notion also in the case of a *triad*, viz., a social group consisting of three persons. In less than the next 8 years, Harary [28] extended this notion to social groups consisting of any number of persons thereby laying the foundation of the theory of balance in *social systems*. To explain this development one would need the notion of a "cycle" and "semicycle" in ambisidigraphs.

Given an integer $k \geq 2$, by a *cycle of length k* in an ambisidigraph S we mean a sequence $C_k = (a_1, a_2, a_3, \ldots, a_k, a_1)$ of vertices of S such that $a_1, a_2, a_3, \ldots, a_k$ are all distinct and (a_i, a_{i+1}) is an arc for each $i, 1 \leq i \leq k$, where indices are reduced modulo k ; if $k \geq 3$ then we call C_k a *significant cycle* in S. Any cycle is said to be *positive* (equivalently, "balanced") or *negative* (equivalently, "imbalanced") according to whether the number of negative arcs in it is even or odd.

Essentially, then, Harary [28, 33] conceived a social group endowed with interpersonal interactions as a *signed digraph* and defined such a *social network* to be *cycle-balanced* if every cycle in the network is balanced. He also defined a signed digraph to be *balanced* if every *semicycle* in it contains an even number of negative arcs, where by a ***semicycle*** one means a closed walk in which all the vertices are distinct and the directions of the arcs are arbitrary; thus, every cycle is a semicycle but, conversely, not every semicycle is a cycle. Thus, it is important to observe right at this stage the following facts: *A signed digraph which is cyclebalanced may not be* *balanced as such whereas every balanced signed digraph must be cycle-balanced.* Furthermore, since in a graph every edge may be regarded as being "undirected" (taken equivalent to the term "symmetric arc" when treated as a digraph), *the notions of cyclebalance and balance coincide on* ***sigraphs*** which are graphs in which every edge is designated to be positive or negative.

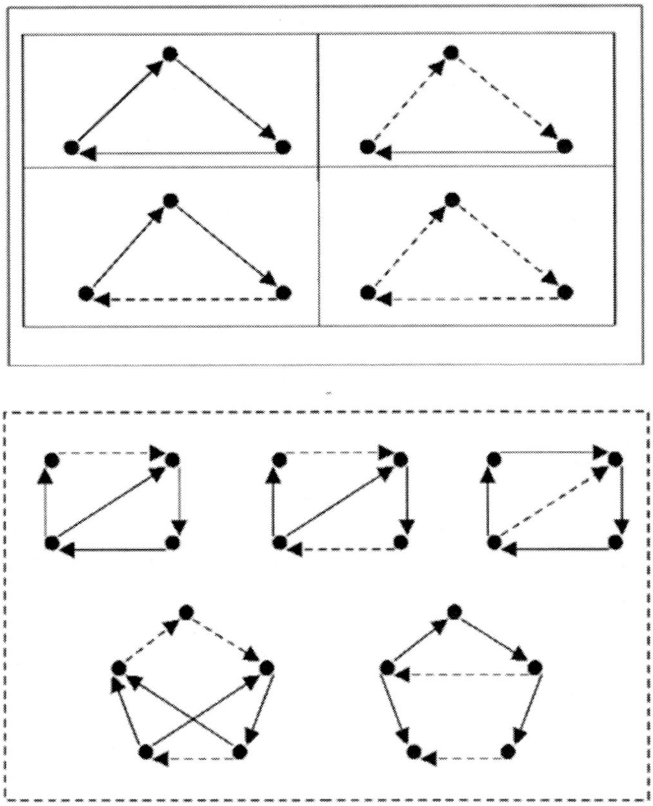

Fig. 16.4 Sociograms of small orders.

Figure 16.4(a) depicts some distinct (up to isomorphism) balanced as well as some unbalanced sociograms in a triad and Fig. 16.4(b) depicts some other such "higher order" sociograms.

The spectrum of a matrix: In this section, we shall explain how cyclebalance is characterized using the notion of the spectrum of a matrix. We have seen that given any $n \times n$ $(-1,0,1)$-matrix, called a *square (1,0,1) matrix of order n)* $\boldsymbol{A} = (a_{ij})$, one can associate with tbiA its sidigraph $S(\boldsymbol{A})$ by taking n vertices $v_1, v_2, v_3, \ldots, v_n$ and joining the vertex v_i to the vertex v_j by a positive (negative or no) arc whenever

$a_{ij} = 1$ (respectively, 1 or 0); in fact, this notion can be extended to be valid for any arbitrary square matrix $\boldsymbol{A} = (a_{ij})$ of order n just by changing the defining condition for $S(\boldsymbol{A})$ to "whenever $a_{ij} > 1$ (respectively, $< l$ or $= 0$)". In this extended sense, $S(\boldsymbol{A})$ is called the *sign pattern* of $\boldsymbol{A}$.

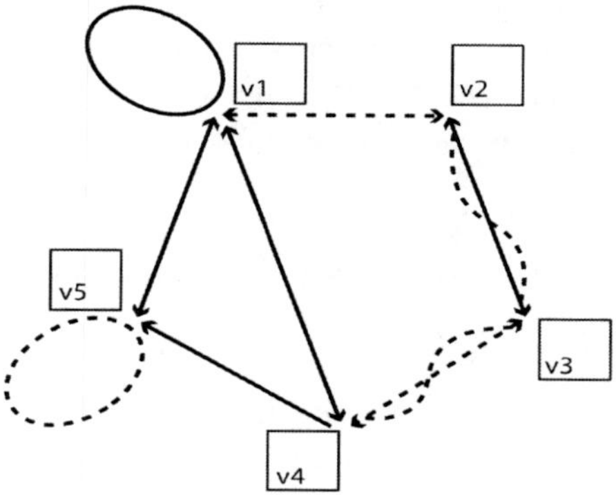

Fig. 16.5 Sidgraph S(A) for the sample 5-5 matrix A.

For instance, if

$$A = \begin{matrix} 0.75 & 0.25 & 0 & 0 & 0 \\ 0 & 0 & 0.07 & 0 & 0 \\ 0 & -0.02 & 0 & -0.22 & 0 \\ 0.21 & 0 & -1.55 & 0 & 1.23 \\ 2.32 & 0 & 0 & 0 & -3.02 \end{matrix}$$

then the sidigraph so associated with $\boldsymbol{A}$ is shown in Fig. 16.5. Note that $S(\boldsymbol{A})$ does not represent the actual values of the entries in the matrix $\boldsymbol{A}$; however, if one wishes, one can assign the *absolute value* $|a_{ij}|$ of the entry a_{ij} on the arc (v_i, v_j) in $S(\boldsymbol{A})$ whence we may regard the so *arc-labeled* sidigraph as a *real network* which we shall denote by $S_l(\boldsymbol{A})$. Clearly, $S_l(\boldsymbol{A})$ represents $\boldsymbol{A}$ completely, in the sense that given one of them the other is completely determined. Hence, the matrix $\boldsymbol{A}^+ = (|a_{ij}|)$ is called the *nonnegative associate* of $\boldsymbol{A}$.

The next notion in matrix theory required for our purpose is that of what is known in that theory as the *determinant* of a given square matrix

$$A = \begin{matrix} a_{11} & a_{12} & a_{13} & \cdots & a_{1n} \\ a_{21} & a_{22} & a_{23} & \cdots & a_{2n} \\ \vdots & \ddots & \vdots & & \\ a_{m1} & a_{m2} & a_{m3} & \cdots & a_{mn} \end{matrix}$$

of order n, which was first defined by the $17th$ century German mathematician Leibnitz in a letter addressed to a French mathematician L'Hospital dated April 28, 1693 but the term "determinant" was coined by another legendary German mathematician C.F. Gauss in 1801. The Leibnitz definition is given by

$$\det \boldsymbol{A} = \prod_{\sigma \in \pi_n (sgn\sigma)} a_{1\sigma(1)}, a_{2\sigma(2)}, a_{3\sigma(3)}, \ldots, a_{n\sigma(n)}$$

where $\prod$ denotes the binary operation of forming the usual product of the terms indicated and π_n denotes the set of all permutations $\sigma : \{1,2,\ldots,n\} \to \{1,2,\ldots,n\}$ which are defined as functions σ that assign $\sigma(i)$ to each $i \in \{1,2,\ldots,n\} := \boldsymbol{n}$ such that $\{\sigma(i) : i \in n\} = \boldsymbol{n}$. Further, the quantity $sgn\sigma$, called the *signature* of σ, is defined as follows: which is defined to be +1 or 1 according to whether the number of *transpositions* (which are *cycles* of length two, viz., tuples of the form $(\sigma(i), \sigma(j)) = (j,i)$, in the unique decomposition of σ into such cycles) in σ is even or odd. For example, let

$$\begin{matrix} & 0 & 0 & -1 \\ A = & -1 & 0 & 0 \\ & 0 & 1 & 0 \end{matrix}$$

Then, $\sigma(1) = (1)(2)(3), \sigma(2) = (1)(23), \sigma(3) = (2)(13), \sigma(4) = (3)(12), \sigma(5) = (13)(12), \sigma(6) = (12)(13)$ whence by the definition of the determinant we get $det\boldsymbol{A} = (+1)(0)(0)(0) + (1)(0)(0)(1) + (1)(1)(0)(0) + (1)(0)(1)(0) + (+1)(1)(1)(1) + (+1)(0)(0)(0) = 1$.

The quantities in the above expression may be put in terms of basic parameters of a certain kind of subgraphs of $S(\boldsymbol{A})$, called "linear subgraphs"; a subgraph L of a (di)graph G in general is called a *linear subgraph* of G if every component of L is a (directed) cycle in G (e.g., see [32]). Some more terminology would be essential to state this fundamental connection between digraphs and matrices.

Let $D = (D^u, w)$ be a *directed network*, with *weight function* w defined on the arcs of its *underlying* digraph D^u, $\zeta_t(D)$ denote the set of all linear subgraphs of order t in D^u, and $c(L)$ denote the number of components in $L \in \zeta_t(D)$; for any subgraph H of D, its weight $w(D)$ is defined as the product of the weights of the arcs in D. The following result has been independently "discovered" by a number of researchers in as diverse fields as molecular chemistry and electrical circuit theory (cf.: Acharya, 1980).

Theorem 16.1 [27, 31] *For any real matrix* $\boldsymbol{A}$ *of order n,*

$$det(A) = (1)^n \sqcap_{L \in \zeta_n(SI(\boldsymbol{A}))} (-1)^{c(L)} w(L)$$

The third idea required for the purpose of expounding what is meant by the *spectrum* of a given square real matrix $\boldsymbol{A}$ is that of its *characteristic polynomial* $\varphi(\boldsymbol{A})$. It is defined by

$$\varphi(\boldsymbol{A}) = det(\boldsymbol{A} - \lambda I)$$

where λ is a real variable and I denotes the *unit matrix* of order n whose entries in the diagonal are all 1s and the rest are 0s. It may be written in the *polynomial form* as

$$\varphi(\boldsymbol{A}) = \Sigma_{i \in \{0\} \cup \boldsymbol{n}} a_i \lambda^i$$

where the coefficients a_i of λ^i can also be expressed in terms of linear subgraphs in the sidigraph $S(\boldsymbol{A})$ of $\boldsymbol{A}$. Since the coefficient $a_i(\boldsymbol{A})$ of λ^i in $\varphi(\boldsymbol{A})$ is $(-1)^j$ times the sum of the determinants of the *principal minors* of order $n-j$ in $\boldsymbol{A}$,

$$a_i(\boldsymbol{A}) = (-1)^n \sqcap_{L \in \zeta_{n-j}(SI(\boldsymbol{A}))} (-1)^c(L)w(L).$$

Two real square matrices $\boldsymbol{A}$ and $\boldsymbol{B}$ are said to be *cospectral* if $\varphi(\boldsymbol{A}) = \varphi(\boldsymbol{B})$. The following important criterion for cospectrality of $\boldsymbol{A}$ with $\boldsymbol{A}^+ = (|a_{ij}|)$ is wellknown.

Theorem 16.2 [1] *Any real $n \times n$ matrix $\boldsymbol{A}$ is cospectral with its nonnegative counterpart $\boldsymbol{A}^+$ if and only if $S(\boldsymbol{A})$ is cyclebalanced.*

This theorem has lead to a number of interesting questions, notions and results in the mathematical theory of sidigraphs; it is beyond the scope of this lecture to enter into those excursions, but interested ones can go through [1, 5, 21, 25]. Its implications to balance theory in social psychology, however, are well enshrined in the seminal paper by Katai and Iwai [39]; it is neither possible to cover development even in that direction in this short lecture.

The following is an important corollary of the above theorem, which is of immediate interest to social balance theorists as such.

Corollary 16.1 [1]*: A sidigraph S is cyclebalanced if and only if S is cospectral to its underlying digraph S^+.*

16.8 Intergroup Relations

Here onwards, we shall treat the terms "individual" and "group" in somewhat equivalent sense, connotation of the term "equivalent" limited to their behaviours and functions, for extending it beyond is fraught with the danger of conveying similarity in their physical characteristics such as, say, both having the same weight. We will see the advantage of this approach in studying some very important socio-psychological phenomena by developing hypothetical models to enrich our conceptual resources to understand them better, even by empirical investigations if required. Towards this end, first we need the following definition of a "group" due to Alderfer [14]:

> A *human group* is a collection of individuals (1) who have significantly interdependent relations with each other, (2) who perceive themselves as a group by reliably distinguishing members from nonmembers, whose group identity is recognized by nonmembers, (4) who, as group members acting alone or in concert, have significantly interdependent relations with other groups, and (5) whose roles in the group are therefore a function of expectations from themselves, from other group members and from non-group members.

This idea of a group begins with individuals who are interdependent, moves to the sense of the group as a significant socio-psychological object whence boundaries are confirmed from inside and outside, recognizes that the group-as-a-whole is an interacting unit through representatives or by collective action, and returns to the individual members whose thoughts, feelings and actions are determined by forces within the individual and from both group and non-group members. This conceptualization of a group makes every *individual* member into a *group* representative whenever he or she deals with members of other groups, and treats every transaction among individuals as at least in part, an intergroup event [15, 57, 64].

In the study of intergroup relations, which in a generic sense can cover interpersonal relations too invoking the above mentioned approach, psychology has fallen into the problem of taking characteristics of interaction patterns and talking about them as though they were attributes of the entities engaged in the interaction [17]. Goffman [26] makes distinction between *characteristics of the actors* in interaction and the *characteristics of interaction patterns*; what we observe in a society is the latter; Smith [65] prefers to analyze the latter since we can only "see" the behaviours of the entities engaged in the interaction whereas we cannot actually "see" interactions. To observe interactions demands, then, a derivative, analytical approach that enables us to focus on what transpires, invisibly, in the space between the interactants. Smith [65] gives the following beautiful metaphor to illustrate his point:

> Focusing on interaction patterns between people or groups are conceptually analogous to noticing the wind as it blows through trees. The wind is recognizable primarily by its impacts on objects that move and behave within the realms of the wind's presence. It would not make sense for us to claim that leaves ripple and trees bend simply because it is their nature to do so. When a tree is bending and twisting vigorously in the middle of a storm, it does make sense to recognize the characteristic of the tree's flexibility, but we are more likely to focus on the intensity and direction of the wind. Just as trees express the nature of the winds blowing against them, a group's behaviour can be thought of as expressing the nature of the prevailing interaction patterns. Just as the nature of the wind is detectable only in the behaviour of the trees, so the nature of the interaction between groups can be derived only from interpreting the group's behaviours.

Secondly, the problem about the patterns that are dynamic is that the moment we attempt to represent fluid, changing phenomena in words, there is a tendency for the pattern to become reified. Then we find the rich, dynamic, movie-like patterns becoming reduced to impoverished, static snapshots. This tendency has been most evident in sociology, which treats dynamic interactions as the focal point of analysis, yet conceptualizes them as crystallized, institutionalized and stabilized properties of social structure. As a result, structured characteristics come to be viewed by sociologists not as crystallized interactions but as entities possessing empirical and conceptual reality themselves [16].

Perhaps, it is because of the complex nature of the sociopsychological situations mentioned above, (i) focus of social network theorists' attention has been mostly on "egocentered" networks, that is, the set of associations radiating from a particular individual, (ii) they have tended to employ a static type of analysis which fails to account for the continuous fluctuations in sociopsychological attitudes, and (iii) they have avoided forming general conclusions about the patterns that they have observed.

A notable exception is one of the earliest network models to represent *behavioural dynamics* due to French [24], in his theory of *social power*. Using directed graphs to represent power structure in a group, he examined the *opinion patterns* that evolved from various *influence configurations*. Rapoport [55] laid the foundation for probabilistic social network models through his investigation of *information transfer* in **random nets**. Lorrain and White [49] gave a rigorous exposition, using the theory of categories, of structural classifications in networks of *binary oppositions*, with their avowed aim to achieve a "global" understanding of social networks by studying the underlying structure. Subsequently, the idea of *cognitive structural balance* first introduced by Cartwright and Harary [20] was placed in a sociometric setting and generalized by Holland and Leinhardt [37], who invoked the concept of a partial order.

Fiksel [23] put a step forward by development of a network model explicitly incorporating the dynamics of interpersonal associations thereby expanding the scope of the model to encompass large groups of interrelated individuals as well as permitting a description of patterns of evolution within such groups. Due to its powerful simplicity, in this lecture, we choose to embark on presenting salient features of this model and point out some new directions for advanced research.

16.9 Societal Networks

Unlike the freely used term "social network", the term "societal network" will be used to denote a mathematical system having certain well-defined properties. In formal terms, it is a *labeled directed network* $(\mathscr{N},\mathscr{A},\mathscr{T})$ where $\mathscr{N}$ is a finite set of nodes, $\mathscr{T}$ is a finite set of relational types, and $\mathscr{A}$ is a set of ordered triples (x,y,u) such that $(x,y) \in \mathscr{N}^2 := \mathscr{N} \times \mathscr{N}$ whence actually $A \in \mathscr{N}^2 \times \mathscr{T} = \{(x,y;u) : (x,y) \in \mathscr{N}^2, u \in \mathscr{T}\}$. The set $\mathscr{A}$ denotes the set of all the *arcs* in the network which link pairs of nodes. Thus, if $(x,y;u) \in \mathscr{A}$ then there is an arc *from* node x to node y bearing the relational type u. In a social context, the nodes would represent individual people while the arcs would represent *directed* relations such as "mother of" or "superior to". We will assume throughout that the network is connected.

Now, let $\mathscr{T}$ have the property that if $u \in \mathscr{T}$, then there exists a *reciprocal* relation $u^{-1} \in \mathscr{T}$; this enables one to consider the arc $(x,y;u)$ as being equivalent to an *oppositely directed* arc $(x,y;u^{-1})$. For example, the reciprocal of "superior to" is "inferior to", so that x is superior to y if and only if y is inferior to x. When $u = u^{-1}$, we are dealing with a *symmetric* relation such as "friend of". Let $N(x)$ be the set of nodes adjacent to x, viz., $N(x) = \{y \in \mathscr{N} : (x,y;u) \in \mathscr{A} \text{ and } u \in \mathscr{T}\}$ and let $\mathscr{R}(x,y)$ be the relation u on the arc $(x,y;u)$. Let us not permit *selfloops*, viz., $x \notin N(x)$.

We can now endow the network with *dynamic* properties. Suppose that each node is always in one of a finite number of possible *states*, denoted by the set $S(x)$. At every instant of discrete time ($t = 0,1,2,\ldots$) a node may undergo a change of state; furthermore, ***let us assume*** for the time being that these state *transitions* are influenced only by the states of adjacent nodes in the network. More precisely, let $\sigma_x(t)$ denote the states of the node x at time t. Then, it is postulated that

$$\sigma_x(t+1) = \psi_x(\sigma_x(t)); \{\sigma_y(t), \mathscr{R}(x,y) : y \in N(x)\}.$$

That is, a state transition for node x is determined according to the *transition rule* ψ_x whose *arguments* are : (i) the *previous state* of node x, (ii) the previous states of all the nodes adjacent to x, and (iii) the relations of these nodes to x.

In a societal interpretation, the states of an individual could represent *political beliefs, social attitudes*, or even *moods*! The model described above simply implies that the major factors determining a change of state are the states of related individuals. Of course, it is possible that *external influences* are present as well, as in Robert's Energy Use Model, but for a wide range of social contexts it is meaningful to focus on the network.

Now, given any initial state at time $t = 0$, the transition rule determines the entire future evolution of all nodes in the network (note that the underlying network itself does not change in its structure). One may not be interested here in the detailed state changes undergone by any one individual, but in the *long-term patterns* and *trends* that emerge when the societal network is viewed as a whole. The key principle underlying this exercise is that *local properties of a societal network affect the global behaviour of the network.* In other words, the minute interactions of related individuals collectively influence the *macroscopic characteristics* of the structure to which they belong.

This model of dynamic societal networks due to Fiksel [23] can be viewed as a generalization of the model by French [24] who postulated a group power structure with a single type of directed relation, denoting influence of one individual over another. Harary [30] later formalized and extended these results, and suggested the consideration of multiple relational types as an open area for research.

In a given state of societal network, if each node x has a different transition rule ψ_x then very little can be predicted about the behaviour of the network. State changes will propagate throughout the network in a chaotic fashion, with no clear pattern emerging. However, Fiksel [23] showed that in a societal network in which nodes may be partitioned into certain equivalence classes, equivalent nodes share certain structural properties. His precise formulation of this seminal notion is given below.

The *circle* of two nodes x and y in a societal network is given by the set $V(x,y) = N(x) \cup N(x) - \{x,y\}$ and the *core* of x and y is the set $\bigwedge(x,y) = N(x) \cap N(x) - \{x,y\}$. A pair of nodes x and y in the network are called *structurally equivalent*, written $x \approx y$, if the following conditions are satisfied:

S1 $V(x,y) = \bigwedge(x,y)$

S2 for any $v \in \bigwedge(x,y), \mathscr{R}(x,v) = \mathscr{R}(y,v)$

S3 if $y \in N(x), \mathscr{R}(x,y) = \mathscr{R}(y,x)$.

This equivalence essentially means that x and y share a common set of relationships with a particular group of other nodes in the network. Note that x and y need not be related directly.

Theorem 16.3 (Fiksel [23]) *The equivalence relation induces $\approx$ a partition of $\mathscr{N}$ into disjoint classes satisfying :*

1. *x and y belong to the same equivalence class if and only if $x \approx y$.*
2. *for $y \in N(x), \mathscr{R}(x,y)$ is uniquely determined by the equivalence classes of x and y.*

The key import of this theorem is that the multitude of relations in the network can be summarized by simply observing the relations that exist among equivalence classes.

Since individuals exposed to the same set of social relationships are likely to display similar behavior, Fiksel [23] postulates that *equivalent nodes have identical transition rules*. One can possibly extend this to a more realistic model by postulating that *equivalent nodes tend to have asymptotically identical transition rules* to accommodate the long-term behaviour of the social system represented by the model. The genesis of this postulation lies in the fact that in reality people exposed to the same set of social relationships are more likely to exhibit varying behaviors *initially* since behavior, being the response of the individual to the situation, depends basically on the personality of the individual which is essentially an evolutionary characteristic; however, their continuous exposure over a long period may be expected to develop in them similar behaviors. This extended postulate leads to an open area for research on the above mentioned model of Fiksel [23].

16.10 Conclusions and Scope

As inquisitive investigators, we are always anxious and excited to know what is appearing or likely to appear, in the horizon we mean, at the frontiers of current knowledge. In the context of social network analysis, we are thrilled to mention the latest from our quarters: The longstanding problem on finding an appropriate definition of *structural complement* of a signed network raised more than 40 years ago by Harary [29] has now been solved [10, 11, 13] conclusively, most interestingly using hints provided in the behavioural science literature! As a result, new insights into *structural stability* analysis of the dynamic social processes are emerging [4, 12, 13, 41–43], providing details of which is beyond the scope of this paper.

There are altogether different areas of social sciences in which network structure plays crucial role. For instance, in management science, functional organization of tasks to accomplish the organizational goals optimally is of paramount importance [30, 67]. In such a context, the notion of domination in a *graph* G (that is, an undirected network or a hypergraph in which every edge has exactly two vertices) comes handy for modeling. Formally, a set D of nodes in G is said to be *dominating* if every node v in G is either in D or is adjacent to some node u in D (c.f.: [19, 35, 52]).

Hence, if we regard G as the interaction network of employees in an organization, then D may be taken as a taskoriented formal "grouping" of employees in the organization represented by G such that D contains the set of Task Group Leaders (TGL) in the organization and every employee of the organization not in D has to interact with exactly one of the Task Group Leaders, since obeying more than one TGL might cause confusion or conflict. Existence of such special groupings, which may be called *perfect command groups*, depends heavily on the architecture of the functional network in the organization's structure. Further, for purposes of optimizing the overall "interaction load" of TGLs, it is desirable to have connectivity within any specific group S of employees through at least one TGL, say u, meaning thereby the requirement that $S \cup \{u\}$ be connected in the organization's functional network G. One may also look for perfect command groups of TGLs along with optimal utilization of the command groups in the above sense. Furthermore, in order to avoid confusion in the command system so set up in the organization, it might be found necessary that no two TGLs should have direct links between them; this condition amounts to requiring that the set of nodes in G that represent TGLs be independent. Thus, the problem reduces to finding what is known in literature as an *independent psd-set* in G that are perfect, or as we may call efficiently interactive command groups. However, very special graphs only can possess independent psd-sets (e.g., see [6–9]); most of these architectures are nonhierarchical which could be quite complex (e.g., see [51]).

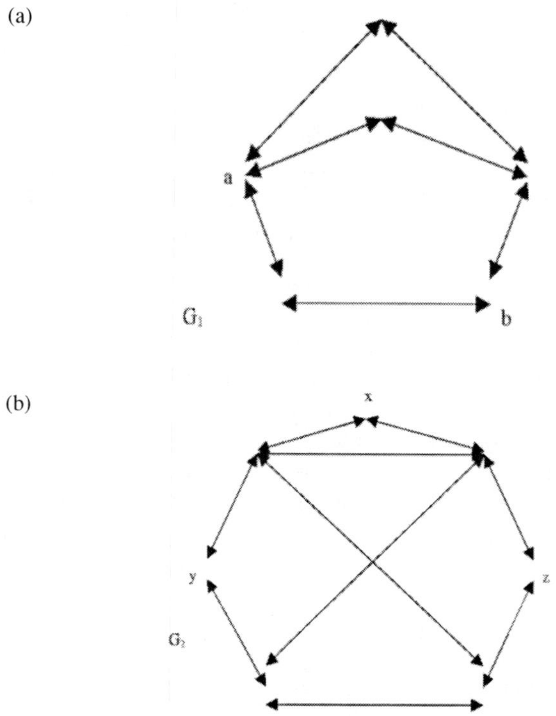

Fig. 16.6 (a) $\{a,b\}$ is an independent psd-set of G_1, (b) $\{x,y,z\}$ is an independent psd-set of G_2.

For example, Fig. 16.6 depicts two such architectures that possess an independent psd-set each (viz., $\{a,b\}$ in the first and $\{x,y,z\}$ in the second).

A graph possessing no independent psd-set is the hexagon C_6. Thus, to configure a functional network containing an independent psd-set in an organization is itself a complex task since the architecture of the organizational network might not possibly admit any!

Hence, if one takes a comprehensive view of social network analysis literature, it would appear largely confined to socio-psychological phenomena as observed or valid in short epochs of time during which the network of attributes and relations involved in the description of particular phenomena remains more or less unaltered; longitudinal studies which deal with evolutional aspects of processes such as *fluctuations* in interpersonal interactions and cooperation/coalition processes amongst the inhabitants of a colony have been relatively less. The national scenario in India has only patches of efforts of either kind; specifically, theories of ***cognitive consistency, sociopsychological balance*** and ***reciprocity*** seem to be dominant amongst discrete mathematicians; yet, amongst social psychologists in the country these topics have not been of any attention thereby creating wide gaps in our understanding of these theories in the context of the country's chequered populations and cultural milieu. We consider these as very important topics for advanced research and rigorous training amongst behavioural/social science community in the country due to bright possibilities of application of the knowledge base in these topics, especially invoking an appropriately suited cybernetic approach, to study a large variety of problems of social relevance such as *conflict resolution, group dynamics, organizational behaviour* and *social choice* in the wake of new national development policies the Government has been continually announcing and pressing for their fast implementation.

At length, we have made amply clear our basic point that even though understanding behavioural or social phenomena by means of any sort of modeling approach can never be perfect, such efforts invoking cybernetic methods can yield better insight into their complexities which perhaps would not be revealed otherwise. The role of modern logic in this enterprise can hardly be underestimated. Better understanding certainly leads to better control and application of the so generated knowledge more gainfully for the human society to pave ways for better living. We believe, this is a positive aspect of the whole endeavour that must be pursued to be cherished by our posterity in the near future.

Acknowledgements The second author wishes to express her thanks to the Department of Science & Technology, Govt. of India for supporting this work under their project SR/S4/MS: 132/2002, especially its Principal Investigator Prof. E. Sampathkumar for bringing to her notice the notion of complement of an arbitrarily edge-colored graph which likely to find its application in addressing the problem of structural stability of a social system endowed with a multitude of interpersonal interaction attributes.

References

1. Acharya, B. D. A graph-theoretical expression for the characteristic polynomial of a matrix, *Proc. Nat. Acad. Sci.*, 50/A III: 169–175, 1980.
2. Acharya, B. D. Spectral criterion for cycle-balance in networks, *J. Graph Theory*, 4: 1–11, 1980.
3. Acharya B. D., and Acharya, M. New algebraic models of a social system, *Indian J. Pure & Appl. Math.*, 17(2): 150–168, 1986.
4. Acharya, B. D., and Acharya, M. A graph-theoretical model for the analysis of intergroup stability in a social system, Manuscript, 1983. In T. Zaslavsky, editor, A mathematical bibliography of signed and gain graphs and allied areas, VII Edition, *Electronic J. Combinatorics*, 8(1): Dynamic Surveys # 8: 124 (Electronic), 1998.
5. Acharya, B. D., Gill, M. K., and Patwardhan, G. A. Quasicospectral graphs. In *Proc. Nat. Symp. On Mathematical Modelling* (MRI, Allahabad: July 19–20), pp. 133–144, 1982. MRI Lecture Notes in Applied Mathematics No.1. In B. D. Acharya, editor, *Mehta Research In-*

stitute of Mathematics & Mathematical Physics (recently, renamed: Harish-chandra Research Institute), Allahabad, 1984.
6. Acharya, B. D., and Gupta, P. On point-set domination in graphs : V. Independent psd-sets, *J. Combin. Info. & Sys. Sci.*, 22(2): 131–146, 1997.
7. Acharya, B. D., and Gupta, P. On point-set domination in graphs : IV. Separable graphs with unique minimum psd-sets, *Discrete Math.*, 195(1–3): 1–13, 1999.
8. Acharya, B. D., and Gupta, P., On point-set domination in graphs : VI. Quest to characterize blocks containing independent psd-sets, *Nat. Acad. Science-Letters*, 23(11–12): 171–176, 2000.
9. Acharya, B. D., and Gupta, P. On point-set domination in graphs : VII. Some reflections, Manuscript, 2002. Extended abstract in: Electronic Notes in Discrete Mathematics, 15 (2003).
10. Acharya, B. D., and Joshi, S. On sociograms to treat social systems endowed with dyadic ambivalence and indifference, Preprint, June 2001. (Invited presentation at the 17th Annual Session of the Ramanujan Mathematical Society held in Banaras Hindu University, Varanasi during June 10–12, 2002.).
11. Acharya, B. D., and Joshi, S. On the complement of an ambisidigraph, R. C. Bose Centenary International Symposium on Discrete Mathematics and Applications: Indian Statistical Institute, Kolkata (December. 20–23, 2002). (See: Electronic Notes on Discrete Mathematics, 2003a).
12. Acharya, B. D., and Joshi, S. Semibalance in signed digraphs. In *Proc. Int. Conf. on Recent Trends and New Directions of Research in Cybernetics & Systems.* IASST, Guwahati: Jan 1–3, 2004.
13. Acharya, B. D., Joshi, S., Rao, A. R., and Rao, S. B., A Ramsey theorem for strongly connected ambisidigraphs, Manuscript, 2003.
14. Alderfer, C. P. Group and intergroup relations. In J. R. Hackman and J. L. Suttle, editors, *Improving Life at Work.* Goodyear, Santa Monica, CA, 1977.
15. Alderfer, C. P., and Smith, K. K. Studying intergroup relations embedded in organizations, *Adm. Sci. Q.*, 27: 35–65, 1982.
16. Bacharach, S. B., and Lawler, E. J. Power and politics in organizations, Jossey-Bass, San Fransisco, CA, 1980.
17. Bateson, G. *Mind and Nature.* Bantam, New York, NY, 1979.
18. Beer, S. *Decision and Control – The Meaning of Operational Research and Management Cybernetics.* Wiley, New York, NY, 1966.
19. Berge, C. *Graphs and Hypergraphs.* North-Holland Elsevier, Amsterdam, 1973.
20. Cartwright, D. W., and Harary, F. Structural balance: A generalization of Heider's theory, *Psychol Rev.*, 63: 277–293, 1956.
21. Cvetkovi , D., Rowlinson, P., and Simi, S. *Eigenspaces of Graphs, Encyclopedia of Mathematics and its Applications,* vol. 66. Cambridge University Press, Cambridge, 1997.
22. Festinger, L. A theory of social comparison process, *Human Rel.*, 7: 117–140, 1954.
23. Fiksel, J. Dynamic evolution in societal networks, *J. Math. Socio.*, 7: 17–46, 1980.
24. French, J. R. P. A formal theory of social power, *Psychol. Rev.*, 63: 181–194, 1956.
25. Gill, M. K. A graph-theoretical recurrence formula for computing the characteristic polynomial of a matrix. In S. B. Rao, editor, *Combinatorics and Graph Theory*, Proceedings, Calcutta, pp. 261–265, 1980. Springer, Berlin, 1981.
26. Goffman, E. *The Presentation of Self in Everyday Life.* Doublday, Garden City, NY, 1959.
27. Greenman, J. V. Graphs and determinants, *Math. Gaz.*, 60(414): 241–246, 1976.
28. Harary, F. On the notion of balance of a signed graph, *Mich. Math. J.*, 2: 143–146, 1953.
29. Harary, F. Structural duality, *Behav. Sci.*, 2(4): 255–265, 1957.
30. Harary, F. Graph-theoretic methods in the management sciences, *Management Sci.*, 5: 387–403, 1959.
31. Harary, F. The determinant of the adjacency matrix of a graph, *SIAM Rev.*, 4: 202–210, 1962.
32. Harary, F. *Graph Theory.* Addison-Wesley, Reading, MA, 1969.
33. Harary, F., Norman, R. Z., and Cartwright, D. *Structural Models: An Introduction to the Theory of Directed Graphs.* Wiley, New York, NY, 1965.

34. Hatcher, Jr., J. H. Arguments for the existence of a general theory of behaviour, *Behav. Sci.*, 32: 179–189, 1987.
35. Haynes, T. W., Hedetniemi, S. T., and Slater, P. J. *Fundamentals of Domination in Graphs*. Marcel Dekker, Inc., New York, NY, 1998.
36. Heider, F., Attitudes and cognitive organization, *J. Psychol.*, 21: 107–112, 1946.
37. Holland, P. W., and Leinhardt, S. A dynamic model for social networks, *J. Math. Sociol.*, 5: 5–20, 1977a.
38. Holland, P. W., and Leinhardt, S. A method for detecting structure in sociometric data. In S. Leinhardt, editor, Social Networks, Academic Press, New York, NY, 1977b.
39. Katai, O., and Iwai, S., Studies on the balancing, the minimal balancing and the minimum balancing processes for social groups with planar and nonplanar graph structures, *J. Math. Psychol.*, 18(2): 140–176, 1978.
40. Knoke, D., and Kuklinski, J. H. *Network Analysis, Ser.: Quantitative Applications in the Social Sciences*, Sage Univ. Paper # 28. Sage Publications Inc., London, 1982.
41. Kovchegov, V. B. A model of dynamics of group structure of human institutions, *J. Math. Sociol.*, 18(4): 315–332, 1994.
42. Kovchegov, V. B. A principle of nonergodicity for modeling of the human groups by nets of probability automata. In *Proc. 14th IMACS World Congress on Computational and Applied Mathematics.* Georgia Institute of Technology, Atlanta, GA, pages 787–790, July 11–15, 1994.
43. Kovchegov, V. B. Application of the theory of locally interacting and product potential networks of automata to modeling balance in social groups, Manuscript, 2004.
44. Krackhardt, D. Simmelian ties: Super, strong and sticky. In R. Kramer and M. Neale, editors, *Power and Influence in Organisations*, Sage, Thousand Oaks, CA, 1998.
45. Krackhardt, D. The ties that torture: Simmelian tie analysis in organisations. In S. B. Bacharach, S. B., Andrews and D. Knoke, editors, *Research in the Sociology of Organisations*, vol. 16, pages 183–210. JAI, Stanford, CT, 1999.
46. Krackhardt, D., and Kilduff, M. Friendship patterns and culture: The control of organisational diversity, *Amer. Anthropo.*, 92: 142–154, 1990.
47. Krackhardt, D., and Kilduff, M., Structure, culture and Simmelian ties in entrepreneurial firms, *Soc. Networks*, 24: 279–290, 2002.
48. Leenders, R. Th. A. J., Modeling social influence through network autocorrelation: Constructing the weighted matrix, *Soc. Networks*, 24: 21–47, 2002.
49. Lorrain, F., and White, H. C., Structural equivalence of individuals in social networks, J. Math. Sociol., 1: 49–80, 1971.
50. Marken, R. S., The nature of behavior: Control as fact and theory, *Behav. Sci.*, 23: 196–206, 1988.
51. O'Neill, B., Structures for nonhierarchical organizations, *Behav. Sci.*, 29: 61–77, 1984.
52. Ore, O. Theory of Graphs, vol. 38. AMS Colloquium Publications, Providence, RI, 1962.
53. Pettigrew, A. M. Foreword. In N. M. Ashkanasy, C. P. M., Wolderom, and M. F. Peterson, editors, *Handbook on Organisational Culture and Climate*, pages 13–15. Sage, Thousand Oaks, CA, 2000. .
54. Rao, A. R., and Bandyopadhyay, S. *Reciprocity in a Village Social Network: Graph-Theoretic Analysis Using Survey Data*. Indian Statistical Institute, Calcutta, 1978.
55. Rapoport, A. An outline of a probabilistic approach to animal sociology, I. *Bull. Math. Biophysics*, 11: 183–196, 1949.
56. Reis, H. T., Collins, W. A., and Berscheid, E. The relationship context of human behaviour and development, *Psychol. Bull.*, 126(3): 844–872, 2000.
57. Rice, A. K. Individual, group and intergroup processes, *Hum. Relations*, 22: 565–586, 1969.
58. Roberts, F. S., Signed digraphs and the growing demand for energy, *Environ. and Planning*, 3: 395–410, 1971.
59. Roberts, F. S., and Brown, T. A., Signed graphs and the energy crisis, *Amer. Math. Monthly*, 82: 577–594, 1975.
60. Romney, A. K., Welter, S., and Batchhelder, W. Culture as consensus: A theory of culture and informant accuracy, *Amer. Anthropo.*, 88: 313–338, 1986.

61. Shepard, R. N., and Arabie, P. Additive clustering: Representation of similarities as combinations of discrete overlapping properties, *Psychol. Rev.*, 86(2): 87–123, 1979.
62. Simmel, G., Individual and Society, In: K.H., Wolf, editors., *The Sociology of Georg Simmel*, Free Press, New York, NY, 1950.
63. Simms, J. R. Quantification of behavior, *Behav. Sci.*, 28: 274–283, 1983.
64. Smith, K. K. An intergroup perspective on individual behavior. In J. R. Hackman, F. E. Lawler and L. W. Porter, editors, *Perspectives on Behavior in Organizations*, McGrawHill, New York, NY, 1977.
65. Smith, K. K. Social comparison processes and dynamic conservatism in intergroup relations. *Research in Organizational Behaviour*, vol. 5, pages 199–233. JAI Press Inc., 1983.
66. Thom, R. *Stabilite structurelle et Morphogenesese Essai d'une Theorie*, Benjamin, New York, NY, 1972 (England Edition: 1975).
67. Walikar, H. B., Acharya, B. D., and Sampathkumar, E. Recent Developments in the Theory of Domination in Graphs, MRI Lecture Notes in Mathematics, No.1. The Mehta Research Institute of Mathematics and Mathematical Physics (currently under the new name, 'Harish-Chandra Research Institute'), Allahabad, 1979. pp. 252.

Part VI
Perspectives from Indian Logic

Chapter 17
History and Development of Indian Logic: An Overview

K. Ramasubramanian

17.1 Introduction

It is the unflinching devotion to the search for truth, from time immemorial, that gave rise to the different schools of philosophy in the Indian tradition as elsewhere in the world. Though the views presented by the different schools largely differ from each other, all the six major philosophical systems – popularly known as *ṣaḍ-darśanas*, namely, *Mīmāṃsā, Vedānta, Sāṅkhya, Yoga, Nyāya and Vaiśeṣika* – believed in "cessation of miseries" referred to in the literature as *mokṣa, apavarga, niśreyasa*, etc., to be the ultimate *puruṣārtha*,[1] wherein all the human pursuits culminate. The six schools mentioned above, are regarded as orthodox schools (*āstika-darśanas*) as opposed to the heterodox schools (*nāstika-darśanas*), which mainly consists of the *Cārvākas* (materialists), the *Bauddhas* and the *Jainas*. In passing it may be mentioned here, that the classification of *āstika* and *nāstika-darśanas* is not based on whether they believe in God or not, but on the fact whether they accept the authority of the Vedas or otherwise.[2]

Of the six *āstika-darśanas*, mentioned above the term "Indian logic" primarily refers to the system of philosophy as propounded in the *Nyāya* and *Vaiśeṣika* schools. Of course, this is not to completely exclude or ignore the contributions of some of the famous buddhist logicians such as Nāgārjuna and Diṅnāga (c. 5th century AD). Even to present a brief history of Indian logic and its development, the span to be covered spreads over a few millenia starting from the time of *Upaniṣads*

K. Ramasubramanian
Cell for Indian Science and Technology in Sanskrit, Department of HSS, IIT Bombay, Mumbai 400 076, India, e-mail: kramas@iitb.ac.in

[1] The word *puruṣārtha* means – that which is sought after by a human being.

[2] In fact, among the six *āstika-darśanas* the schools of *Mīmāṃsā* and *Sāṅkhya* (a group of them known as *Nirīśvara-sāṅkhyas*) do not accept *Īśvara* as the creator of the world. However, they still believe in the authority of the Vedas and hence are grouped under *āstikas*.

J. van Benthem et al. (eds.), *Proof, Computation and Agency*, Synthese Library 352, DOI 10.1007/978-94-007-0080-2_17,

– passing through different phases, that includes origin the evolution of *Navya-nyāya* – till date. The literature to be considered into account is quite enormous. This comprises some published and huge amount of unpublished works in the form of *vivṛtis* (explanations) and *kroḍapatras*[3] which are mostly found in the repositories of the traditional scholars. Particularly with the advent of commentaries and super-commentaries on the *Tattvacintāmaṇi* of Gaṅgeśa, during 17th century, the growth in the *Navya-nyāya* literature has been prodigious.

Hence, it would be a mammoth task to venture into providing a comprehensive account of the growth and development of Indian logic in general. Our aim in the present paper is very modest. After briefly tracing the history and mentioning the purpose of this discipline, we move on to highlight the development of Indian logic in different phases. Here we present a couple of tables that gives a list of some important logicians and their major contributions. Then we proceed to discuss the motivation behind the development of logic and its nature. Here we present a few observations of the contemporary scholars and finally conclude the paper with a few remarks.

17.2 Place of Logic in the Scheme of Inquiry and Its Purpose

Kauṭilya in his *Arthaśāstra* ($\sim$ 400 BC) broadly dividing the different spheres of knowledge, recommends human endeavour towards the following branches of learning and inquiry :

Name	Subject matter dealt with includes
Ānvīkṣikī	Analytical philosophy that includes logic, epistemology and so on.[4]
Trayī	*Vedas, Vedāṅgas* & other *śāstras* like *Dharma, Mīmāṃsā, Vedānta* etc.
Vārtā	News to be made available for public related to – Agriculture, Economics, Trade and so on.
Daṇḍanīti	Study of Polity, Governance and State-craft

Table 17.1 Different branches of knowledge as given by Kauṭilya.

Thus *Arthaśāstra* of Kauṭilya is one of earliest works from which we come to know that *ānvīkṣikī* was the name by which *Nyāya-śāstra* was popularly referred

[3] The term *kroḍapatram* is used to refer to a piece of literature that is very deep and highly technical, which could be understood only by the experts who have specialized on that topic. In the text *Amarakośa* the word *kroḍa* is defined to be – नअनाक्रोडं भुजान्तरम् । Literally the word *bhujāntaraṃ* means – the portion inbetween the hands, which could be taken as chest. However, in the present context we simply take it to be referring to the middle portion. Thus *kroḍapatraṃ* means – a scholarly note or paper written choosing some section from the "middle" of a text and lead the discussion of the topic to much deeper levels.

[4] To quote – साङ्ख्यं योगो लोकायतं चेत्यान्वीक्षिकी।

to in those days. Commenting upon the origin of the word *ānvīkṣikī*, Vātsyāyana observes:

प्रत्यक्षागमाभ्यां ईक्षितस्य अन्वीक्षणम्...आन्वीक्षिकी ।[5]

Logic is that (discipline) with which one investigates (by supplying proper reason) what is already known through perception or verbal testimony.

The etymological derivation of the word *ānvīkṣikī* is as follows:

अनु + ईक्षा + ठञ् + ङीप् = आन्वीक्षिकी।
श्रवणादनु ईक्षा, पर्यालोचना प्रयोजनं अस्याः ॥[6]

The purpose of which is to generate contemplation after listening.

Perhaps, having this etymological derivation of the word *ānvīkṣikī* in mind, Kauṭilya in his *Arthaśāstra* while classifying and describing the scope of the different branches of learning observes:

....धर्माधर्मौ त्रय्याम् । अर्थानर्थौ वार्तायाम् । बलाबले चैतासां हेतुभिरन्वीक्षमाणा आन्वीक्षिकी लोकस्योपकरोति । ....व्यसने, अभ्युदये च बुद्धिमवस्थापयति । प्रज्ञावाक्यक्रियावैशद्यं च करोति ।[7]

Ānvīkṣikī renders great help, by proper reasoning, in distinguishing between right and wrong in the Vedic disciplines; between profits and non-profits in agro-economy between correct and the incorrect decisions in the science of polity; determining comparative validity and invalidity of all these disciplines (under special circumstances). It renders great help in also keeping one's mind steady in woe and weal, and produces clarity/dexterity in understanding, speech and action.

Capturing the essense of the above, he succinctly puts forth the purpose served by *ānvīkṣikī* in relation to all spheres of human activity, in the form of a short verse:

प्रदीपः सर्वविद्यानां उपायः सर्वकर्मणाम् ।
आश्रयः सर्वधर्माणां शश्वदान्वीक्षिकी मता ॥

It acts as a lamp to all disciplines of learning, serves as a tool to perform all actions and provides perpetual support for all the activities carried out at all times. Hence (it is rightly) described as *ānvīkṣikī*.

Besides being a philosophical system in its own right, the central idea behind providing a training in *ānvīkṣikī* is to enhance the analytical skills which a human being is inherently endowed with. In other words, it tries to equip the student with the necessary capacity to methodically analyse the knowledge that he receives through perception (*pratyakṣa*), inference (*anumāna*) or verbal testimony (*śabda*). In fact,

[5] Commentary on the *Nyāya-sūtra* of Gotama (I.1.1).

[6] *Śabdakalpadruma*, vol I, Reprinted by Rashtriya Sanskrit Samsthan, 2002, p. 177.

[7] *The Kauṭlīya Arthaśāstra, Adhikaraṇa 1, Adhyāya 2* Ed. by R. P. Kangle, II Edition, Bombay University, 1969. Reprinted Motilal Banarasidass, Delhi, 2000, p. 4

it is precisely to highlight this aspect of *Nyāyaśāstra*, it is called *pramāṇaśāstra* in contrast with the *padaśāstra* (science of grammar, *Vyākaraṇa*), and *vākyaśāstra* (the exegetics, *Mīmāmsā*). An oft-quoted maxim that aptly putsforth the idea, is as follows:

काणादं पाणिनीयं च सर्वशास्त्रोपकारकम्।

[A training in] the discipline whose foundation were laid by Kaṇāda (logic)[8] and Pāṇini (grammar) will be useful in all other disciplines.

17.3 Rise and Development of Logic in India

When we speak of rise and development of logic, we do not mean the logical thinking that human being is endowed with – which is as old as the human species itself – but the earliest explicit mention or usage of the term logic in the literature and the literature exclusively devoted to the study and analysis of thinking process. From this perspective, the growth and development of logic in India can be broadly classified into three phases.

- **Phase 1** : *Ārṣa-Nyāya* – (before 4th century BC)
- **Phase 2** : *Nyāya-Vaiśeṣika* – (4th century BC–12th century AD)
- **Phase 3** : *Navya-Nyāya* – (13th century AD – till date)

17.3.1 Phase 1

A class of texts known as *Upaniṣads* form a part of the Vedas which are considered to be the oldest corpus of literature extant today. In one of the *Upaniṣads* we find an explicit use of the term *tarka* to refer to the process of ratiocination which sometimes leads to futile argumentation.

नैषा तर्केण मतिरापनेया।[9]

This knowledge (regarding the nature of the Self) is not attainable by ratiocination.

The above statement is made by *Yama* (the God of Death) to Naciketa, in the context of explaining the nature of the Self (*Ātmā*). Here Yama wants to convey the message that understanding the nature of *Ātmā* is beyond the frontiers of the mind

[8] Kaṇāda is the author of *Vaiśeṣika-sūtras*. Reference to discipline of logic through the name of Kaṇāda is because he preceded Gotama, the author of *Nyāya-sūtras*. It is preceisely an amalgamation of these two, at a much later date, that came to be declared commonly as *Nyāyaśāstra* in India.

[9] *Kaṭhopaniṣad*, I.2.9. In this *mantra*, the word *matiḥ* refers to the knowledge of the self.

and intellect. In other words, the domain of intellect is confined to the frontiers of the world of objects and cannot be extended to the world of "Truth", which can be comprehended only by following the foot-paths of the seers as laid down in the scriptures. This statement should not be misunderstood to mean that reasoning is undermined or ignored in the Vedic literature. In fact, in another *Upaniṣad*, the process of reasoning (in a channeled way) is recommended for the proper understanding of the nature of the Self.

आत्मा वा अरे द्रष्टव्यः श्रोतव्यो मन्तव्यो निदिध्यासितव्यः ।[10]

Indeed the Self is to be understood, heard, reasoned and contemplated upon.

From these prefigurations appearing in the *Upaniṣads*, either recommending the exercise of reasoning or suggesting that mere ratiocination may not help, and also considering fact that a lot of philosophical debates have been described in the *Upaniṣads* – for instance, between Yājñavalkya and other sages in the court of the King Janaka as portrayed in the *Bṛhadāraṇyakopaniṣad* – one is led to conclude that some kind of logical doctrines must have been present even during the Upaniṣadic age in some form or other (which is not extant today), if not in a crystalline form.

Further there are ample evidences to show that there have been discussions carried at great length in order to resolve the meaning of a Vedic texts. These analysis and interpretations of the Vedic passages – which at a later period gained the status of a separate branch of study in the hands of Jaimini known as *Mīmāṃsā-śāstra* – laid solid foundations to the development of logic in the Indian context. Moreover, around 4th century BC, Jain author Bhadrabāhu talks about 10 step argument[11] which is listed in the form of a table below.

1. General thesis	6. Denial of counter-argument
2. Particular thesis	7. Quoting example
3. General reason	8. Doubting example
4. Particular reason	9. Allaying the doubt
5. Counter argument	10. Conclusion

Table 17.2 The ten-step given by Bhadrabāhu around 4th century.

17.3.2 Phase 2

Starting from *Kaṇāda* (4th century BC) and soon followed by Gotama (3rd century BC), the *Nyāya-Vaiśeṣika* tradition in India seems to have produced an array of illustrious scholars, commentators and exponents, if not in continuous succession at

[10] *Bṛhadāraṇyakopaniṣad*, IV.5.

[11] See for instance, Ingalls, *Logic in India*, Encyclopaedia Britannica, p. 225.

least intermittently as evidenced by the works extant today. The works of Kaṇāda and Gotama form the basic foundation on which the whole edifice of what is known as *Nyāyaśāstra* (Indian logic) stands today.

Among the literature on *Vaiśeṣika* philosophy extant today, the earliest one is by Kaṇāda known as *Vaiśeṣikasūtras.* As the name of the work itself suggests, it has been composed in *sūtra* style, a style that was commonly employed in those days, as evidenced from the works of Paṇini and Bādarāyana.[12] The *Vaiśeṣikasūtras* are 368 in number and they are divided in 10 chapters. Briefly indicating the purpose of the work at the beginning of Chapter 1, the author quickly moves on to categorize *padārthas.* Any entity that becomes an object of knowledge is a *padārtha.* Kaṇāda classifies substances (*dravyāṇi*) into 9 types; qualities (*guṇāḥ*) into 17 types and actions (*karmāṇi*) into 5 types.

After this classification (*sūtras* 4–6), in the very next *sūtra* he mentions – among other things – that "generality" and 'particularity' (*sāmānya* and *viśeṣa*) is something which is commonly present in the above three *padārthas*, namely substances, qualities and actions. The rest of the text primarily deals with the definitions and details of what has been covered in *sūtras* 1–7. The last three chapters touch upon the means of knowledge such as perception and inference. There is also discussion about causality. The name *Vaiśeṣika* to the system stems from the belief held by the philosophers of the school, that every substance has its unique essence or particularity (*viśeṣa*) – right from the atomic to the cosmic level – because of which it is distinguishable from the other.

Nyāya-sūtras of Gotama on the other hand are ≈535 in number – barring minor variations among different editions – and are distributed in 10 chapters. While most of the texts belonging to the *sūtra* literature commence with the word *atha* (then), the *Nyāyasūtras* form an exception. It directly commences with the listing of 16 objects, whose "true" knowledge is supposed to fetch "supreme bliss" (*niḥśreyasa*). One of the important discussions that has been carried out by Gotama that merits mentioning here is – the refutation of the contention that verbal testimony, *śabda*, cannot be considered as an independent *pramāṇa*. Gotama, also enlists certain other *pramāṇas* such as *aitihya, arthāpatti*, etc., and adduces reason as to why they cannot be considered as independent means of knowledge (*svatantra pramāṇas*). Essentially the entire text is an elaboration on the first couple of *sūtras* that defines the means for *summum bonum*, which is what every living being seeks after.

The most prominent logicians, belonging to phase 2 besides Kaṇāda and Gotama are, Vātsyāyana, Uddyotakara, Śivāditya Miśra, Vācaspati Miśra and Udayanācārya. The greatest contribution of *Vātsyāyana* lies in the form of his commentary to the *Nyāyasūtras*, which is generally known after his name as *Vātsyāyanabhāṣya.* It is the earliest commentary on *Nyāyasūtra* that is extant today. Here Vātsyāyana presents a detailed exposition of the *sūtras* with plenty of references here and there from the *Mīmāṃsā-śāstra.* The language and the style of

[12] Paṇini is the renowned author of *Aṣṭādhyāyī* and Bādarāyana of the *Brahmasūtras.*

presentation is somewhat comparable to the *Mahābhāṣya* of Patañjali. To give a flavor here we present excerpts from the commentary of the first *sūtra* followed by translation.

अथ प्रयोजनम् – येन प्रयुक्तः प्रवर्तते तत् प्रयोजनम्; यमर्थमभीप्सन् जिहासन् वा कर्मारभते; तेनानेन सर्वे प्राणिनः, सर्वाणि कर्माणि, सर्वाश्च विद्याः व्याप्ताः। तदाश्रयः न्यायः प्रवर्तते। कः पुनरयं न्यायः? – प्रमाणैः अर्थपरीक्षणं न्यायः।

Now [we define] *Prayojanam* – *Prayojana* is that, induced by which a man sets into action. Or it is that, desiring to accomplish or avoid which a man involves into activity. It is by this (*prayojana*), all the living beings, all actions, and all the body of knowledge is pervaded. [Naturally] the *Nyāya* (*śāstra*) too proceeds having it base on that (*prayojana*). What is this *Nyāya*? Examination of things by employing *pramāṇa* (right means of knowledge) is *Nyāya*.

Having defined the term *Nyāya* succinctly in the context of defining *prayojanam* – the fourth of the 16 *padarthas* listed by Gotama in the first *Nyāyasūtra* – Vātsyāyana proceeds to present the etymological derivation of the term *ānvīkṣikī*. As mentioned earlier, *ānvīkṣikī* is this term which has been in vogue during the early periods to refer to the discipline of *Nyāyaśāstra*, as evidenced by the usage of *Kauṭilya* in his *Arthaśāstra*. The derivation of the term presented here by Vātsyāyana is quite instructive. Incidentally he also defines what *Nyāyābhāsa*, a pseudo-*Nyāya* or an impostor of *Nyāya*, is.

प्रत्यक्षागमाश्रितं अनुमानं, सा अन्वीक्षा, प्रत्यक्षागमाभ्यां ईक्षितस्य अन्वीक्षणं अन्वीक्षा। तया प्रवर्तते इत्यान्वीक्षिकी, न्यायविद्या, न्यायशास्त्रम्। यत्पुनरनुमानं प्रत्यक्षागमविरुद्धं न्यायाभासः स इति ॥

Inference is based on perception or verbal testimony; it is (also called) *anvīkṣā*. [This is so because] Assessing whatever that has been gathered/assessed by perception and/or verbal testimony is *anvīkṣā*. Since this discipline proceeds with this, it is (called) *Ānvīkṣikī*, also known as *Nyāyavidyā*, *Nyāyaśāstram*. That inference which is opposed to the (right knowledge obtained through) perception or verbal testimony is an impostor of *Nyāya*.

We know from our own experiences that any process of investigation carried out to arrive at the "truth", more often than not, involves making a lot inferences. The inferences made are necessarily based on the information that has been gathered either through perception (*pratyakṣa*) or verbal testimony (*āgama*). Hence, *Vātsyāyana* at the very onset of the text categorically states that – any inference that leads to a conclusion which is in contradiction to the perception or verbal testimony must be considered as *Nyāyābhāsa*. This statement incidentally throws light on one of the fundamental principles – the "truth value" of the conclusion arrived through the process of inference, is not to be determined merely by looking at the formal structure of the propositions, but also by considering the "truth value" of the propositions themselves – on which the entire edifice of the theory of inference (*anumāna*) rests upon.

The commentary of Vātsyāyana underwent adverse criticism on certain topics by the famous buddhist logician Diṅnāga who lived around the beginning of 5th century AD. One of the grounds for the criticism being the acceptance of the Vedic authority by the *Nyāya* school, which was not acceptable to the Buddhists. The Buddhist school around that period was very strong and were producing hyper-critical polemical works. It is in this atmosphere that Uddyotakara's *Nyāya-vārtika* – the earliest extant commentary on *Nyāya-bhāṣya* of Vātsyāyana – came into being. Here Uddyotakara refutes the objections raised by Diṅnāga and thereby establishes the *Nyāya* stand point on firm grounds. This important contribution of Uddyotakara has been aptly putforth by *Mahāmahopādhyaya* Kuppuswāmi Śāstrī, as follows:

> Uddyotakara's great service to *Nyāya* consists in his successful endeavour to lift it up from the slough into which it was thrown by Diṅnāga's confutation on Vātsyāyana's *Bhāṣya*.[13]

Śivāditya Miśra, more commonly known as Śivāditya is well known for his *Saptapadārthī*. The title of the text stems from the fact that the author classified the *padārthas* (the objects of knowledge)[14] into seven categories. The number "seven" because Śivāditya explicitly included *abhāva* into the category of *padārthas*, which were known to be six right from the time of Kaṇāda.[15] It is perhaps for the first time that such a classification is made in the history of Indian logic, as one does not find any other text available hitherto – that is extant today – taking this approach. Śivāditya Miśra concludes the work with the following interesting verse:

> सप्तद्वीपा धरा यावत् यावत् सप्त धराधराः ।
> तावत् सप्तपदार्थीयं अस्तु वस्तुप्रकाशिनी ॥
>
> May this (treatise) *Saptapadārthī*, be the torch-bearer on the nature of things, as long as the earth has seven islands (continents)[16] and possesses seven mountains.[17]

The hall mark of *Saptapadārthī* lies in presenting the entire *Vaiśeṣika* system in a short and simple manner. Another significant feature and perhaps the most important one about this text that merits mentioning here is the fact that almost all the introductory texts – such as *Tarkasaṅgraha* of Annambhaṭṭa or *Bhāṣāpariccheda* of Viśvanātha – composed more than five centuries later have adopted more or less the same framework as laid down by Śivāditya.

Among philosophers, Vācaspati Miśra is well known for his highly erudite gloss on the *Brahmasūtrabhāṣya* of Ādi Śaṅkara called *Bhāmatī*.[18] Most of the treatises of Vācaspati, though considered authoritative, are primarily in the form of commentaries or expository glosses. One of the outstanding features of Vācaspati is that, his erudite compositions spans over almost all systems of Indian philosophy. It is for this reason, many scholarly works refer to him with great reverence

[13] S. Kuppuswami Sastri, *A Primer of Indian Logic*, KSRI, Chennai, 1998, p. 31.

[14] Technically a *padārtha* is defined to be – that which is knowable or nameable.

[15] The six leaving *abhāva* are – *dravya, guṇa, karma, sāmānya, viśeṣa* and *samavāya*.

[16] The seven continents are – Jambū, Plakṣa, Śāka, Śālmali, Kuśa, Krauñca and Puṣkara.

[17] They are – *Himavat, Vindhya, Malaya, Mahendra, Sahya, Ṛkṣa* and *Pariyātra*.

[18] As per the legend, the name *Bhāmatī* was given by Vācaspati Miśra to his work, as a token of recognition of the selfless, dedicated services rendered by his wife (called *Bhāmatī*) while he got himself completely absorbed in the preparation of the gloss.

as *sarvatantrasvatantra* – "one who has achieved par excellence in all branches of learning". The two important works by Vācaspati in the field of *Nyāya* is listed in Table 17.3. For his monumental contributions in the form of *Nyāya-tātparyatīkā*, with due respect, he has been referred to as *Tātparyācārya* in the *Nyāya* literature.

Name	Period (Generally accepted)	Important contributions
Kaṇāda	4th century BC	*Vaiśeṣika-sūtras**
Gotama	3rd century BC	*Nyāya-sūtras**
Vātsyāyana	3rd century AD	*Nyāyasūtra-bhāṣya* - Earliest extant commentary on the *Nyāyasūtras*
Praśastapāda [19]	(?)	*Vaiśeṣikasūtra-bhāṣya*
Vasubandhu	4th century AD	1. *Abhidharamakośa* 2. Mentoring Diṅnāga
Diṅnāga	c. 400 AD	1. *Hetucakra* (Wheel of Reason) 2. *Pramāṇa-samuccaya* 3. *Nyāya-dvāra* & *Nyāya-praveśa*
Uddyotakara	6th century AD (later part)	*Nyāya-vārtika* – Earliest extant commentary on *Nyāya-bhāṣya*
Dharmakīrti[20]	7th century AD	*Nyāyabindu* – Explains the validity of an inference by resorting to a novel way of classifying *hetus*[21]

* - Foundational text

Table 17.3a Prominent logicians of Phase 2 and their major contributions.

Among the logicians till 10th century AD, Udayanācārya is considered to be the greatest next only to Gotama. Besides erudite commentaries, Udayana has written

[19] Randle – one of the western scholars of the early part of the 20th century who has made a serious study of Indian logic in the early schools – adduces arguments to place Praśastapāda after Diṅnāga. Certain other scholars like Stcherbatsky have found reasons to believe that Praśastapāda could be a contemporary of *Vasubandhu*, the *guru* of Diṅnāga. Weighing the evidences on both the sides, certian other scholars have opined that the balance swings in favour of placing Praśastapāda prior to Diṅnāga. See for instance : A. B. Dhruva, *Nyāyapraveśa* of Diṅnāga, Bibliotheca Indo Buddhica No. 41, II Edition, Sri Satguru Publications, Delhi, 1987.

[20] A famous Buddhist logician who has been referred to by I-tsing – a Chinese Buddhist pilgrim who left for India in 671 AD and returned to China in 695 AD. During I-tsing's stay in India he seems to have resided at the great Nalanda monastery for about 10 years (676–685 AD).

[21] Dharmakīrti says – There are three and only three types of *hetus*. In his own words – *trirūpāṇi ca trīṇyeva liṅgāni*. They are : (i) *Anupalabdhi-hetu*, (ii) *Svabhāva-hetu* and (iii) *Kārya-hetu*. For details refer to – Rajendra Prasad, *Dharmakirti's Theory of Inference*, OUP, 2002.

three important works (see Table 17.3). Of the three, *Nyāya-kusumāñjali* is a brilliant piece of work primarily aimed at the refutation of the anti-theistic theories. In this work which is considered as the *magnum opus* of Udayana, he not only provides a large number of inferential proofs to establish the reality of god, but also builds the necessary *Nyāya* doctrines that would enable the reader to understand/appreciate the proofs presented in the text later.

Name	Period (Generally accepted)	Important contributions
Vācaspati Miśra	9th century AD	1. *Nyāya-sūcī-nibandhana* 2. *Nyāya-vārtika-tātparya-ṭīkā* [22]
Jayantabhaṭṭa	9th century AD	*Nyāyamañjarī* – Expository gloss on select *sūtras* of Gotama
Śivāditya	c. 950 AD	*Saptapadārthī* [23]
Bhāsarvajña	10th century AD	*Nyāya-sāra* – Book marked with certain unique features [24]
Udayanācārya	10th century AD	1. *Prabodhasiddhi* 2. *Ātmatattvaviveka* 3. *Nyāyakusumāñjali* 4. *Kiraṇāvali* (984 AD) 5. *Tātparyapariśuddhi* [25]
Śrīdhara Miśra	10th century AD	*Nyāya-kandalī* – (A masterpiece in *Vaiśeṣika-darśana*)

Table 17.3b Prominent logicians of Phase 2 and their major contributions.

In the other voluminous treatise *Ātmatattvaviveka*, Udayana critically examines the Buddhist doctrine of momentariness (*kṣaṇikavāda*), the unreality of the objective world (*asadvāda*) and so on and refutes their arguments strictly on log-

[22] An erudite work intended to primarily explain certain difficult sections of the *Nyāyavārtika* of Uddyotakara and also clarify certian obscure portions of the *Vātsyāyana-bhāṣya*.

[23] This work in all consists of 199 aphorisms. The commentaries that are extant today are the ones that were written about 5 centuries later. The earliest one by Jinavardhana Sūri around 1415 AD. This was followed by two more commentaries namely *Padārthacandrikā* (c. 1459 AD) and *Mitabhāṣinī* (c. 1523 AD).

[24] The uniquessness of this work stems primarily from two distinct features : (i) the author does not accept *Upamāna* as a separate *pramāṇa* and (ii) while presenting the theory of *hetvābhāsa*, besides the accepted five types, he includes a sixth one called *anadhyavasita*.

[25] Of the five texts enlisted, the first three are independent works and the last two are commentaries on *Praśastapāda-bhāṣya* and *Tātparyaṭīkā* of Vācaspati Miśra respectively.

ical grounds thereby defending the *Nyāya-Vaiśeṣika* position on the Nature of Self (*Ātmā*). Either from the view point of language or the display of unrelenting polemics, *Ātmatattvaviveka* seems to stand matchless till date. It is said that the process of criticism and counter-criticism of each other's views in the *Nyāya-Vaiśeṣika* and the Buddhist traditions – which continued unabated for centuries – almost came to a halt when *Ātmatattvaviveka* made its appearance on the stage. Another significant contribution of Udayana lies in his efforts to syncretize the *Nyāya* and *Vaiśeṣika* schools which had already made its beginnings in the works like *Saptapadārthī*.

Name	Period (Generally accepted)	Important contributions
Varadarāja	c. 1100 AD	*Tārkikarakṣa*
Vallabhācārya	12th century AD	*Nyāyalīlāvatī* – on which a number of commentaries were written during 13–15th century [26]

Table 17.3c Prominent logicians of Phase 2 and their major contributions.

17.3.3 Phase 3

This phase of Indian logic, which gave birth to a huge amount of literature – that is distinct both in the nature and structure that was available hitherto – commences with the monumental work of Gaṅgeśa, known as *Tattvacintāmaṇi*. This is considered to be one of the most creative works of the highest order on logic and epistemology. Though the exact date of composition of this classic is not known, it is believed to have been authored around the beginning of 13th century.

One of the various reasons for which Gaṅgeśa's *Tattvacintāmaṇi*, more popularly known as *Maṇi*, is considered to be an epoch-making treatise, is the introduction of several new terminologies and concepts such as *anuyogin* (possessing correlate), *pratiyogin* (counter-correlate), *avacchedaka* (delimitor), *avacchedya*, (delimited) and so on in a very systematic manner. The organization of the text is also quite different from the ones produced hitherto. Besides other special features, *Maṇi* presents a beautiful synthesis of both the *Nyāya* and *Vaiśeṣika* schools. Though the

[26] A dissertation – aiming at an understanding of the discourse in *Nyāya* literature of the post-Udayana and pre-Gaṅgeśa period – carrying the title "The Happening of *Nyāya* tradition: Vallabha on *Anumāna* in *Nyāyalīlāvatī*", has been submitted for the award of the degree of Doctor of Philosophy in January 2007 in the Department of Linguistics and Philosophy, Uppsala University, Sweden.

germination of this synthesis seems to have begun a couple of centuries earlier during the period of Udayana himself, the culmination of it had to wait till Gaṅgeśa came up with his brilliant contribution in the form of *Tattvacintāmaṇi.* In fact, this seminal work marks the beginning of a new era in the history of Indian logic known as *Navya-nyāya.*

Gaṅgeśa's singular contribution lies in shifting the emphasis from the *categoristic* approach to *epistemological* approach. In other words, by laying more emphasis on the means of knowledge (*pramāṇa*) than categorizing the *padārthas*, Gaṅgeśa turned the course of the *śāstra* from primarily being a *padārtha-śāstra* to *prāmāṇa-śāstra.*[27] The work of Gaṅgeśa had such a profound and ever-lasting impact on the logical tradition of India that almost all the later literature produced in logic followed the same pattern of organizing the chapters of their works as noted in the *Tattvacintāmaṇi.* The sphere of influence of this work did not remain confined to the walls of literature in logic. Scholarly literature in all disciplines which includes *Vedānta, Mīmāṃsā, Vyākaraṇa* and even *Sāhitya* started using the new concepts and terminologies introduced by Gaṅgeśa in order to bring in high degree of precision, accuracy and brevity in their own writings. In fact, it would be hard to find even a single polemical text, composed at a later period, which does not employ these concepts introduced by Gaṅgeśa.

Gaṅgeśa was followed by a galaxy of logicians, which includes his own son Vardhamānopādhyāya, who wrote several illuminatory expositions generally known by the name *prakāśa* (see Table 17.4). Commentaries and illuminatory expositions kept coming in quick succession both on Gaṅgeśa's *Tattvacintāmaṇi* and the glosses that were written on them. While tracing the development of logic in India, one would invariably notice a surge in the production of classics during the 16th and 17th century, which in turn laid the foundation for the generation of enormous amount of literature, called *Navya-nyāya.* It is during this period the most outstanding work *Dīdhiti* was composed by Raghunātha Śiromaṇi. In fact, the origins of the upsurge can be traced back to one of the greatest *gurus*, Vāsudeva Sārvabhauma whose singular contribution lies in training four brilliant disciples namely Caitanya, Raghunātha Śiromaṇi, Raghunandana and Kṛṣṇānanda. Of the four disciples Raghunātha Śiromaṇi was endowed with exceptional capacities both as a teacher as well as a writer. This made his contemporaries refer to him as *Tārkika-Śiromaṇi* – "The crest-jewel among the logicians."

Raghunātha was followed by a series of illustrious scholars such as Mathurānātha, Jagadīśa, Gadādhara, Śaṅkara Miśra, Viśvanātha Pa ncānana, Annambhaṭṭa and so on. Among them Mathurānātha – most famous for his splendid expositions on *Tattvacintāmaṇi* and *Dīdhiti* – was a direct pupil of Raghunātha. The commentary on *Dīdhiti* by Jagadīśa and Gadādhara are popu-

[27] The very fact that Gaṅgeśa has divided his *Tattvacintāmaṇi* into four chapters named after the four means of knowledge (*pramāṇas*) – namely *pratyakṣa, anumāna, upamāna* and *śabda* – stands testimony to this. However, it is not be misunderstood here, that Gaṅgeśa's work ceased to be a *padārtha-śāstra.* It still continued to be one, with the tools of *prāmaṇa* being much sharpened by introduction of a technical and more formal language.

Name	Period (Generally accepted)	Important contributions
Gaṅgeśopādhyāya	13th century AD	*Tattvacintāmaṇi* – An epoch-making work that influenced almost all the later works on *Nyāya*
Keśava Miśra	c. 1275 AD	*Tarkabhāṣā*
Vardhamāna	c. 1300 AD	Commentaries called *prakāśa* on – 1. *Kiraṇāvali* of Udayana 2. *Maṇi* of Gaṅgeśa 3. *Nyāyalīlāvatī* of Vallabhācārya
Jayasiṃhasūri	c. 1350 AD	*Tātparyadīpikā* on *Nyāyasāra*
Bhāvanātha Miśra	c. 1410 AD	Father and teacher of Śaṅkara Miśra
Vāsudevasūri	15th century AD	*Pañcapādikā* on *Nyāyasāra*
Śaṅkara Miśra	c. 1440 AD	1. *Kalpalatā* on *Ātmatattvaviveka* 2. *Āmoda* on *Nyāyakusumāñjali* 3. *Kaṇṭhābharaṇa* on *Nyāyalīlāvatī* 4. *Kaṇādarahasya* on *Padārtha-dharmasaṅgraha* of Praśastapāda 5. *Upaskāra* on *Vaiśeṣikasūtra*
Yajñapati Upādhyāya	c. 1450 AD	*Prabhā* on *Tattvacintāmaṇi* and teacher of Pakṣadhara Miśra
Pragalbha Miśra	c. 1455 AD	*Pragalbhī* on *Maṇi* and a few other works – but not available in full

Table 17.4a Prominent logicians of Phase 3 and their major contributions.

larly known after them as *Jāgadīśī* and *Gādādharī*. Till date these are considered to be most important texts that need to be mastered in order to establish oneself as a scholar in *Navya-nyāya*.

The other three scholars mentioned above, namely Śaṅkara Miśra, Viśvanātha Pañc-ānana and Annambhaṭṭa lived in the later half of the 17th century. While *Śaṅkara* is famous for his commentary on *Jāgadīśī*, the latter two are well known for their introductory hand-books *Bhāṣāpariccheda* and *Tarkasaṅgraha*. These primers written by Viśvanātha and Annambhaṭṭa also have glosses written on them by the authors themselves that go by the name *Muktāvali* and *Dīpikā* respectively. The reception of the work *Tarkasaṅgraha* of Annaṃbhaṭṭa has been so much so, that today every student of Indian logic (traditional or otherwise) gets introduced into the subject with this work its gloss *Dīpikā*. Though the popularity of

Name	Period (Generally accepted)	Important contributions
Pakṣadhara Miśra	c. 1460 AD	Commentaries titled *viveka* on – 1. *Kiraṇāvalīprakāśa* of Vardhamāna 2. *Tattvacintāmaṇi* of Gaṅgeśa 3. *Nyāyalīlāvatī* of Vallabhācārya
Vāsudeva Sārvabhauma	c. 1480 AD	Teacher of stalwarts like Raghunātha Śiromaṇi and Caitanya[28]
Raghunātha Śiromaṇi	c. 1500 AD	*Dīdhiti* on *Maṇi*[29] *Padārthatattva-nirūpaṇa*
Keśava Miśra Tarkācārya	c. 1525 AD	*Nyāyacandrikā* on *Tarkabhāṣā* *Prakāśa* on *Nyāya-sūtras*
Mathurānātha	c. 1570 AD	Commentaries entitled *Rahasya* on *Kiraṇāvali, Maṇi* and *Dīdhiti*
Jagadīśa		*Jāgadīśī* and *Setu*[30]
Tarkālaṅkāra	c. 1625 AD	*Śabdaśaktiprakāśikā, Tarkāmṛtam, Nyāyadarśanam* – own works
Viśvanātha Pañcānana	c. 1634 AD	*Bhāṣāpariccheda* with an auto-commentary, *Siddhānta-muktāvali*

Table 17.4b Prominent logicians of Phase 3 and their major contributions.

Annambhaṭṭa reached great heights through this introductory hand-book, his profound scholarship in other disciplines of *śāstra* such as *Vedānta, Mimāṃsā* and *Vyākaraṇa* should not be underestimated (see Table 17.4). To summarize, the 16th and 17th century seems to have been a golden age in the history of Indian logic, which wittnessed a galaxy of extraordinarily brilliant scholars producing a series of splendid works on logic and epistemology.

In the centuries that followed, commentaries and super-commentaries got emerged in prodigious measure, and it would almost be next to impossible to provide even a partial list of them in an article of this nature. However, in Table 17.4, we have made an attempt to provide aglimpse of some of the most important works,

[28] Chaitanya, a great saint of the early 16th century Bengal, played a major role in cultivating *Bhakti-yoga* among the masses. His line of followers, known as *Gauḍīya Vaiṣnavas*, consider him as an *avatar* of Lord Kṛṣṇa himself.

[29] This is such an erudite commentary on Gaṅgeśa's *Tattvacintāmaṇi* that it eclipsed almost all the earlier existing commentaries and also greatly influenced the later ones.

[30] These are commentaries on *Didhiti* and *Praśastapāda-bhāṣya* respectively.

Name	Period (Generally accepted)	Important contributions
Gadhādhara Bhaṭṭācārya	c. 1650 AD	*Gādādharī* on *Dīdhiti* *Vyutpattivāda*, *Śaktivāda*, etc. *Tarkasaṅgraha* with *Dīpikā* and a comm. *Siddhāñjana* on *Maṇyāloka*
Annambhaṭṭa	c. 17th century AD	Commentaries on *Nyāya-sudhā*, *Aṣṭādhyāyī*, *Brahmasūtra*, etc.
⋮	⋮	⋮ [31]
Kālī Śaṅkara Bhaṭṭācārya	c. 1810 AD	*Kroḍapatra* on *Gādādharī* *Kroḍapatra* on *Jāgadīśī* & several other works
⋮	⋮	⋮
Dharmadatta Jhā [32]	c. 1910 AD	1. *Vivṛti* on *Gādādharī* 2. *Vivṛti* on *Jāgadīśī* 3. *Tippaṇi* on *Nyāyakumāñjali* 4. *Gūḍhārthatattvāloka*
Vāmācaraṇa Bhaṭṭācārya	c. 1940 AD	Several *Kroḍapatras* on *Jāgadīśī* of Jagadīśa Bhaṭṭācārya
Varadācārya K S	c. 1959 AD	*Nyāyasaurabha* on *Nyāyamañjarī*
Rāmānuja Tātācārya	c. 1980 AD	1. *Bālapriyā* 2. Series titled *Bālabodhinī* [33]

Table 17.4c Prominent logicians of Phase 3 and their major contributions.

that got evolved during phase 3. It must be admitted that this table is neither complete,[34] nor comprehensive i.e., trying to include all the authors. For a more detailed account, the readers may refer to the remarkable work of Potter[35] and his team venturing to provide list of authors and their works (not merely confining to logic) chronologically beginning from the 15th century to the present date.

[31] The "vertical dots" here and below is used to indicate the enormous amount the literature – which may take dozens of pages to list – produced during the 18th and 19th centuries.

[32] Dharmadatta Jhā was more popularly known as Bacchā Jhā.

[33] Of the two, the former is a commentary on *Tarkasaṅgrahadīpikā* and *Nīlakaṇṭhī* (a commentary on *Dīpikā*), while the latter is on different chapters of *Gādādharī* such as *Pañcalakṣaṇī*, *Caturdaśalakṣaṇī* etc. Besides this, Tātācārya has authored a few works in *Vyakaraṇa* also.

[34] Giving a list of all the contributions of a particular author.

[35] http://faculty.washington.edu/kpotter/ckeyt/txt4.htm

17.4 Motivation Behind the Development of Logic in India

Towards the end of *Ṛgveda* one finds the following passage –

सङ्गच्छध्वं संवदध्वं सं वो मनांसि जानताम्।[36]

Go and meet each other, speak well, thoroughly understand thy minds.

It is quite evident from the above passage that the Vedic seers, knew that the most crucial factor in having a congenial relationship has to do with having a clear understanding between one another. The latter generally has it roots in having meaningful discussions with each other, with a clearly defined purpose. For a discussion to be meaningful and purposeful it needs to fulfill two aspects – (i) there must be clarity in expression and (ii) it must be free from prejudices. It is precisely these two aspects that are being conveyed by the Veda when it says "speak well" (*saṃvadadhvaṃ*).

The desired outcome of any discussion – though more often than not is not achieved – is to arrive at a conclusion that is "valid". Validity in the Indian logic is not merely decided based upon the structure of the propositions, but is also dependent upon the material content of it. This may be contrasted with the development of logic in the west, which primarily focussed on the structure of propositions – as distinguished from their content. For instance, the argument –

Whosoever gets admission in IIT is of age > 25.
Rama has got admission into IIT.
Hence, Rama's age is > 25,

would be accepted as a valid argument in the western formal logic, as there is no logical fallacy in it, whereas it would not be accepted so in the Indian logic unless the invariable concomittance (*vyāpti*) between admission into IIT and age > 25 has been well established. In other words, the truth value of the content of the argument is also considered into account in order to decide upon the validity of argument.

The motivation behind the development of logic in India as a separate discipline, stems from two distinct traditions –

- *Vāda* tradition – Debate/discussion leading to valid knowledge
- *Pramāṇa* tradition – Inquiry into the accredited source of knowledge

In Indian tradition, a debate (*vāda*) would be considered meaningful only when it is based on things known through a certain *pramāṇa* – and the theoretical foundations of both got developed hand in hand like the theory and experiment (in any scientific discipline). Gotama in his *Nyāya-sūtra*[37] classifies debates into three types, which are summarized below:

[36] *Ṛgveda*, X.191.2.

[37] Gotama, *Nyāya-sūtra*, I.2.1–3.

1. वादः – तत्त्वबुभुत्सुकथा

- schema for proper argumentation among disputants who are engaged in an honest, non-eristic, and balanced debate.
- aim of each participant is not victory, but a fair assessment of the best arguments for and against the thesis.

2. जल्पः – विजिगीषुकथा

- schema for conducting a sophistical debate employing tricks, underhanded debating tactics.
- aim is to win at all costs and by any means necessary.

3. वितण्डा – स्वपक्षस्थापनहीना, परपक्षमात्रनिराकरणपरा

- schema adopted by sceptics mererly to refute. (refutation-only debate)
- aim is to argue against a thesis without commitment to any counter-thesis.

It goes without saying that it is only *Vāda*, (Type (1) above) that is most desirable form of debate, wherein, arguments putforth are equally weighed and fairly assessed without prejudice or bias. As a result, the debate becomes purposeful and based on sound logical reasoning one arrives at valid knowledge. In fact, according to Gotama, "obtaining valid knowledge was the only means to the supreme good in which all moral values are included or from which they are derived" (*summum bonum*) – which he calls as *niḥśreyasa*.

प्रमाणप्रमेयसंशयप्रयोजन-दृष्टान्तसिद्धान्तावयवतर्कनिर्णयवादजल्पवितण्डा-हेत्वाभासछलजातिनिग्रहस्थानानां तत्त्वज्ञानान्निःश्रेयसाधिगमः ।[38]

It is only by having a true knowledge of – the means of knowledge, the object of knowledge, [source of] doubt, purpose, example, thesis to be established, members [of *anumānaprayoga*], confutation (reductio ad absurdum), ascertainment, discussion, wrangling, cavil, fallacy, quibble, analogue and the point of defeat that a person can have supreme felicity.

It is this belief that the true knowledge is the source of supreme felicity, which the Indian logicians held, made them clearly identify the different means of knowledge (*pramāṇa*) and also build the necessary theory to demonstrate that one *pramāṇa* does not reduce into the other. This theory is based upon identifying the aggregate causal conditions that give rise to a certain *pramāṇa*. The Indian logicians have also developed the theory of inference at great length, which includes methods of identifying false *reason* adduced in an argument.[39]

---

[38] Gotama, *Nyāya-sūtra* I.1.1.

[39] Further details on this can be found in the article – "The concept of Hetvābhāsa", published in the present volume.

17.5 Nature of Indian Logic

On account of the nature of genesis of Indian logic, explained in the previous section, it imbibed an epistemological character, that it has retained throughout the history. Though the focus may vary from text to text, the fact that both logic and epistemology forms an integral part of single discipline of study known as *Nyāyaśāstra* remains invariant. Actually,

Indian Logic = *Vaiśeṣika* + *Nyāya*

Here it may be mentioned that the classification – that the *Vaiśeṣika* system represents the analytic or inductive form of reasoning and that the *Nyāya* represents the synthetic or deductive form of reasoning is all based on superficial analysis. In fact, the *Vaiśeṣika* philosophy is based on the maxim

साधर्म्यवैधर्म्याभ्यां तत्त्वविज्ञानम् ।

The comprehension of 'Truth' is through (the comprehension of) similarities and dissimilarities.[40]

It is impossible to comprehend similarities and dissimilarities, without employing both the inductive and deductive techniques. Thus it is obvious that the classification as to *Vaiśeṣika* represents inductive system and *Nyāya* the deductive has no real basis and does not stand scrutiny. Actually, the two together forms an integrated system, which later simply came to be known as *Nyāyaśāstra*. One more important thing that deserves to be mentioned here, is that the so called *Nyāyaśāstra* also combines to some extent, psychology, ethics, ontology and religion. Mixed composition is *not an outcome* of ability to appreciate differences, but due to the *cultural ethos* of India. To drive the point, as it were, here it may be appropriate to quote to the first two *sūtras* of Kaṇāda

अथातो धर्मं व्याख्यास्यामः ॥ १ ॥
यतोऽभ्युदयनिःश्रेयससिद्धिः स धर्मः ॥ २ ॥

Now (after hearing the queries), therefore (and having confirmed that the student is seriously interested), we proceed to explain *Dharma*. (1)

That entity which is responsible for attaining *abhyudaya* (material prosperity for enjoying this world) and *niḥśreyasa* (tranquility that gives supreme felicity), is *Dharma*. (2)

Those who are naive to Indian ethos, may find the commencement of a work on *Vaiśeṣika* philosophy with the above two *sūtras* quite odd, as the subject matter

---

[40] This has been aptly putforth by one of the earliest commentators Candrānanda when he introduces the seventh *sūtra* of the text – विज्ञातसाधर्म्यवैधर्म्याणां च द्रव्यादीनां अभ्युदय-निःश्रेयसहेतुत्वात् साधर्म्यं तावत् कथयति ।

of the text – a clear understanding of the *padārthas*, based on the principle understanding of "generality" and "particularity" – seems to be quite different from what has been mentioned in the above two *sūtras*. But the moment we realize that all the systems of philosophy have been constructed with the fundamental principle of specifying the means for "supreme bliss" called *niḥśreyasa*, the apparent contradiction noted evaporates into thin air. The ultimate bliss that is sought after is referred to as *Kaivalya* by Sāṅkhyās; *Mokṣa* or *Brahman* by the Vedāntins; and *Samādhi* by the Yogins.

Moreover, a philosopher, logician or a scientist does not form an isolated system. He very much forms a part of the mixed fabric with which the society in which he lives is weaved. As a result, there is bound to be a certian influence on the values and norms observed by the society on his thought process. This is something that is unavoidable. Hence, while reading the literature – philosophical or otherwise – produced in a society it would be good to keep this ideas as the bottom line in order to have a better appreciation and judgement. Just to place this thesis on a stronger footing, we may cite another example (concluding verse *Nyāyakusumāñjali* by Udayanācārya):

इत्येवं श्रुतिनीतिसंप्लवजलैः भूयोभिराक्षालिते
येषां नास्पदमादधासि हृदये ते शैलसाराशयाः ।
किन्तु प्रस्तुतविप्रतीपविधयोप्युच्चैर्भवच्चिन्तकाः
काले कारुणिक त्वयैव कृपया ते रक्षणीया नराः ॥

In spite of cleansing their (atheist's) minds with huge amounts of water – like the ones flowing during floods – in the form of *śruti, smṛti, nyāya*, etc., if You (*Īśvara*) are not able to take a seat in their hearts, then surely, their hearts are made of the essence of rocks. Though, in this matter (existence of *Īśvara*), they (atheists) maintain a diametrically opposite view, they indeed keep contemplating on You (perhaps more than what I do to refute *Īśvara*). Hence, when the time is ripe, they also need to be protected only by you, the all-compassionate one!

The above prayer of Udayanācārya, clearly reflects the philosophy held by him, that it is only through the grace of *Īśvara* that supreme felicity can be achieved. Earning grace – according to Udayana's philosophy – presupposes belief in *Īśvara*. As Udayana does not want any one to suffer, he pleads to the Lord, the all-compassionate one, to shower mercy and thereby save the the non-believers too. This incidentally mirrors how accommodating Udayana's heart has been to include even the rivals,[41] into the family of devotees. Udayana's monumental contribution to the theistic philosophy in the form *Nyāyakusumāñjali* and *Ātmatattvaviveka* – both of them composed with atmost logical rigor – made the later scholars to place him on a high pedestal and describe him as *Nyāyācārya*.

In fact, in the Indian tradition right from its inception, *Nyāya-śāstra* was never construed to be merely a "grammar of reasoning". It was, and till date it is, con-

[41] Rivals only in the academic sense of holding diametrically opposite view and not otherwise.

sidered as one of the six orthodox schools (*āstika-darśanas*) of Hindu philosophy, whose purpose is defined to be *summum bonum*. Here a quotation from Ganeri may be in order. Towards the end of his introduction to a recently published work Jonardon Ganeri observes:

> The effort must continually be made to explain the distinctiveness in the goals, methods and techniques of Indian logic, in order that we might better understand the nature of the challenge that this alternative tradition of inquiry into the basis of reasoned thought presents, most particularly when we assume that our own theories are free from commitments specific to our own history.[42]

In yet another interesting paper, it has been well-argued by Claus Oetke[43] that Ancient Indian Logic can be construed as a Theory of Non-monotonic reasoning (NMR). He observes:

> The principal aim of this paper is to establish the thesis that a significant relationship exists between Indian theories of proof and inference (*anumāna*), in particular the most ancient varieties of what is commonly called 'Indian Logic,' and a number of quite recent developments in the theory of "commonsense-inference" which are often subsumed under the term 'Non-monotonic Logic(s).' ...
>
> It is claimed that there is a connection both on the, so to speak, "object level" as well as on the "theoretical level"[44]

17.6 Concluding Remarks

Indian logic, commonly referred to as, *Nyāya-śāstra* is not merely a "grammar of reasoning" but is considered as one of the six orthodox schools (*āstika-darśanas*) of Hindu philosophy. The *Nyāya* school which has its foundational basis on the *Nyāya-sūtras* of Gotama, which are believed to have been composed at least by the end of the 3rd century BC,[45] if not earlier has a very long lineage, which continues till today.

As we turn the pages of the history of Indian logic, it becomes quite evident that Gaṅgeśa's *Tattvacintāmaṇi* – more popularly known as *Maṇi* composed at the interface of the 12th and the 13th century[46] – has been considered to be a major landmark, at which the emphasis got shifted from study of *prameya* to *pramāṇa*. A lot of new expressions and concepts, that could be considered as thought-measuring devices, got introduced to achieve more precision in arguments. The successors of

---

42 J. Ganeri, (ed), *Indian Logic. A Reader*, Richmond, Surrey: Curzon Press, 2001.

43 Professor, University of Stolkholm.

44 *Journal of Indian Philosophy*, 24: 447–539, 1996, Kluwer Academic Pub.

45 This is the opinion maintained by the majority of well-informed and unprejudiced Indologists.

46 This is the period agreed upon by a majority of scholars.

Gaṅgeśa, including his own son, took forward the lineage to great heights. Even as late as 18th century, it has been recorded that Navadveep – the birth place of *Navya-nyāya* – used to be a great center of learning, wherein thousands of students used to come from distant parts and have rigorous training in *śāstra*. Incidentally the record indicates the teacher-student ratio to be less than 1 : 8 in the University of Nuddeah (Navadveep) that flourished during the 17–18th century. Here is an excerpt from the records that provides an interesting reading.

> The grandeur of the foundation of the Nuddeah university is generally acknowledged. It consists of three colleges–Nuddeah, Santipore and Gopulparrah. ... And their resources are so ample, there are at present about eleven hundred students, and one hundred and fifty masters. Their numbers, it is true fall very short of those in former days. In Rajah Rooddre's time, there were at Nuddeah, no less than four thousands students, and masters in proportion. ... Shankar pundit, is head of the college of Nuddeah, and allowed to be the first philosopher and scholar in the whole university. His name inspires the youth with the love of virtue – the pundits with the love of learning – and the greatest Rajahs, with its own veneration. The students that come from distant parts, are generally of a maturity in years, and proficiency in learning, to qualify them for beginning the study of philosophy, immediately on their admission; but yet they say, that to become a real pundit, a man ought to spend 20 years at Nuddeah, in close application.[47]

It is remarkable to note that this lineage continued to be a vibrant tradition not merely critically examining the earlier works but also evolving new concepts and taking the discussion to much deeper levels in a specific direction. Notwithstanding the series of onslaughts that took place in India, which had been greatly detrimental in many ways in preserving and passing on the traditional knowledge systems, that *Nyāya* tradition is still a living tradition producing commentaries and *kroḍapatras* is indeed a spectacular achievement that is quite gratifying. However laudable the efforts and the achievements of the Indian tradition in logic may be, they would only remain confined within the walls of traditional scholarship, unless its relevance is established for the contemporary times. Here, it may be in order to cite an observation from one of the young, energetic, modern scholars.

> In conclusion, while the history of logic in India shows a strong tendency towards formalisation, the logic of ancient India tried to model informal patterns of case-based reasoning, patterns that are increasingly becoming recognised as widespread and representative of the way much actual reasoning takes place.[48]

The above quotation clearly showcases the relevance of the studies of Indian logic as presented in the ancient texts. Hence, it seems that it would be a useful exercise to carry out a thorough analysis as to how the methods and techniques employed in the texts of Indian logic would be useful – if at all – for the furtherance of the studies at in logic at levels far deeper than the present, and also explore its application to other disciplines even in modern times.

---

[47] The Calcutta Monthly Register and India Repository, January 1791, pp. 136–139.

[48] Jonardon G., Ancient Indian Logic as a theory of Case Based Reasoning, *Journal of Indian Philosophy*, 31: 3345, 2003.

Chapter 18

Indian Logic and Philosophy of Science: The Logic-Epistemology Link

Sundar Sarukkai

18.1 The Logical and the Empirical in Indian Logic

There are many interesting themes in Indian logic which illustrate not just philosophical complexity and rigour but also their potential use in philosophy of science. Matilal describes Indian logic as the "systematic study of informal inference-patterns, the rules of debate, the identification of sound inference *vis-à-vis* sophistical argument, and similar topics." An important task for Indian logicians was to critically understand which inferences are valid and what conditions they should obey in order to have certainty in inference. Thus, Indian logicians were deeply concerned about establishing a theory to know which inferential statements one could be certain about and the methodology to decide on their validity. The early *Nyāya* logic is exemplified by the five-step argument and there has been much discussion on whether it is equivalent to a syllogistic form.

Perhaps the most common representative example of inference in Indian logic is that of inferring the presence of fire from seeing smoke. The analysis of this example illustrates a deep and complex engagement with the idea of inference and various related philosophical themes. Some of the questions concerning this example are the following. How do we infer that there is fire given that we only see the smoke? How can we establish the validity of this conclusion? What are the criteria for knowing that our inference is right, that we are justified in making it? What kind of inferences can give us certainty? There are two kinds of inferences described by the Naiyāyikās and the Buddhist logicians. One is the inference one makes for oneself. This occurs when one sees smoke on a hill and then infers to oneself that there is fire on the hill. There is no further communication or convincing somebody else about this inference. The other type of inference is called inference-for-another, which is a

Sundar Sarukkai
National Institute of Advanced Studies, Indian Institute of Science Campus, Bangalore - 560012, India, e-mail: sarukkai@nias.iisc.ernet.in

J. van Benthem et al. (eds.), *Proof, Computation and Agency*, Synthese Library 352, DOI 10.1007/978-94-007-0080-2_18,

demonstration of the inferential process and conclusion for another person. Through this rational demonstration, one can convince another about the inference that one has made and explication of the reasons one holds for making a particular inference. *Nyāya* has a five-step process to accomplish this. The five parts are (1) the statement of the thesis of inference, (2) stating the reason or evidence for this thesis, (3) citing an example of a generalisation supporting the reason and this example is such that it is well accepted by others, (4) application of the present observation to the generalisation and (5) conclusion or assertion that the statement of the thesis has been proven. The well-known illustration of these five steps is as follows.[1]

1. Proposition: There is fire on the hill.
2. Reason: For there is smoke
3. Example: (Wherever there is smoke, there is fire), as in the kitchen
4. Application: This is such a case (smoke on the hill)
5. Conclusion: Therefore it is so, i.e., there is fire on the hill.

The structure of argument sets out the rational, logical way of convincing others that the inference one makes is perfectly valid. The reasoning is that we make an inference of fire when we see smoke because we know, from previous experiences, that wherever smoke is present so is fire. The second step states the reason, which is smoke, for making the inference. The reason in conjunction with a more general principle or natural law that we know allows us to make the inference. But given *Nyāya*'s commitment to empiricism, any inference we make has to be grounded in some observations, as given by examples. These examples should be commonly accepted and can be both positive and negative examples. Positive examples are those that support the inference, such as a kitchen because in the kitchen we know that smoke and fire are seen to occur together. Negative example, such as a lake, supports the inference by looking at cases where there is no smoke in places where there is no fire. The fourth step applies this general principle to the case at hand and finally, the fifth step, states the conclusion.

This five-step process is the Indian logician's equivalent of Aristotelian syllogisms. Thus, it is the most common example seen in a discussion of Indian logic. In the way it is presented, it seems clear that this process is not what was understood as logic in the West, largely because of the use of examples, inference which is of a singular case and so on.

The *Nyāya* formulation of inference was modified and replaced by an influential formalism by *Dignāga*, the great Buddhist logician. *Dignāga* reformulated the question of logic into a question of semiotics. Inference by its very nature is related to signs. Therefore, smoke, first and foremost, has to be considered as a sign which is in some sense related to what we infer. *Dignāga*'s logic is primarily an attempt to

---

[1] Matilal 1999, p. 4. See also Vidyabushan 1920, p. 61.

clarify what kinds of valid signs are possible and how we can make justified inferences from these signs. In this sense, *Dignāga* was concerned with justification of an inductive statement such as, "Wherever there is smoke, there is fire". The aim is to know how an inductive cognition can be absolutely certain.

There is yet another peculiarity in his formulation, and this has to do with the synonymic usage of sign, reason and evidence. These are terms that are often used interchangeably in Indian logic. *Dignāga*'s theory of inference sets out a structure of inference based on the nature of the sign, thereby defining when a sign can properly stand for another. He formulated the "triple nature of the sign", three conditions which a sign must fulfil in order that it leads to valid inference.

1. It should be present in the case (object) under consideration.
2. It should be present in a *similar* case or a homologue.
3. It should not be present in any *dissimilar* case, any heterologue.[2]

These conditions can be used to check whether any sign is a logical sign of another. These conditions are also conditions to check whether a reason is a valid reason or not. For example, in the thesis, "Sound is impermanent, because it is audible", the reason is audibility and the inference is the impermanence of sound. If we want to know if this reason is a valid reason for the conclusion, then we have to check if the reason satisfies the three conditions.

Dharmakīrti suggests three broad divisions into which inferences can be grouped: inference based on (1) own-nature, (2) causal relation and (3) non-perception. All three are not only interesting in their own right but are also kinds that can be discerned in scientific inference. The relation to science is further illustrated when we understand that the two kinds of inference based on own-nature and causality are based on the notion of natural relations and the presence of "natural" properties. The first kind, namely, inference based on *own-nature* is illustrated with the common example, "This is a tree because it is an oak tree". Considering that *Dharmak īrti* also used the term identity to describe this relation, it seems clear that this kind of inference is based on identifying the oak tree *necessarily* with a tree. This example also suggests, as others have pointed out earlier, that the inference is "based upon the relation of class inclusion" and can be seen as an analytical statement.[3] What is interesting here is the relation of analyticity with inference. The idea that there is inference in analytic statements is an interesting one, especially given the fundamental idea that inductive inference increases knowledge content. The second kind which deals with causal relations describes examples such as the smoke-fire inference.

The third kind of inference is a unique one based upon non-perception. A commonly cited example of this type is "There is no pot here because no pot is perceived here." The basic idea here is that we also make inferences based on non-perception

---

[2] Matilal 1999, p. 6.

[3] See Prasad 2002.

of something. This means that not all our inferences arise through perceiving something positively. For example, the very fact of not seeing smoke (in a situation when you expect to see it) makes us infer something. So, non-perception of various factors is also a cause for inference. *Dharmakīrti* goes on to give eleven varieties of inference from non-perception. Thus, we see the complex ways of formulating inferences in Indian logical systems.[4]

18.2 Logic-Epistemology Link in Indian Logic

Matilal points out that logic in India arose out of two different traditions – one, the tradition of debate and dialectics, and the other, the epistemological, empirical tradition, because of which the distinction between logic and epistemology, as in Western logic, is not made in Indian schools of logic. He identifies three essential differences between Indian and Western logic.[5] First, the lack of a clear distinction between epistemology and logic; second, the apparent presence of psychological elements in logic; and third, the lack of deduction-induction distinction.

The relation between epistemology and logic, so integral to Indian logic, is what is erased in Western logic, based on the belief that epistemology which is about knowledge, empiricism and truth must have no relation to logical structures. In the Indian case, inference is actually a *pramana*, a valid means and instrument of knowledge. It is one of the sources of knowledge in all Indian traditions except the materialist *Cārvākas*. This is also the reason why the notion of evidence is so important in Indian logic. As we saw earlier, evidence and reason are often synonymously used in analysing valid inference. In the Western tradition, it is well known that deductive structures are independent of the truth of the premises. Thus, it is commonly observed that logic is not about truth, especially truths of the world, the set of empirical truths. In fact, what sets logic apart is this indifference to the dictates of the nature of our world. But this is where Indian logic significantly departs from Western logic. The use of the example in the *nyāya* five-step process and the similar and dissimilar conditions in *Dignāga*'s formulation illustrate the importance of factoring in empirical truth as part of logical process. This does not mean that there is no formal argumentative structure which can be discerned in Indian logic and such analysis are available in the literature.[6]

The rejection of empty referring terms is another indication of the essential connection between epistemology and logic. For example, one of the premises in Aristotelian syllogism could well be "All Martians are blue". This would be a valid statement in the syllogism independent of the question whether there are Martians

---

[4] See Matilal 1999. Also R. Prasad 2002.

[5] Matilal 1999, pp. 14–18.

[6] For example, see S. Bhattacharyya 1987. Also Matilal 1999, Chapter 7.

and whether they are really blue. Now, one might argue that this is exactly the reason why we have logic so as to be able to transcend the empirical restrictions arising from the real world. Logic helps us to order empirical facts and so has to be beyond these facts. Even if this is so, we can understand Indian logic as showing how there can be logical structures without letting go of a commitment to truth and knowledge as part of logical argumentation. The real problem would arise only if it is claimed that empirical truths *dictate* the logical structure. This is a view that is not found in the dominant traditions of logic in India. As we have seen, they acknowledge the logical structure present in moving from some premises to a conclusion but they do not say that these empirical truths dictate the kind of arguments that are possible. In fact, it is useful to see the empirical and epistemological content in Indian logic only in terms of a regulative factor and not as a determining factor. Thus among all possible logical statements, some are disallowed. Thus, empiricism and issues of epistemology are ingrained into logic in so far as they are used to delimit the possible range of logical arguments. The various attempts, with varying degrees of success, of rewriting Indian logic in terms of formal and modern logic only points to the essential logical equivalence in the arguments and the use of empiricism and epistemology to cut down the possible logical structures which are allowed. Finally, it would be useful to understand the nature of Indian logic as reflecting a worldview which is not only suspicious of perception but also of reason.

The lack of a clear distinction between epistemology and logic arises because of the empirical foundations demanded of Indian logical arguments. While formal deductive structures are present in the *Nyāya* and Buddhist schemes, the emphasis on the premises being empirically true, or at least available for empirical judgement, brings logic and epistemology together. However, the use of *tarka* or suppositional/reductio arguments is prevalent in Indian logic. The use of *tarka* is indicative of the conceptualisation of counterfactuals and the use of deductive arguments. While *tarka* was seen as an important instrument of reason, it was not seen as a means to knowledge like ordinary inference was, because of the lack of empirical validity in counterfactual statements. Such reasoning comes into play when we want to argue for one position against another by showing the absurd consequences of holding the opposed position. This is "negative" argumentation in the sense that a particular thesis is not proved by this reasoning but only that another thesis is shown to have undesirable consequences. Positive argumentation in this sense is based on empirical evidence in contrast to *tarka*. It has also been argued that *tarka* is more broad-based than reductio arguments since the undesirable consequences are more than just contradictions. This larger canvas of *tarka* is captured by Udayana's classification of *tarka* into five types: self-dependence, mutual dependence, cyclical dependence, lack of foundation and undesirable consequence.

The second difference which Matilal notes has to do with the presence of psychological elements in Indian logic. It is precisely the removal of psychology from logic that distinguishes logic in the modern logical tradition. So retaining these elements in logic is either a fallback to ancient logical systems or worst, not being aware of something essential to logic. Matilal counters this by saying that the Indians did not commit the bigger blunder of emphasising psychology in logic thereby leading to psychologism and/or that one could understand the Indian "psychologised" logic as a different kind of logic. The latter point is reasonable to hold when we see the kinds of logic present today. In particular, there would be similarities in Indian logic and in the approach of intuitionist logic and the presence of psychological elements in it. But there is another argument that we can invoke here and this has to do with the relation between logic and psychology.

It is well accepted that Indian logic shows explicit psychological elements. Let me discuss Mohanty's detailed argument in this context.[7] First of all, there is an explicit psychological description of the process of inference, starting from perceiving a sign (say smoke), then remembering the connection between smoke and fire, which then leads to an awareness that this connection is applicable to the present case, finally leading to the inferential cognition. Further the psychological dimension is explicit in the invocation of desire (either absence or presence of) as a condition for inference. However, the inferential schema which is used to explain the inference-for-another does not carry over the consequences of the psychological description. In some sense, this is like the description of any rational, objective act, which is dependent on some mental process but whose content is independent of the consequences of a subjective, psychological act. For example, even in the creation of scientific theories, ideas arise through a cognitive process which can be described in terms of psychological terms but the content of the theory is objectively accessible. Inference-for-oneself is only based on the recollection of the connection between smoke and fire whereas inferencefor-another needs the five-step process described earlier. The argument that Indian logic is psychological also depends on what is meant by the psychological. If being psychological implies validity only to subjective experiences then Indian logic is definitely not psychological, especially because while using the language of psychology it still deals with and exhibits universal structures of human thought. The greater challenge to the logical character of Indian logic actually would come from the entanglement of the empirical within the logical, but as we have seen above this too is not a significant problem. As Mohanty also notes, *tarka* is a good example of a formal logical structure and has no mention or use of examples. So it is not that formal logical structures were unknown to Indian logic. The use of examples as part of inferential reasoning does not negate its formal character. We can understand the use of example along the following lines suggested by Mohanty: one, because the example should be one that is acceptable to the opponent, it not only dealt with true premises but it also served an important role in the "dialogical-disputational context"; two, because of the insistence on the truth of the premises, example preserves the "epistemic purity" of the inference; and three, an example succeeds in introducing "an individual by the rule of existential instantiation."[8]

[7] Mohanty 1992, pp. 100–132.

[8] Ibid., pp. 117–118.

There are other important points of difference, details of which I will not go into here. These include the idea of the necessary in Indian logic and whether this logic is intensional or extensional. We also have to remember that whichever type of logic we choose, we are making commitments to some underlying worldview; in the case of Western logic, there is a commitment to the existence of abstract entities such as propositions. Now whether it is better to make this assumption as against what Indian logic does is a question that cannot be resolved by logical arguments alone! Mohanty in the conclusion of his essay on Indian logic notes that Western logic is framed by oppositions between logical and the empirical, necessary and contingent and so on, and that Indian logic, taking a middle path, shows that "we need not treat the logic and the psychological, the epistemic and the causal, the extensional and the intensional as though they are irreconcilable opposites."[9]

Thirdly, Matilal notes that mathematics profoundly influenced Western logic whereas in India, it was grammar that was most influential. There are two points we can note about this brief observation. Firstly, in what sense was mathematics a model for Western logic? It is commonly observed that Plato's forms, inspired as they were with geometrical forms, also influenced the importance of formal structures in logic. In fact, the extreme step of considering premises without any empirical or truth content as part of valid logical reasoning can also be seen as a consequence of privileging the notion of form over content. Thus, while geometrical forms illustrate the importance of forms in the Greek imagination, and thus might suggest the influence of mathematics on logic, there is also the added reason that mathematics was associated with deductive certainty, universal truth and so on. In a sense, all these characteristics of mathematics (actually of geometry and of early arithmetic) get transplanted to logic.

What could be the equivalent status in the Indian system? Matilal notes that grammar was the dominant influence on Indian logic. There are two pertinent comments that are relevant here. Firstly, Indian mathematics was itself quite different in character compared to ancient Greek mathematics. In particular, the emphasis was not on models and axiomatic formulation but on algorithms and pragmatic use. Even in the case of geometric forms, the circle was not the ideal form for Indians but the square. Mathematics, especially as used in astronomy, was built not from axioms but from algorithmic rules of manipulation to get the best results for practical use. Therefore, this pragmatic, result-oriented character of Indian mathematics is part of a larger worldview which it shares with the pragmatic, empirically oriented character of Indian logic. Secondly, Matilal's point that Sanskrit was the dominant influence in Indian logic also indicates a relation between Indian logic and mathematics. Sanskrit is a highly structured language and it is more like a formal language like mathematics especially in comparison to many natural languages like English. Therefore, even being influenced by Sanskrit instead of mathematics (of the Greek variety!) might not indicate an essential difference between Indian and Western logic. Given these two arguments about the nature of mathematics and of the special nature of

[9] Ibid., p. 131.

Sanskrit (in opposition to English for example) it seems reasonable to believe that looking to explain the difference between Indian and Western logic as one between mathematics and grammar might not be too illuminating. However, this does not mean that the special characteristics of Sanskrit are not reflected in Indian logic. The language of Indian logic itself is conditioned by the peculiarities of Sanskrit. In fact, some of the unique themes in Indian logic arise from the demands of Sanskrit.

18.3 Indian Logic and Philosophy of Science[10]

It is now possible to see the close relation between philosophy of science and Indian logic. Philosophy of science has drawn upon various philosophical and logical categories from the Western tradition in order to reflect upon the nature of science. One essential part of this approach was to clarify the relation between science and logic. Science's relation with logic is most clearly exemplified in its use of mathematics. While the mathematical component of science is seen to reflect the deductive structure of argument in science, there is much in science which draws upon other forms of argumentation, particularly inductive inferences. Many of these inferences in science follow the same structure as the common inferences we make, such as the inference of fire from smoke. Thus, the philosophical issues that arise in analysing inferences in science share some common conceptual ground with "ordinary" inferences. Paradoxically, the strong empirical grounding of science places it in potential conflict with logic, understood in the Western way. Science draws upon observations and reason, and using them together constructs its narrative of the world.

The relationship between science and logic is fraught with various problems. I want to suggest two metaphoric images that capture two different ways of analysing this relation. One is along the lines which philosophy of science has done, and done so admirably. This I call the *logic in science view*, which describes the attempt to fit logical categories from the Western tradition into the practice and discourse of science. Thus, this becomes a search for elements of logic in science. A consequence of this approach is that sometimes science is moulded to fit logic and this most often leads to a belief among many scientists that philosophy of science not only misunderstands science but also misrepresents it.

However, there is another way of understanding this relation, which is well exemplified by Indian logic. I would like to characterise this approach as *science in logic*. As we saw earlier, many conceptual ideas that arise in Indian logic resonate strongly with scientific methodology and praxis. This does not in any way imply that Indian logic is "doing" science. *What it is doing however, is expecting logic itself to be scientific, in contrast to the position that science be logical.*

[10] For a detailed description of this link, see Sarukkai 2005.

18.4 Indian Logic, Science and Semiotics

Looking at science through the rubric of conceptual themes in Indian logic allows a reformulation of some standard problems in philosophy of science. For example, some new trajectories include analysis of the relation between logic and semiotics, the origin of theory through the use of signs and symbols, the natural relation between sign and signified in applied mathematics (in contrast to the conventional understanding of mathematical symbols as being arbitrary), the structure of Indian logic as manifesting an explanatory structure, particularly the deductive-nomological model of scientific explanation and so on.

The American philosopher Charles Sanders Peirce made seminal contributions in the areas of semiotics, logic, pragmatism and on scientific method. In many of his ideas, he shares an affinity with Indian philosophers and could have passed off as a member of some of these Indian traditions!

I begin with what Peirce writes about logic. "Logic, in its general sense is . . . only another name for semiotic (. . .), the quasi-necessary, or formal, doctrine of signs." By this he means that "we observe the characters of such signs as we know, and from such an observation, by a process which I will not object to naming Abstraction, we are led to statements, eminently fallible, and therefore in one sense by no means necessary, as to what *must* be the characters of all signs used by a 'scientific' intelligence, that is to say, by an intelligence capable of learning by experience."[11] Note the use of certain important concepts, which are also those which we discussed earlier in the context of Indian logic: observation, fallibility, and "scientific" intelligence, which he sees as an intelligence which learns from experience.

Peirce begins by defining sign, in a way similar to *Dignāga*, as "something which stands to somebody or something in some respect or capacity." There are three elements to a sign: the creation of another sign in the mind, the sign standing for an object, and the presence of an idea in reference to which the sign stands for the object. The second element of the sign, namely its capacity to stand for some other object, is what Peirce calls logic. Thus, "logic proper is the formal science of the conditions of the truth of representations."[12]

Buchler in his introduction notes that Peirce's path-breaking contribution is his "conception of logic as the philosophy of communication, or theory of signs."[13] He also says that the "conception of logic as semiotic opens broad, new possibilities". Arguably, we can best understand the aims of Indian logic alongside the approach towards logic and signs by Peirce. Peirce's ideas about signs share a common conceptual space with *Dignāga* and other Indian logicians. If, therefore, Peirce is an important figure in logic, then it seems reasonable to also expect a similar acknowl-

[11] Peirce 1955, p. 98.

[12] Ibid., p. 99.

[13] Ibid., p. 12.

edgement for the ancient and medieval Indian logicians who based their logic on the nature of signs and therefore first understood logic as semiotics.

Is *Dignāga* formulation of the triple condition of logical sign a precursor to logic as semiotics? Firstly,*Dignāga* and Peirce are both interested in valid logical conclusions and judgements. Both of them see the sign as the path towards valid judgements of the logical kind. Both of them have a broad view of signs, ranging from material signifiers such as smoke to nonmaterial such as words. As a theory of signs, Peirce's classification is much more detailed whereas as a theory of logical sign, *Dignāga*'s conditions do more than Peirce's analysis. For example, Peirce's idea of similarity as used in iconic signs has no definition of what similarity is, which is exactly what the similarity condition in *Dignāga*'s theory tries to answer. In Indian logic, the sign is synonymously used with reason and evidence. The idea of reason is already inherent in the meaning of a sign since a sign is a sign of something else and it is conceivable that there is a reason for this connection. The question for the Indian logicians consisting in knowing that the sign "really" stood for the thing which it referred to. Therefore, this involved understanding the reason why the sign comes to stand for another. This could be psychological (by seeing concomitance for example) or social (linguistic conventions for instance) but still part of the doubt about inference comes from doubt about the origins of the relation between sign and the signified object. The conditions of similarity and dissimilarity are also attempts to clarify this relation as also the origins for making the connection in the first place.

Dharmakīrti's three types of inference, the ones based on identity, causality and nonperception, can be seen as a classification that answers as to why some signs come to stand for another. Identity is based on similarity and is actually a more comprehensive definition of icon as compared to Peirce. When we say that an oak tree is a tree our inference is based on a perception of some similarities. In general, it would seem that our "perception" or inference of universals, such as say cowhood, is based on recognition of similar characteristics and therefore signs in such an inference function as icons.

The causal type, with smoke and fire as example, is illustrative of indexical signs. Smoke is an index, which refers to a fire in that location. Smoke and fire are associated through contiguity and smoke obeys all the characteristics of an index which Peirce describes as follows: "First, that they have no significant resemblance to their objects; second, that they refer to individuals, single units, single collections of units, or single continua; third, that they direct the attention to their objects by blind compulsion."[14] Causal signs are just one type of index signs. The interesting question is whether *Dharmakīrti*'s second type can be extended to indexicals in general and not be restricted only to causal signs? Or equivalently whether Peirce's large set of indexical signs need to be pared only to causal ones?

Given the insistence of Indian logicians on reason for a sign, it is natural that the arbitrary nature of sign is not available in the logical formulations discussed earlier.

[14] Ibid., 108.

But it would be wrong to say that the arbitrary nature of sign was not known to them, since *Dignāga*'s formulation draws upon his *apoha* doctrine of language. It was also very clear for Indian philosophers that words function as signs standing for something else. Moreover, it was clear to them that this relation between words and things, for example, was arbitrary. In fact, I think it is reasonable to argue that the stringent conditions on a valid sign may actually be a reflection of the problems of arbitrary connection between words and what they stand for. Since philosophy of language was one of the pillars of ancient Indian thought, the influence of these philosophies on logic might have succeeded in making the conditions on signs more rigorous than it perhaps ought to have been!

Therefore, if in the first place, arbitrary relation of sign-signified is what is being sought to be eliminated in the three conditions of *Dignāga*, then it is no surprise that the kinds of valid signs in Indian logic is highly restricted. On the other hand, we can see that the use of symbolic notation in Western logic, including in the representation of terms in Aristotelian syllogisms, is itself a use of signs. As is commonly done, in the premise "All Greeks are mortals", we can let A stands for Greeks and B for mortals and this sentence becomes "All A is B". However, note that this move of letting A and B stand for something else, is actually a semiotic one, where A is a sign which stands for a set consisting of Greeks. Symbolising thus is arbitrary and therefore there can be no necessary pervasion relation between the symbol A and the set of Greeks. This use of sign is just like the use of language where words are in arbitrary relation with things. Using a term to stand for something is actually to create a sign-signified relation.

So once we use arbitrary symbols, we are not inferring but dealing with issues of language. If we are dealing with issues related to language, then the kinds of questions that arise are very different than if we are dealing with cognitive inferences. Use of arbitrary symbols for Indian logicians would be a movement into the domain of language and thus perhaps outside inferential cognition. This implies that even the attempts to symbolise Indian logic is to misunderstand something essential about it!

18.5 Signs, Symbols and Theory

The basic question that is common to both philosophy and science, and not just Indian philosophy and modern science, is the problem of understanding how we move from an empirical observation to a hypothesis or theory about that observation. We might begin a move of this kind by first describing the observation. Even a mere description needs language and concepts that belong to that language help us to describe that observation in a particular way. We move from observation to theory even in this first step of describing that particular observation in words. Talking about observations is already taking us into the activity of theory-making. But why so? What is that language does? First of all, the use of language to describe what we observe takes us into a different level, to that of language and not of the phenomenal event. The use of language that marks a distinct qualitative shift from the universe of observations to that of expression suggests a character of theory that has been very influential, namely, the idea that the use of signs and symbols make possible the shift from the empirical to the theoretical.

This should not be too surprising. After all, the moment we use language to describe an observation we are already in the world of signs. Words in a language are signs. The most important and fundamental characteristic of a sign is that a sign stands for something else. A word "chair" stands for the thing chair. So words stand for things, properties, relations and so on. How words stand for the things they stand for is quite a different topic altogether and philosophy of language and semiotics deal with this issue in great detail. In the case of language it is often the case that words are arbitrary and are a matter of conventions of a particular linguistic community. There is in general no "natural" association between the word and the phenomenal world the word describes and talks about. But verbal languages are not the only examples of signs. There are many different types of signs as we have already encountered in the discussion on Peirce. In fact, the use and theory of signs have a long tradition within Western philosophy. In particular, many influential thinkers have formulated various theories about signs and their relation to knowledge. Since I am focussing on the philosophy of science, I will summarize a few of these approaches to an understanding of signs and their role in the creation of theories. This also illustrates one fundamental similarity in the *concerns* of the Indian logicians and philosophers in Western traditions, namely, an understanding of the relation between sign and signified, and the role of signs in theory-making.

The issues about the meaning and role of signs can also be placed within the larger distinction between empiricism and rationalism. Empiricism holds that knowledge of the world needs empirical, observation inputs whereas pure rationalism holds that all knowledge can be guaranteed by human reason alone. Leibniz is an influential philosopher of the rationalist tradition. In rationalism, the problematical issues have to do with the nature of reason, what is actually meant by reasoning, how do we reason, what guarantees the truth of reason and so on. Therefore, reason is enmeshed with mental states, thoughts and so on. Leibniz held the view that we can think and even know only by using signs. These signs could be natural, in the sense that we make "natural" associations and the relation between sign and signified is "natural" or the signs could be artificial, in the sense that we create as a community a set of arbitrary signs to stand for various other things. For example, a red light at a signal is a sign indicating stop. We could conceivably have used any other colour to indicate this so the choice of red is just a convention and arbitrary in this sense. Now, if we can think or know only through the use of such signs, as Leibniz would have us believe, then the real fundamental entities of human knowledge and thought are only signs. The consequent problem of this view would then be to explain how the sign is related to the signified.

Signs don't grow on trees (although sometimes "natural" signs do) and the mind has a constitutive role in creating a whole domain of signs through which we think, express and know. Since there is no knowledge without the mediating role of signs, this leads us to a "constitutive" theory of knowledge. Theory, enmeshed as it is in a world of signs, is therefore possible only by adding something to the empirical world and this addition consists of the human activity of signcreation and sign-meaning.

Although there is a fundamental role to signs in human thought, signs and symbols take on a primary role in certain disciplines, particularly logic and mathematics. Frege considered the idea of the sign as a "great discovery". He also held the position that ideas and concepts are possible only through the creation and use of signs. Further, an added advantage of particular symbolic notations was that they did not manifest some common problems associated with verbal languages. But sign here refers largely to an arbitrary set of symbols which are created by us with no specific natural meaning associated to them. In the same logical tradition, George Boole understands the discipline of logic as expressed entirely by the world of signs and symbols, and that the "laws of signs are a visible expression of the formal laws of thought."[15]

Cassirer, much influenced by Leibniz, analyses in detail the central importance of symbols and science. For him, the structure of science rests on the "logic of things", namely, "the material concepts and relations" and this logic of things cannot be separated from the logic of signs.[16] An important consequence of this approach is his position that "concepts of science are no more imitations of existing things, but only symbols ordering and connecting the reality in a functional way". The importance of this view needs to be stressed again, especially the argument that concepts are also semiotic. This is important because of the often held belief that the transition from observation to theory occurs through concepts and that concepts are neither linguistic nor part of a system of signs. The characterisation of observation by concepts is indeed central to the activity of science. The ability to create new concepts is at the heart of the scientific and mathematical enterprise but is not restricted to them. Even philosophical theories generate new concepts. The belief that concepts mediate observation and theory tend to see concepts as being beyond the symbolic domain that constitute theories. Concepts, in one view, are objective realities which are already present in the phenomena and thus they are discovered. But if concepts themselves belong to the world of symbols they cannot play the mediating factor between phenomena and description of these phenomena.

The emphasis on signs and symbols leading to their essential role in disciplines such as logic and mathematics is initiated by the first move of creating signs to stand for various elements of an observation. But this emphasis on sign and symbol is not special to the Western tradition. Cassirer observes that, thanks to Leibniz, "at one stroke the concept of the symbol has become the actual focus of the intellec-

[15] Ferrari 2002, p. 15.

[16] Ibid., 25.

tual world."[17] Alas, Cassirer was over 1,000 years off the mark. As we saw earlier, *Dignāga*'s logic was actually all about the nature of the sign and Indian philosophers for centuries after debated and fine-tuned these ideas on signs.

In the context of scientific theories, Helmholtz, Hertz and Duhem considered the essential role of symbols in theories. Helmholtz, influenced by Kant, believed in the constitutive view that knowledge about the world cannot be independent of the organisation of the mind and that signs play this mediating role.[18] In particular, Helmholtz was concerned with the difference between a property that causes a sensation and the nature of that sensation in us. For example, the colour red induces the quality of redness but this quality is dependent on our constitutive apparatus and thus is only a sign of the colour that is present in the red object. Thus sensations only indicate the effect an object or a particular property of the object has on human subjects. The similarity with Peirce's classification of qualisin may be noted here. But Helmholtz also holds a view of natural connection for he sees these signs as "natural signs" in that the "same sign must be always assigned to the same object." An important point in such formulations is that a sign of an object is not the object and therefore signs do not "capture" the reality of the world but they can be faithful to the lawful relations that are present in the world.

In many formulations of the sign, we tend to come across such a view that while signs may be quite different from the object, they more faithfully represent relations if not the objects. This kind of relational realism is also to be found in some theories of mathematics, which is after all immersed in the symbolic universe. This relational aspect is clearly articulated by Heinrich Hertz, another important scientist, who saw symbols as images that mirror the relations in and of objects. Although he too sees knowledge as being constitutive, he privileges the importance of the objects which actually determine the nature of the symbols that are created in the mind, although what symbols capture are the relations and not the objects themselves. (The problem in such approaches is that the nature of the mind and its "necessary" structures are seen to be real on a order different from that of the reality of an object. This is problematical since it defines reality of the mind in a different way and also does not expect the mind to be understood symbolically like external objects.)

Finally, it will be useful to consider Pierre Duhem's conception of a symbol, particularly since there are some elements in it that are quite different from the Indian view of sign. Ihmig summarizes five general features of a symbol for Duhem.[19] Firstly, signs are arbitrary and conventional. They do not possess any natural relation and therefore a sign does not have a natural connection with the signified. He also considers signs to be at a different level than phenomena. Thus, for Duhem, smoke cannot be a sign for fire since both smoke and fire belong to the same phenomenal level whereas if smoke were to be a sign it has to be qualitatively different from fire. I think we can see that this difference which Duhem posits is similar to the

[17] Cassirer 1957, p. 57.

[18] Ibid., p. 17.

[19] Ihmig 2002. p. 79.

semiotic character of language, where a word is an arbitrary sign but is also qualitatively different from the phenomenon/object it refers to. Further, for Duhem, signs are always part of a larger connected universe of symbols and therefore cannot be understood in isolation. The arbitrary nature of the symbol implies that there cannot be any truth value associated with them. Since they also represent something else instead of the truth or falsity of a symbol we can only ask whether they are appropriate or inappropriate. Finally, Duhem makes an important distinction between symbols associated with scientific formulation and those that arise in ordinary generalisations. This point is similar to the ones made by others about the difference between concepts in science and those in everyday life. Concepts in science undergo constant test, modification and rectification, and symbols in science, for Duhem, are similarly open to complex processes of creation and modification. Given Duhem's belief in the intrinsic relation between mathematics and theories, one can see that his formulation of symbols is very close to the mathematical use of symbols.

With this background, it may be a useful exercise to explore the meaning and role of signs in Indian logic and where necessary compare with the Western semiotic tradition. It is clear that there is fruitful comparison at the junction of Indian logic, Western semiotics and the nature of scientific theorising. Similarities and dissimilarities between these fields can illustrate some fundamental lessons about the nature of these philosophies.

Here is an interesting paradox. The constitutive role of the mind, in the Kantian sense or otherwise, is a common theme in many of the Western theories that discuss the relation between symbol and knowledge. Cassirer goes to the extent of remarking that physics does not constitute of "signs of something objective" but only of "objective signs". The paradox is this: Indian logic and epistemology, which are seen to be essentially "psychological", actually does not offer any simple constitutive pictures. In fact, the sign is not described as arising from an activity of mind and hence the enormously complicated attempts to define the pervasion-pervaded relationship.

18.6 Science and Semiotics

Signs are ubiquitous. They are everywhere and occur in all human thought and actions. Peirce's classification of sign illustrates the different kinds of signs and the ways in which they could be classified. The Indian logicians understood inference through the medium of sign. Science too constructs a universe of signs and its discourse is influenced by the ways it uses and interprets signs. Is there a special nature to signs as used in science?

There is indeed a special relation between signs and science, and this has to do with how science understands the world and perception. One of the most important contributions of science has been its capacity to open up a new world of entities and processes which are unobservable by our normal perception. The use of instruments extends our perceptual limits and allows us a glimpse of a world hidden directly from us. Thus, for science, there are at least two modes of perception, the direct and the indirect. Not that these modes are special to science; we need only to remember the Buddhist argument for inference where they argue that while direct perception is possible, inference functions as an indirect perception. The strict meaning of "indirect" in these cases is slightly different. For the Buddhists, indirect involved reason, a capacity to infer since the domain was beyond that of perception. For science, perception itself is possible indirectly and instruments extend the range of perception in order for us to be able to perceive things which we cannot do with our unaided senses. However, there is an important question we need to consider about this instrumental perception: is this perception an inference? Do we really see electrons or do we infer their presence? And if we infer them, can we develop a method by which our certainty of this inference increases?

Obviously, there is no one simple answer to these questions. The range of instruments that are used in science are remarkably vast and the meaning of perception itself is quite different in them. For example, when we use a telescope or microscope we are able to "directly see" distant or tiny objects. When I see germs in a microscope, I believe that I am really seeing them, just as I see a tree. For science, there is really no distinction between seeing a tree in front of me and seeing germs through a microscope because the eye itself is an instrument. Seeing through a microscope is only an extension of the capabilities of this instrument called the eye.

Also, sceptical questions about these types of "direct" instrumental perception are answered in the same way that sceptical questions about ordinary perception are answered. Science understands and accepts to a large extent the appearance/reality distinction and understands that perception has to do with appearance whereas its task is to understand the "reality" behind the appearance. So the kind of manipulations it does with our ordinary observations is what it does to instrumental observation. But it is not as if there is no inference even in direct instrumental perception. For example, while we may be able to see germs directly through a microscope there is a systematic reasoning process to conclude that these are microscopic entities which are magnified by a certain amount and therefore their "real" size etc., are also correspondingly small. So, information about and meaning of these entities are often found by methods of scientific reason, in which inference plays an important role. But at the level of perception, instruments such as telescope and microscope yield direct perception. The fact that the size of what we see does not correspond to the "real" size is not important because even in ordinary perception we see objects having different sizes at different distances.

There is also indirect instrumental perception. In these cases, what we see directly through instruments is not indicative of the nature of the entity which is perceived unless we make appropriate interpretation. In the case of a microscope, if I see an object that is circular then it is highly probable that the object is indeed circular (given that the microscope does not cause distortion etc.), whereas for instrumental detection of various sub-atomic particles, my perceptual data would be very different perceptually from what the object would "really" be like. (Sub-atomic particles are of course a bad example since it is anyway difficult to say what they really look like!) The cloud chamber experiment is one good example. In this experiment, we can detect electrons and even use the data to calculate the ratio of charge to mass of an electron. The cloud chamber has an electric and magnetic field. The experimental output, the observational data, can be generated as a photograph which shows a lot of lines on it. One of these lines (which look more like scratches on the photograph) corresponds to an electron. There is no direct perceptual content in the scratchy line that we see and we really cannot infer how an electron "looks" based on this line. However, this line indicates the presence of an electron. If we do not hold a theory that can tell us how an electron would behave under the action of electric and magnetic fields, how to isolate an electron which can first of all be placed under the effect of these fields and so on, we would not be able to interpret a particular line as being indicative of the motion of an electron. But given a theory we can make some inference from the perceptual data.

This process of inferring the existence and motion of an electron from a line in a photograph is similar to the inference of an object from a sign. The line on the photograph is like smoke at a locus. Our perceptual data is only the line/smoke from which we infer that there is an electron/fire. The line/smoke is the sign. Following *Dignāga*, we can ask how can we be certain that this sign is indeed referring to an object? The line to an electron and smoke to the fire.

Let us first do a *Dignāgian* analysis on this sign. We want to know if the particular line in a photograph can be a valid sign that indicates with certainty the presence of an electron. The first condition is satisfied as long as there is at least one case. The second condition is that there should be some similar cases at the least. An example of similar case would imply that we should look for similar situations where we see both the electron and the photographic line. Obviously in an experimental situation such as this, the similarity case is equivalent to reproducibility of a phenomenon. What does the condition of reproducibility do? It is equivalent to the homologue condition because for controlled experiments two properties arise together not in natural situations but under particular experimental conditions. Why is reproducibility of an experiment important? We think reproducibility verifies a conclusion, makes it more certain that the relation we see in an experiment indeed occurs over and over again. In other words, we create homologues by reproducing the event. Each of these events are different because each time we reproduce the experiment we are doing it at least at a different time, if not at a different place. Thus each one of the experimental repetitions is a potential homologue. Reproducibility is weaker than the condition of similar cases but the condition that signs should be present in similar cases is also very much a part of scientific process. The priority given to experiments that reach the same conclusion through different processes is indicative of this move. There are many examples of this, including the well-known one of finding Avogadro's number by different means. The fact that all these different means of finding this number yield the same result implies that this number is an objective fact.

Is it enough just to have similar cases? How would the third condition of heterologue play out in this example? The basic point about dissimilar case is that if there is a similar line in the photograph even when an electron is not present then we cannot be sure that the line is a mark of an electron. Since the line occurs under the influence of an electric field the contrary would be a positive charged particle, say a proton. Now, if a proton's motion generated the same line in the photograph then we can say that the presence of the line in the photo is not a valid inference to an electron. Once again, since these are experimental setups and not natural cases, if dissimilarity is part of the inferential argument then we can see its presence in some experimental methodology, just like similar case is part of replicability. The importance of null experiments in science is well documented. Not finding expected results after performing an experiment gives an indirect proof of some aspects related to these results.

The question is this: how can an experimenter be sure that the mark of an electron is indeed a mark of the electron and not something else? Reproducibility can only yield the equivalence of similar cases but can reproducibility give us certain knowledge that the mark is indeed that of an electron?

To answer this, let us look at how this inference is generated in science. First of all, there is no language of signs and relation of one sign-property to another. The standard description is as follows. There are two possible general ways of understanding observational facts. One is that experiments are independent of theory which they may be trying to prove or disprove and the other is that many experimental results are discovered independent of any theory that one holds while doing the experiment. The consequence of this is that experimental observations either make meaning within the context of a particular theory or that these observations are independent of any theory. The former view is called the theory-laden view of observations and claims that observations make meaning only when interpreted according to the theory one holds. This problem is compounded by the observation that for a given set of observational facts we can have more than one equivalent theory, in the sense that we can have theories that postulates very different kinds of entities and principles and yet give an effective explanation of the facts. This view is enshrined as the Quine-Duhem thesis. The basic idea here is that observations are theoryladen and thus doing experiments to test a theory is to be involved in a circle of validation. In the example of detecting electrons, one can argue that unless we knew how electrons behaved under an electric field and how electric force modifies its "trajectory" there is no way of interpreting the line as a trace of an electron's motion.

18.7 Sign-Signified Relation in Applied Mathematics

There is one interesting aspect of the use of symbols in mathematics that is worth mentioning here. The motivation for doing this is to indicate a formulation that takes arbitrary symbols and creates signs which have a necessary relation with some signified. The implication of this is that a sign-signified relation, which is necessary, as in Indian logic, is closely aligned with formal structures in mathematics.

Symbols in mathematics are arbitrary. We can choose to let "m" stand for mass. There is no natural relation between "m" and the real property of mass. The implication of this is that in one problem I can choose "m" for mass and in another I can choose "s" for mass if I like. Similarly for the velocity, which although is often symbolised by "v" is in principle independent of the symbol I choose to represent it. This much is obvious but the next step adds an interesting twist. Given m and v representing mass and velocity of an object, I construct a new symbol, mv^2. Now, if the earlier symbols were arbitrary, then it should follow that any complex symbol formed from the simpler ones must also be arbitrary. However, the term $\frac{1}{2}mv^2$ is the "sign" for another property of the moving object, its kinetic energy. Interestingly, this sign for kinetic energy is the valid sign for it; no other combination of the simpler symbols can stand for kinetic energy. Therefore, kinetic energy of any object (in classical physics) will be represented only by this sign and no other. Thus, miraculously, from a combination of arbitrary symbols we seem to have constructed a sign which is in a natural relation with the signified!

Note that we do not assign an arbitrary symbol for the kinetic energy, which we could have done if all we wanted to do was to assign symbols to all the properties. But if we did assign an arbitrary symbol for all properties, including kinetic energy, we would have missed this interesting occurrence of one property being associated in an essential sense with one "sign". Therefore, this suggests that while it is possible to associate an arbitrary symbol with every property, we might be missing some important information if we focus only on the arbitrary nature of signs. In Western philosophy, there is a strong move towards privileging the arbitrary nature of sign. Ironically, one of the reasons for privileging the arbitrary nature is that it enables construction of a rich domain of symbols, one that is very important both for mathematics and logic.

This shift from the arbitrary to the necessary is not restricted to kinetic energy but is found in all physical concepts that have a mathematical sign representing them. In fact, this natural association is a very important methodological tool for theoretical research since it allows us to detect where physical concepts could lie hidden in some mathematical description.

What is interesting is that this is not necessarily of that much importance in pure mathematics as it is in applied mathematics. One of the reasons why mathematics is so effective in describing the real world lies in its use of this matching between signs in its discourse and some physical concepts that these signs stand for. It is clear that there is no causal link between these signs and the appropriate physical concepts. But then what gives the notion of necessity in this relation? We can recollect *Dharmakīrti*'s classification of the types of inference as being of own-nature along with causal relation and non-perception. It has been suggested that own-nature inferences are like analytical statements. Associating specific complex signs for specific physical concepts is made possible through definition. Thus I define kinetic energy to be $\frac{1}{2}mv2$ but this is not an arbitrary definition. I cannot define kinetic energy in any other way! And in representing kinetic energy mathematically, I can only use this sign or its symbolic equivalents (like $\frac{p^2}{2m}$, where p is the momentum, which classically is equal to mv).

This suggests that the analysis of the sign-signified relation in the use of mathematical symbols for physical concepts falls under a particular analysis of signs. And this analysis is well described by *Dignāga*'s basic question of asking when is a sign a logical sign? To paraphrase it for the example discussed above, we can ask when is the sign for kinetic energy a valid sign which really stands for the concept kinetic energy? More detailed analysis of this issue will take me too far away from what I want to do in this book. I mention it only to emphasise that the notion of natural related signs is of central importance to both science and Indian logic.

Mathematics is the best example of a discipline that essentially depends on the power of symbolisation. However, the notion of arbitrary symbols has been given undue importance in understanding the nature of this symbolisation. Applied mathematics, mathematics that is used in the sciences, poses a challenge to the arbitrary nature of symbols that occur in "pure" mathematics. That meaning accrue to symbols is a possibility that mathematics has to accept. This is manifested in the practice of applied mathematics in many different ways. I will briefly discuss two uses of symbols in mathematics that demand a more sophisticated interpretation of signs in mathematics, which includes the possibility of certain signs capturing a special relation to the signified.

The first is the mode of what I have elsewhere called "alphabetisation" in mathematics.[20] This is a description of how mathematics creates its "alphabets". Mathematical discourse has an interesting way of representing various mathematical entities. The patterns of the symbols actually communicate information. If symbols are seen as arbitrary it is equivalent to saying that symbols do not carry any prior meaning. But if we look at how symbols are created, especially complex patterns representing various mathematical objects, then we can easily see how meaning is encrypted into the patterns of the symbols. Therefore, to look upon mathematical symbols as being arbitrary is to misunderstand the creative use of symbols in mathematics. There is an intriguing possibility that by looking at symbols we can know what it signifies by unpácking the meaning of its pattern. That is, the relation between sign and signified could be hidden in the way the symbol is written! One illustration of this possibility lies in the importance of visual form and its role in applied science. From mathematical patterns, physicists can deduce what mass a

[20] Sarukkai 2002.

particle has or what physical concept occurs where in a mathematical description. A discussion of this will take me too afar but I mention this here to motivate the notion of the importance of "natural" relations which are captured in particular signs. These are the kinds of signs that are privileged in Indian logic and we thus see new ways of drawing upon this tradition to reflect on contemporary science and mathematics.

Indian logic is an exemplar of scientific method. The strict conditions on confirmation, verification and falsification are methodological constraints in any inference that is accepted as valid. The Indian philosopher's search for necessary relations as invariable concomitance is similar to the modern day scientist's search for laws. The excessive dependence of Indian logicians on causal relations brings their concerns closer to the revivalist theories of causal description of scientific processes. It is, therefore, a cause of great surprise that philosophy of science has been immune to the intellectual charms of Indian logic![21]

References

1. Bhattacharyya, S. *Doubt, Belief and Knowledge*. New Delhi, ICPR, 1987.
2. Buchler, J., editor. *Philosophical Writings of Peirce*. New York, NY: Dover, 1955.
3. Cassirer, E. *The Philosophy of Symbolic Forms, The phenomenology of Knowledge (1929)*, Vol. 3. New Haven, Yale University Press, 1957.
4. Ferrari, M. The concept of symbol from Leibniz to Cassirer. In M. Ferrari and I.O. Stamatescu, editors, *Symbol and Physical Knowledge*. Berlin, Verlag, 2002.
5. Ihmig, K. The symbol in the theory of science: Duhem's alleged instrumentalism or conventionalism and the continuity of scientific development. In M. Ferrari and I. O. Stamatescu, editors, *Symbol and Physical Knowledge*, Berlin, Springer, 2002.
6. Matilal, B. K. *Character of Logic in India*. New Delhi, Oxford University Press, 1999.
7. Mohanty, J. N. *Reason and Tradition in Indian Thought*. Oxford, Clarendon Press, 1992.
8. Prasad, R. *Dharmakīrti's Theory of Inference*. New Delhi, Oxford University Press, 2002.
9. Sarukkai, S. *Translating the World: Science and Language*. Lanham, MD, University Press of America, 2002.
10. Sarukkai, S. *Indian Philosophy and Philosophy of Science*. Delhi, CSC/Motilal Banarsidass, 2005.
11. Vidyabhusana, S. C. *A History of Indian Logic*. Delhi, Motilal Banarsidass, 1920.

[21] For a more detailed analysis of these relations, including that between Indian theories of language and philosophy of science, see Sarukkai 2005.

Chapter 19
The Concept of Hetvābhāsa in Nyāya-śāstra

K. Ramasubramanian

19.1 Introduction

The singular and unprecedented contribution of the *Nyāya* school[1] that greatly influenced all other branches of learning in the Indian tradition is the development of a full-fledged *pramāṇa* theory. The word *pramāṇa* refers to the different means of knowledge, such as perception, inference, analogy and verbal testimony. Of the different *pramāṇas* – whose foundational basis, relative strengths and the domain of applicability have been thoroughly analysed and discussed at great length by philosophers belonging to different schools – it is the *anumāna-pramāṇa* (theory of inference) that has received greatest attention amongst the *Naiyāyikas*[2] belonging to the ancient, [3] medieval as well as the recent past.

The influence of the *pramāṇa* theory both on the dialogical and philosophical discourse has been, and still today is, primarily three fold : (i) it furnishes the required grounding for the thesis to be backed up by a certain *pramāṇa*, (ii) provides the necessary platform to incisively weigh/analyse the proposed thesis, and (iii) enables to succinctly summarise the observations, that stem out of the analysis, in the form of an *anumāna* or *nyāya-prayoga*[4]. The advantages that could be had – in terms of increased precision, brevity and clarity – either in formulating the problem or in expressing one's thoughts in a concise form using the *anumāna-pramāṇa*,

K. Ramasubramanian
Cell for Indian Science and Technology in Sanskrit, Department of HSS, IIT Bombay, Mumbai 400 076, India, e-mail: kramas@iitb.ac.in

[1] One of the major philosophical systems that flourished in ancient India which was given a solid shape by Gotama (c. 3rd century BC) in the form of *Nyāya-sūtras*.

[2] Those who have committed themselves to the study of *Nyāya-śāstra*.

[3] Bhadrabāhu, a Jaina logician of the 4th century BC – considered to be the eighth successor of Mahaveer – recommends a ten-step argument to ensure that the inference made is a valid one.

[4] Discussion on *nyāya-prayoga* is to be found in Section 19.3.

J. van Benthem et al. (eds.), *Proof, Computation and Agency*, Synthese Library 352, DOI 10.1007/978-94-007-0080-2_19, © Springer Science+Business Media B.V. 2011

gave the logicians the necessary impetus for advancing their theory of inference to much greater depths. Consequentially, an enormous amount of literature in the form of textbooks, glosses and commentaries, came to be produced continuously in quick succession. Particularly, this is true of the post-Gaṅgeśa (13th century AD) period during which the growth was indeed exponential.

The works in *Nyāya-śāstra* not only discuss the aggregate causal conditions and the process involved in getting an inferential cognition, but also thoroughly analyse the factors that help to assess and decide whether a given inferential cognition is valid or otherwise. It is precisely in this context, that the theory of *hetvābhāsa* got germinated and later developed into a rich discipline of study in itself. In fact, there are numerous source works in the *Nyāya* literature extant today, that are exclusively devoted to discuss the theory of *hetvābhāsa* and its classifications, besides the secondary ones – which are practically innumerable – in its multifarious forms.

The aim of the present paper is to explain the concept of *hetvābhāsa* – primarily making use of the definition provided by Gaṅgeśa in his monumental work *Tattvacintāmaṇi* – with an emphasis to bring out the distinction it bears with what is generally known as *logical fallacy* or *fallacious reasoning* in the western logic. We also briefly describe the classification of *hetvābhāsa*, as presented in any of the typical introductory texts on Indian logic, with a few illustrative examples.

19.2 Means and Types of Cognition in Nyāyaśāstra

The present section, and the one that follows – though may seem to be, of course deceptively, not linked with primary theme of the article – are meant to provide the necessary background to facilitate the reader develop a fuller and proper understanding of the concept of *hetvābhāsa* to be presented in the later sections.

In *Nyāyaśāstra* – based upon the temporal aspect of the event that is being cognized – cognition is broadly classified into two types, namely (i) *Smṛti* (Recollection) and (ii) *Anubhava* (Experience). From a completely different view point, both the above mentioned types can be further divided into:

1. Valid cognition (*yathārthānubhavaḥ*) – A cognition *that cognizes an attribute as belonging to a thing which really possesses it* is valid.[5] It is also referred to as *pramā* in the literature; and
2. Invalid cognition (*ayathārthānubhavaḥ*) – A cognition *that cognizes an attribute as belonging to a thing which does not really possess it* is invalid.[6] It is also referred to as *apramā/bhramaḥ* in the literature.

As an illustration of the latter classification, let us consider the experience that most of us would have had – as passengers seated in a train – at sometime or the

[5] तअद्वति तत्प्रकारकोऽनुभवः यथार्थः – *Tarkasaṅgraha* of Annambhaṭṭa.

[6] तदभाववति तत्प्रकारकोऽनुभवः अयथार्थः – *Tarkasaṅgraha* of Annambhaṭṭa.

other. For a moment, let's assume that the train in which we travel has come to a halt in a railway platform, just next to another stationary train. The moment the other train sets into motion, we would feel as if our train has set into motion. This type of cognition – in which the attribute (motion) belonging to something else (the other train) has been attributed to something else (our train) which does not possess it – that is called "invalid", as against the "valid" cognition of recognizing the motion of the other train to be its own, which anyway occurs very soon as the other train picks up speed and/or leaves the platform. Without getting into the details of whether the above definition is complete and comprehensive to take into account all the different possible illusions that can arise – such a disucssion being out of the scope of the present paper – we quickly move on to briefly outline the different means of knowledge accepted in the *Nyāya* school.

19.2.1 Means of Knowledge

As mentioned earlier, in Sanskrit literature, the sources or means through which a cognition arises is called *pramāṇa*. In the *Nyāya* school, the different means that contribute to the generation of cognition has been broadly grouped into four categories as indicated in Table 19.1. This division of *pramāṇa*-s – which must have been there, in some form or other from time immemorial, nevertheless which is certainly found in the Vedas[7] – into exactly four types is noted in the *Nyāya* tradition atleast from the time of Gotama (3rd century BC), if not earlier. The third *sūtra* of Gotama's *Nyāyadarśana* presents the list of the four *pramāṇas* as follows :

प्रत्यक्षानुमानोपमानशब्दाः 'प्रमाणानि'।[8]

Perception, inference, assimilation and verbal testimony are the four means of knowledge.

1.	*Pratyakṣam*	Perception
2.	*Anumānam*	Inference
3.	*Upamānam*	Analogy
4.	*Śabdaḥ*	Verbal testimony

Table 19.1 The different means of knowledge accepted in the *Nyāya* tradition.

Cognition of certain things can happen by employing all the four *pramāṇas* listed in Table 19.1. For instance, consider an "iron ball" that has been just heated and kept aside for annealing, and whose temperature is slightly below the red hot temperature. The heat present in the iron ball can be cognized by all the four *pramāṇas*:

[7] In *Kṛṣṇa-yajurveda-Āraṇyaka*, Chapter 1, we have the following lines – स्मृतिः प्रत्यक्षं ऐतिह्यम्। अनुमानश्चतुष्टयम् – which speaks of four means of knowledge.

[8] Gotama's *Nyāya-sūtras*, I.1.3.

- The father (a blacksmith) advises his little son – "Don't go near it. It will be very hot". If the son understands, then the cognition of heat that arose is through *Śabda.*
- If the son does not know how hot the iron ball would be and the father explains – The ball will be as hot as burning fire – then the assimilation of heat is through simile called *Upamāna.*
- Because of enthusiasm, the little son goes near the ball, feels the radiation, infers that it should be quite hot and returns back without physically touching the ball – Here the cognition is through *Anumāna.*
- If the boy is over enthusiastic/mischievous, he may even touch the ball to burn his fingers – It is *Pratyakṣa,* direct experience.

While the above example illustrates that certain things can be cognized by employing all the four *pramāṇas,* there are instances wherein the cognition can take place only by means of a single *pramāṇa.* This has been explained succinctly by Vātsyāyana, with aptly chosen examples, while commenting upon the third *sūtra* of *Nyāya-darśana.*

> 'अग्निहोत्रं जुहुयात् स्वर्गकामः' इति लौकिकस्य स्वर्गे – न लिङ्गदर्शनं न प्रत्यक्षं, स्तनयित्नुशब्दे श्रूयमाणे शब्दहेतोरनुमानं – न प्रत्यक्षं नागमः, पाणौ प्रत्यक्षतः उपलभ्यमाने – नानुमानं नागम इति, ...[9]
>
> "One who is desirous of heaven may perform *agnihotra*".[10] From this [injunction found in the Vedas], an ordinary human being comes to know of "heaven" [by verbal testimony and] – the means is not the sight of a "*liṅga*"[11] or perception. On hearing the sound of a thunder, he infers of thunder and – the means here is neither perception nor verbal testimony. Looking at an entity in one's own palm, [the object is cognized by perception and] – the means here is neither inference nor verbal testimony.

Another feature that merits mentioning here is that the *pramāṇa* theory as presented by the Indian logicians is "not" a near-kin of the Western epistemology in the sense that in the latter, the epistemological issue which primarily determined the course of western philosophy is the fundamental question – Does knowledge arise from reason or experience? However, in Indian philosophy there has been no such debate, regarding *pratyakṣam* or *anumānam.* In fact, all schools of philosophy have unequivocally accepted *pratyakṣam* as the primary means of knowledge. However, in Indian philosophical systems, *pratyakṣam* is not restricted to sensory perception alone. It includes, cognition of the Self (*ātmāparokṣatvaṃ*), the Universals like cowness *jātipratyakṣatvaṃ,* and so on.

[9] *Ibid .*

[10] *Agnihotra* is a ritual and this sentence is an injunction found in the Vedas.

[11] *Liṅga* is a technical term used to refer to something (say A), the knowledge of which makes us immediately infer something else (say B), because of the knowledge of the invariable concomitance (*vyāpti*) between A and B, that we already possess.

19.2.2 Irreducibility of Knowledge

It was explained in the previous section with an illustration that the same object may be cognized using different *pramāṇas*. When different *pramāṇās* cognize a given *prameya*[12] the cognition that arises – which is termed as *phalam* in Table 19.2 – would naturally be of the same *prameya*. However, notwithstanding the fact that different *pramāṇas* give rise to *pramā* (a valid knowledge) of the same *prameya*, it doesn't mean that the *pramās* themselves reduce/collapse to one another. It would still be possible to distinguish one *pramā* from the other, because the first two of the three factors listed in Table 19.2 differ from *pramāṇa* to *pramāṇa*.

1. KARAṆAM	– Special or efficient cause.
2. VYĀPĀRA	– Intermediate cause.
3. PHALAM	– Final result (cognition)

Table 19.2 Factors that distinguish the different *pramāṇas*.

In fact, a salient feature of *pramāṇa* theory is that a *pramā* that is generated by one *pramāṇa* "cannot" be reduced to the one generated by the other, even if it is of the "same" entity, as each of them is caused by a *unique* aggregate of causal conditions.

It is precisely these unique set of causal conditions that give rise to different *pramāṇas*. Having identified the aggregate causal conditions, Indian logicians have grouped them with three names/factors as indicated in Table 19.2. As an illustration, in Table 19.3 we indicate the three factors that distinguish the two *pramaṇā*-s *pratyakṣam* and *anumānam*.

	PRAMĀṆAM	
Factors	*Pratyakṣam* (Perception)	*Anumānam* (Inference)
KARAṆAM	Sensory organs	Universal correlation called the *Vyāpti*
VYĀPĀRA	Link between the sensory organ and the object	Subsumptive correlation called the *Parāmarśa*
PHALAM	Perceptive knowledge called *Pratyakṣam*	Inferential knowledge called *Anumiti*

Table 19.3 Aggregate causal conditions that generate perceptual and inferential cognition.

It may be mentioned here that the concept of *karaṇam* – that happens to be *vyāpti* in the case of *anumānam* – introduced here, plays a key role in the discussion of *hetvābhāsa* to be taken up in the following sections.

[12] An entity or event that comes to the light of knowledge through a certain *pramāṇa*.

19.3 Anumāna and Nyāya-prayoga

The etymological derivation of the word *anumāna* is –

ANUMĀNA =	ANU	+	MĀNA
	(that which follows *or* becomes operative after)		(means of knowledge)

From the above derivation it is clear that *anumāna-pramāṇa* come into play only when it is preceded by a cognition generated through certain other *pramāṇa*. The *pramāṇa* that should precede *anumāna* – or after which *anumāna* becomes operative, if it can – could either be (i) *pratyakṣa* or (ii) *śabda*. Vātsyāyana in his commentary to *Nyāya-sūtra* observes:

प्रत्यक्षागमाश्रितमनुमानम् । सा अन्वीक्षा।
तया प्रवर्तत इत्यान्वीक्षिकी, न्यायविद्या, न्यायशास्त्रम् ।

Anumāna is dependent on (takes place only after) *pratyaksa* or *āgama*. [Thus] it is the process of looking over again. Since this (discipline of logic) springs out from that (*anvīkṣā*) it is called *ānvīkṣikī*, also known as *nyāyavidyā* or *nyāyaśāstram*.

A generic *anumāna-prayoga*[13] will be of the form – "*S* is *p* because of *m*."

पर्वतो	वह्निमान्	धूमात्
(The mountain)	(has fire)	(because of smoke)

पक्षः	साध्यं	हेतुः
(Minor term)	(major term)	(middle term)

Note: Here it needs to be mentioned that the English translations such as minor term, major term and the middle term, do not convey the meanings of their Sanskrit terminologies. However, in order to facilitate the understanding of those who are completely alien to the Indian terminologies, they have been provided here.

A renowned scholar of modern times, J. N. Mohanty, commenting upon the theory of *anumāna* observes:

The word logic is of western origin and has gathered a connotation of its own. The *anumāna* theory is a system by its own right. It is not the same as either Aristotelian syllogistic or modern predicate calculus, it is not for that reason illogical.[14]

A complete syllogistic expression called *Nyāya-prayoga* consists of five parts –

[13] The term *prayoga*, employed here as well as in other contexts later, means – usage.

[14] J.N.Mohanty *Reason & Tradition in Indian Thought*, Claredon Press, Oxford 1992, p.106.

1. The thesis (*pratijñā*)
2. The reason (*hetuḥ*)
3. The exemplification (*udāharaṇam*)
4. The subsumptive correlation (*upanayaḥ*)
5. The conclusion (*nigamanam*)

As an illustration of the above we may consider the standard example of the smoke and the fire:

1. पर्वतो वह्निमान् ।
   The mountain possesses fire.
2. धूमात् ।
   Because it possesses smoke.
3. यो यो धूमवान्, स स वह्निमान्; यथा महानसः ।
   Whatever possesses smoke possesses fire as does the kitchen stove.
4. तथा च अयम् ।
   The mountain is like that.[15]
5. तस्मात् तथा ।
   Therefore, the mountain is like that (i.e. possesses fire).

This scheme of presenting an argument or proving the existense of something based on the knowledge of something else is at least as old as Gotama. The concept of invariable concomitance (*vyāpti*) plays a crucial role in *nyāya-prayoga*.

19.3.1 Justification for Using Five-Membered Structure

In the scheme of five-membered structure described above, it may seem as though (1) and (5) are same; But strictly speaking they are not so. This could be recognized once we consider the fact that it is in the context of *parārthānumāna* (dialogical-disputational discourse) that this five-membered structure is recommended. It must be noted that:

(1) – is just a thesis (to be proved) and
(5) – is conclusion (asserted statement).

Further, (1) may be true for the I cognizer (utterer), but not for the hearer in whom the inferential cognition is desired. Similarly (2) and (4) are not the same. The smoke in (4) is no more the smoke *per se*, but smoke *qua* with which the fire is known to be invariably concommitant. Once this step is taken, i.e. the cognition content of the *upanaya-vākya* is aroused in the hearer – true to the nomenclature which means

---

[15] That is, possesses smoke with which fire is invariably concommitant.

"taking close" – the final cognition is almost unavoidable. This is indicated by the use of the word *tasmāt.* Thus we see that in the inferential process described above, if the first four cognitions (1) – (4) occur, then the set of all the necessary and sufficient conditions for the final inferential cognition to arise being satisfied, it is bound to occur and the inferential process is said to be complete.

In the section that follows we will explain the key role played by *hetvābhāsa* either in allowing the inferential process to go thorough and thereby ensure that the inferential cognition that finally arises is both valid and sound, or in stalling the inferential process somewhere inbetween – which can happen anywhere between stages (1)– (4) listed above – and thereby terminate the process before it gets successfully completed.

19.4 Hetvābhāsa

The word *hetvābhāsa* is actually a compound word. By compound word we mean a word – which is to be considered as "one" unit as per the rules of grammar – formed by grouping a few nouns together. The process of grouping is technically called *samasanaṃ* or more commonly as *samāsa*[16] in the Sanskrit literature. The word *hetvābhāsa* serves as an example of a compound word formed out of two nouns put together as indicated below.

HETVĀBHĀSA =	HETU	+	ĀBHĀSA
	Mark/Reason		Defect/Impostor

It does not require much expertise in Sanskrit to get a hunch of the word *hetvābhāsa* to be a compound word. But beyond that, to know as to what type of compound (*samāsa*) it belongs to, is not quite a trivial thing to be guessed and requires a much deeper analysis. However, for our present discussions – without getting much into the details of the various possible derivations of this compound[17] and the etymological derivation of its consitutents – it would suffice to simply take it as *hetu-doṣa.* That is,

HETVĀBHĀSA = Defect in the *hetu*/Impostor of the *hetu*.

It would be best to proceed with further discussions on *hetvābhāsa* with a word of caution here. It is not uncommon to find secondary literature translating *hetvābhāsa*

[16] In making *samāsās* (forming compound words), while there is a lower bound – which conceivably is two – there is no upper bound on the number of nouns that can be grouped together.

[17] Whether it has to be considered as a *ṣaṣṭhī-tatpuruṣa* compound or as a *upamāna-pūrvapada-karmadhāraya* compound is a matter of detail and depends on how one would like to view it. In the former case, we have to derive the compound as – *hetoḥ ābhāsaḥ (doṣāḥ) = hetudoṣāḥ*, which means a "faults in *hetu*", while in the latter the compound has to be derived as – *hetuvat ābhāsante = hetvābhāsāḥ*, which means "semblance of *hetus*".

as "fallacious reasoning" or "fallacy in reasoning". Such a translation is simply wrong – and is quite likely to be misleading from having a proper understanding of the significant role played by *hetvābhāsa* in the *anumāna* theory – because in the western tradition the term fallacy is used to connote either a defective process of thinking or a violation of some formal structure of argument or rule that has been agreed upon. The term *hetvābhāsa* refers to neither of them and in fact, the word literally means – *what appears to be a hetu while it is not so*. Strictly speaking, the concept of *hetvābhāsa* does not have a parallel in the western logic at all.

As far as the studies in Indian logic – or for that matter any other discipline of *śāstra* – is concerned, a proper understanding of the concept of *hetvābhāsa* and its classification is very essential. On may go to the extent of saying that it is a must without which it may not even be possible to have a purposeful and/or meanigful discussion. This is because, in dialogical debates, more often than not, *anumāna* or *nyāya-prayoga* would be made use of, to establish one's own thesis. When an *anumāna* is employed, and if the opponent is able to point out a *hetvābhāsa*, the *anumāna* becomes invalid and one may have to retract back embracing defeat, unless one could strongly defend his thesis against the *hetvābhāsa* that has been spelled out. It is precisely in the context of dialogical debates, that a sound knowledge of *hetvābhāsa* and its classification comes quite handy. Further, it must be noted that when there is an *ābhāsa* in *hetu*, there is bound to be an *ābhāsa* in the inferential cognition that would be arising shortly. Thus the notion of invalid/valid inference (*sadanumāna/anumānā-bhāsa*) depends crucially upon the recognition of the presence/absence of *hetvābhāsa*.

19.4.1 Definition of Hetvābhāsa

Though Gotama himself has used the word *hetvābhāsa*,[18] and has classified it into five categories, the first and most comprehensive general definition of *hetvābhāsa* is given by Gaṅgeśa (14th century) in his *Tattva-cintāmaṇi* –

अनुमिति-कारणीभूताभावप्रतियोगियथार्थज्ञानविषयत्वम् - हेत्वाभासत्वम् ।

For the purposes of convenience, this definition of Gaṅgeśa has been rephrased by the later authors such as Annambhaṭṭa in their introductory texts – to avoid certain technical jargon, while retaining the purport in tact – as follows:

अनुमितिप्रतिबन्धकयथार्थज्ञानविषयत्वं - हेत्वाभासत्वम् ।

The defect of *hetu* is its being the object of contradictory valid knowledge, which inhibits the inferential cognition.

[18] Of the sixteen *padārthas* listed by Gotama in the very first *sūtra* of the *Nyāya-śāstra*, *hetvābhāsa* happens to the 13th in the listing.

While explaining the above definition, it is further mentioned by the commentators that the word *anumiti* occuring in the above definition should be taken as *upalakṣaṇa*. That is, the word *anumiti* should be understood to be referring to not only *anumiti* per se, but also to the factors that contribute to *anumiti*. Among the various factors[19] two most important ones are – *vyāpti* and *parāmarśa* (refer Table 19.3). Referring to these factors are *karaṇas* a more refined definitions can also be found in various texts.[20].

The central idea behind the above definition is – An inferential cognition of the form "*S* is *p* because of *m*" would be prevented from occuring if the hearer had a valid knowledge of a situation, which has in fact "a defect of *m* as a mark of *p* in *S*". As illustration of this, let us consider the following examples –

- Example 1: Directly *anumiti* is prevented.

  ह्रदो वह्निमान् धूमात्
  The lake possesses fire because it possesses smoke.

- Example 2: Here *anumiti-karaṇa*, namely *vyāpti* is prevented.

  पर्वतो वह्निमान् गर्दभात्
  The mountain possesses fire because it possesses donkey.

In Example 1, the intended inferential cognition (lake possessing fire) namely *anumiti* does not take place, because the hearer already possesses a contradictory valid knowledge (that the lake does not possess fire), that is born out of his visual perception. In Example 2, the *anumiti* does not take place because the required *vyāpti* of the form, *wherever there is donkey there is fire* is prevented by a contradictory valid knowledge.

19.4.2 Import of the Word yathārtha in the Definition

In order to ensure that a valid inference does not stand inhibited – by a false contradiction that is entertained because of the non-apprehension of its falsity – the term *yathārtha* has been employed as an adjective to the word *jñāna* in the definition of *hetvābhāsa*. We illustrate this idea directly by citing an example –

- Suppose someone entertains the thought –

  – उद्यानं न वह्निमत् हरितात् ।

---

[19] Some of them quite peripheral such as the environment, one's own status of mind and so on.

[20] One of the commonly used ones, that is obtained by replacing the word *anumiti* by *anumiti-tatkaraṇa-anyatara*. Thus the full form would be – अनुमितितत्करणान्यतरप्रतिबन्धकयथार्थ-ज्ञानविष्यत्वम् – हेत्वाभासत्वम् ।

   - The garden has no fire because it is green.

- Such a thought would contradict the right conclusion one may draw based on the *anumāna* –

   - उद्यानं वह्निमत् धूमात् ।
   - The garden has fire because of smoke.

If we were to omit the word *yathārtha* in the definition of *hetvābhāsa*, then there is a possibility that the valid conclusion (that the garden has fire) may be pronounced invalid, because there is a contradictory proposition (that the garden is green), which stifles the conclusion. It is precisely to circumvent this problem, the word *yathārtha* has been rightly introduced. Since there is no invariable concomitance (*vyāpti*) between greenery and absence of fire, a mere contradictory proposition does not become an object of a valid knowledge *yathārtha-jñāna*, and hence does not inhibit the right conclusion. In other words, inclusion of the term *yathārtha* guarantees that what is sought after is the "factual inhibition" and not inhibition based on the thoughts entertained.

For factual inhibition, the contradiction must be valid; needless to say that the logical validity of the contradictory proposition is to be determined by its correspondence to fact. In the language employed by the *Nyāya* school, this correspondence to the fact is called *artha-sārūpya* or *viṣaya-sārūpya*.

19.4.3 Falsifier is Valid Knowledge and Not Fact

A proposition can be declared false or a knowledge invalid only if there exists a contradictory valid knowledge corresponding to a contradictory true proposition. A proposition is true provided it corresponds to a fact. This correspondence called *artha-sārūpya* may be stated as – *Structural semblance between propositional knowledge and its factual object.* Now it obviously follows that, it is the valid knowledge of the true proposition which possesses this *artha-sārūpya* and not the invalid knowledge of the false proposition. Thus it is the presence or absence of *artha-sārūpya* that makes a true proposition true and its contradictory proposition false.

Here, it must be noted that the false proposition does not directly correspond to the fact. It has only indirect reference to the fact through the valid knowledge of the true proposition which corresponds to the fact and contradicts the false proposition. Thus the fact falsifies a false proposition only through the valid knowledge and not directly. The direct and immediate falsifier of a false proposition is the valid knowledge of an acceptedly true contradictory proposition which corresponds to the fact. A careful study of the definition of the *hetvābhāsa*, reveals that, it is only after considering all this into account that Gaṅgeśa comes up with his brilliant definition.

19.5 Classification of Hetvābhāsa

For the purposes of convenience, *hetvābhāsa*-s are primarily classified into the following categories :

1. सव्यभिचार (*Savyabhicāra*)	Straying reason
2. विरुद्ध (*Viruddha*)	Contradicting reason
3. सत्प्रतिपक्ष (*Satpratipakṣa*)	Opposable reason
4. असिद्ध (*Asiddha*)	Unestablished reason
5. बाधित (*Bādhita*)	Stultified reason,

The list above is the set of defective reason. The defects are :

1. व्यभिचार (*Vyabhicāra*)
2. विरोध (*Virodha*)
3. सत्प्रतिपक्ष (*Satpratipakṣa*)
4. असिद्धि (*Asiddhi*)
5. बाध (*Bādha*)

These defects when identified in the *hetu* prevent it OR make it impotent to give rise to the inferential cognition. In the following, we provide an overview of these defects – along with their subdivisions – with a few illustrative examples of the same without getting deep into the technical details of the basis and soundness of their classification.

19.5.1 Savyabhicāra (The Straying Reason)

The first of the five *hetvābhāsa*-s listed above is further divided into three types –

1. *Sādhāraṇa* (Common strayer)
2. *Asādhāraṇa* (Uncommon strayer)
3. *Anupasamhārin* (Non-conclusive strayer)

The *Nyāyasūtra* (1.2.5) reads – *anaikāntaḥ savyabhicāraḥ*. The derivation of the word *anaikāntaḥ* is as follows:

Ekānta	One extreme
Ikāntika	Confined to one extreme
Anaikāntika	Not confined to one extreme

The concepts of *Ekānta* and *Anekānta* as defined by the commentators Vātsyāyana and Uddyotakara are similar to the exclusive and non-exclusive disjunction which are quite familiar in modern logic. A valid generalisation (*vyāpti*) of the form *that which has smoke has fire* is a member of exclusive disjunction. There are two extreme cases possible – (i) Any object having smoke has fire. (ii) Any object having smoke does not have fire.

Here the truth of the first excludes the truth of the other. Thus the two members are two *ekānta*-s (exclusive extremes). The *hetu* cannot go with both. On the other hand, the term *Anaikāntika* gives the idea of non-exclusive disjunction. Of the three varieties of *savyabhicāra* we consider only the first one for a detailed explanation below.

Sādhāraṇa **(Common Strayer)**

We explain this by considering the example provided by Vātsyāyana in his commentary on *Nyāya-sūtra*:

शब्दो नित्यः अस्पर्शत्वात् ।
Sound is eternal because it is non-tactile object.

For the inferential cognition to arise we need to have *vyāpti* of the form – *That which is a non-tactile object is eternal.* Now we can think of two extreme possibilities:

1. A non-tactile object is eternal. (Ex: *Ātmā*, Self)
2. A non-tactile object is not eternal.(Ex: *Buddhi*, Mind)

Here the truth of the first DOES NOT EXCLUDE the truth of the other. That is the two members are not two *ekānta*-s (non-exclusive extremes). The *hetu* can go with eternality as well as non eternality. Thus we have a situation where the *hetu Anaikāntika* (i.e. not exclusively attached to one member of the disjunction). Hence, NO *vyāpti* and NO *anumiti*. This is succinctly described in a captive style by Vātsyāyana –

नित्यत्वमपि एकः अन्तः; अनित्यत्वमपि एकः अन्तः; एकस्मिन्नन्ते विद्यते इति ऐकान्तिकः, विपर्ययात् अनैकान्तिकः, उभयान्तव्यापकत्वात् ।

This defect is called *sādhāraṇa* because the *hetu* is common to both the presence and absence of *sādhya*.

19.5.2 Viruddha (Contradicting Reason)

This refers to defect in proposing the *hetu* which is in direct contradiction to *vyāpti*. Consider the example

शब्दो नित्यः कार्यत्वात् ।
Sound is eternal because it is produced.

In this instance, for the inferential cognition to arise we need to have *vyāpti* : *That which is produced is eternal* (V_1).

It is well known that anything that is produced is non-eternal. If it were eternal it cannot be produced (a self-contradiction). In fact, there is *vyāpti* between producibility and non-eternality (V_2). This contradicting valid knowledge of the V_2 will inhibit V_1, that is required for the inferential cognition to take place. Such an instance is called *viruddha* and the definition given by Gaṅgeśa is:

साध्यव्यापकाभाव-प्रतियोगित्वं विरुद्धत्वम् ।

19.5.3 Satpratipakṣa (**Opposable Reason**)

Consider the following inferences:

1. शब्दः अनित्यः कार्यत्वात्, घटवत् ।– (A1)
   Sound is not eternal because it is produced like a pot.
2. शब्दो नित्यः श्रावणत्वात्, शब्दत्ववत् ।– (A2)
   Sound is eternal because it is audible like the soundness.[21]

For the inferential cognition to arise – in the case of (A1) we need to have *vyāpti* : *That which is produced is not eternal*, example "pot" and – in the case of (A2) we need to have *vyāpti* : *That which is audible is eternal* – example "soundness". Now there is a crying contradiction:

From A1 – A is B; and From A2 – A is not-B,
where A is *śabda* and B is non-eternality.

If we represent the *hetu*-s in A1 & A2 by C and D, then the *vyāpti*-s are:

In A1	: C is B
In A2	: D is not-B

It may be noted that the two *hetu*-s C and D try to counter each other, by trying to prove the presence and absence of B in A. Logically, either of them may be right or both could be wrong. But the situation is dicey. Y makes a counter proposition; X gets confused. He is unable to fix the falsity in either of the propositions. This situation which is referred as – *agṛhītāprāmāṇyaka* that leads to mutual inhibition.

[21] A maxim in Indian logic – that goes as, यो गुणो यदिन्द्रियग्राह्यः, तन्निष्ठा जातिः तदभावश्च तदिन्द्रियग्राह्यौ – states that, a sense organ that perceives a property also perceives the "propertyness" as well as the absence of that property.

19.5.4 *Asiddha* (Unestablished Reason)

This refers to defect in proposing the *hetu* whose existence has not yet been established with reference to the *pakṣa* (minor term). This is primarily of three types:

1. *Āśrayāsiddha* (in respect of locus)
2. *Svarūpāsiddha* (in respect of itself)
3. *Vyāpyatvāsiddha* (in respect of concommitance)

Now let us consider the example of 2:

शब्दो गुणः चाक्षुषत्वात् ।
Sound is quality because it is visible.

For the inferential cognition to arise we need to have *vyāpti* : *That which is visible is a quality* – example "color".

But the problem here is – the existence of visibility in sound is not yet established. Only when its presence is settled, can we think of proceeding further to prove whether sound is quality or otherwise. This idea is succinctly putforth by Uddyotakara in his commentary to *Nyāyasūtra* :

प्रज्ञापनीयधर्मसमानः ।
This has the same status of the thing to be established.

Here the term *prajñāpanīya* means the thing to be shown/proved (*sādhya*). The problem of *Āśrayāsiddha* may be stated as follows:

Let's assume "A is B" is the minor premise & there exists no known entity in the world corresponding to A. In such a case, obviously B cannot be predicated of A.

We do not venture to explain the third variety of *asiddha* as it is more subtler and needs an elaborate explanation, which is out of scope of the present paper.

19.5.5 *Bādhita* (Stultified Reason)

This defect is called *kālātīta* in earlier texts. For instance, the definition in the *Nyāyasūtra* (1.2.9):

कालात्ययापदिष्टः कालातीतः ।
That which has been proposed beyond the time limit.

To understand why such a nomenclature is given, consider the example

वह्निः अनुष्णः पदार्थत्वात् ।
Fire is not hot because it is a substance.

It is generally accepted that *hetu* is posed to establish the *sādhya* in the *anumāna* at that point of time when there is a *doubt* about its existence. If there is no doubt then, posing *hetu* is useless. The uselessness can arise in two contexts.

1. *Siddha-sādhana* – the existence of probandum already known.
2. *Bādha* – the existence is stultified by other means.

Now, consider the example given above. Here, even before the *hetu* is uttered, the proposition is contradicted and dismissed by the valid knowledge that the FIRE IS HOT, which is obtained through perception. The proposition is dismissed beyond doubt. Therefore, posing *hetu* is fruitless because its employment has crossed the time-limit which is determined by doubt, which in turn determines the fruitfulness of its posing. This idea is clearly brought out by Vācaspati:

हेतोरपदेशस्य हि साध्यसन्देहविशिष्टः कालः । यथाहुः – नानुपलब्धे, न निर्णीते न्यायः प्रवर्तते; अपि तु सन्दिग्धे।

The right time for supplying *hetu* is when there is a doubt in the presence of *sādhya* (and not otherwise). As is stated – An argument does not proceed on those issues that are completely unknown, or those that are well established, but only where there is a doubt.

19.6 Summary and Conclusion

The scholars belonging to *Nyāya* tradition – apart from discussing various philosophical, epistemological and ontological issues – took great pains to analyze, classify, define and systematize the different *pramāṇa*-s. They also constructed the necessary theory to identify a valid cognition called *pramā*, in contrast to the invalid ones called *apramā* or *bhrama*. Particularly, great attention was paid to develop the theory of inferences in order to distinguish valid inferences against the invalid ones and it is precisely in this context that the concept of *hetvābhāsa* got germinated.

In the case of formal logic developed in the west, for an argument to be considered valid, neither the premises have to be true nor the conclusion has to be true. Validity is purely a conditional notion: For instance,

Whereever there is smoke there is donkey. – (A)
The mountain possesses smoke. – (B)
Therefore, the mountain possesses donkey. – (C)

is a perfectly valid argument because there is no logical fallacy. All that needs to be looked for validity is – if the premises happen to be true, then the conclusion has to be true. But for students trained in Indian logic, the above argument will not be accepted a valid argument. The *anumāna* is wrong because of presence of *hetvābhāsa*. *Bādha* will be pointed out even in Step (A) – as it is disproved by perceptual experience – and hence the argument cannot be carried any further. The point we try to drive home is, (i) there is no parallel to the concept of *hetvābhāsa* in the western logic and (ii) notion of *hetvābhāsa* should not be confused with the notion of "logical fallacy" or fallacious reasoning.

Moreover, till recent times while the logicians in the west were generally concerned with the formal validity or consistency in a string of the so called propositions, in the Indian context, the logicians were trying to develop *anumāna* theory – which includes the theory of *hetvābhāsa* as an integral part of it – whose purpose was *not merely* to help in deciding whether the argument is valid, but to ensure that it is also "sound". In other words, the aim of the *Naiyāyikas* in formulating the *anumāna* theory is to obtain *pramā* ("right" knowledge), and not merely verifying formal validity. This being the objective, their theory of *hetvābhāsa* explains how this is accomplished, by giving the epistemic conditions that prevent the inferential cognitions from arising, in those instances wherein *pramā* cannot arise from the *nyāya-prayoga* that is being made.

CPSIA information can be obtained at www.ICGtesting.com
Printed in the USA
LVOW102312170912

299227LV00007B/49/P

9 789400 700796